TECHNICAL WRITING

FRANCES B. EMERSON

Honeywell Information Systems Inc.

HOUGHTON MIFFLIN COMPANY
BOSTON

Dallas Geneva, Illinois
Lawrenceville, New Jersey Palo Alto

Book and cover design Daniel Earl Thaxton.

Cover credit "Magnetic Force," original painting by Brian Magel, 1985. Private collection.

Line art credit Line art created by ANCO, Boston.

Photo credits
180, 181 Figure 1 reproduced from *The New Science of Skin and Scuba Diving*, 6th Edition, copyright © 1985, Council for National Cooperation in Aquatics. By permission of New Century Publishers, Piscataway, NJ.

262 Figure 11-16 photo courtesy of Honeywell, Inc.

376, 382, 383 Figures 15-1 and 15-4 reproduced with the permission of Sharp Electronics Corporation.

379, 391, 396 Figures 15-2, 15-7, and 15-10 from *Commodore 64 User's Guide*, courtesy of Commodore Business Machines, Inc.

381, 384, 398 Figures 15-3, 15-5, and 15-11 taken from *Robert Bentley's Volkswagen Rabbit/Jetta Diesel Service Manual: 1977–1982.*

414 Figure 16-1 courtesy of Apple Computer, Inc.

Credits continue on page 545.

Printed in the U.S.A.

Library of Congress Catalog Card Number: 85-80764

ISBN: 0-395-35503-6

BCDEFGHIJ-H-8987

CONTENTS

TO THE STUDENT

Good technical writing skills can be a key to your success in business, research, or industry. The more clearly and articulately you communicate your ideas, the more likely they will be accepted and acted upon.

Technical Writing is designed to reflect the workplace, creating real writing situations that resemble the ones you may soon face. Throughout the book, you will be asked to use information from your major field of study as the foundation for writing assignments; in this way, you will gain practical experience in communicating the technology of most interest and concern to you. You will learn to accommodate that information to a variety of audiences and purposes, thereby learning how to meet the requirements of any writing situation. You will also learn about the various contingencies that any writer on the job must consider—deadlines, review cycles, budget constraints, and format specifications.

The broad objective of this book is the development of technical writing skills. To that end, topics are arranged incrementally so that you move gradually from basic skills to more sophisticated analyses and presentations. Part I, "The Writing Process," guides you through the initial evaluation of the writing situation to the research, planning, writing, and revising stages. Part II, "Techniques," presents particular writing strategies that will enable you to organize materials logically, explain key concepts in your field, describe processes and mechanisms, and argue your case convincingly; it also instructs you on the best use of graphic aids. Part III, "Applications," samples important writing formats such as letters and memos, proposals, reports, and manuals. It also covers the topics of product promotion, oral presentation, and job seeking. Part III will give you the opportunity to apply all of the processes and strategies you have studied. Finally, the appendix sections present a formal report example and a succinct handbook and style guide.

The many examples throughout the book are drawn from such diverse disciplines as mechanical engineering, computer science, veterinary medicine, oceanography, petroleum engineering, scuba diving, and paleontology. Both professional and student samples appear and range from very simple expositions to quite elegant arguments.

Exercises and writing assignments will help you practice what you have learned. A set of editing criteria appears at the end of each part. By consistently applying these handy check lists, you will be able to gauge your own progress as you move through the book and gain confidence in your writing ability. Your instructor may use these same editing criteria to evaluate your assignments.

Perhaps most importantly, this book demonstrates pragmatic concerns. If you are like most people working in scientific and engineering disciplines, you want to know two things about any writing situation: what needs to be done and the most efficient way to do it. Therefore, rhetorical theories appear only as they are immediately applicable to a specific writing strategy or situation.

Ultimately, I hope that *Technical Writing* will help you develop good technical writing skills through advice that is realistic, comprehensive, informative, and enjoyable.

F.B.E.

ACKNOWLEDGMENTS

I would like to thank my many reviewers for their contributions:

W. Steve Anderson, University of Arkansas at Little Rock
Roger Bacon, Northern Arizona University
Penny Bladon, Florence Darlington Technical College
Rebecca Burnett Carosso, University of Lowell
Joseph F. Ceccio, The University of Akron
Michael Connaughton, St. Cloud State University
Rick A. Eden, The RAND Corporation
Roy Flannagan, Ohio University
Robert Gieselman, University of Illinois
John S. Harris, Brigham Young University
Paul Killorin, Portland Community College
Lorita Langdon, Columbus Technical Institute
Sherry Burgus Little, San Diego State University
Victoria M. Mikelonis, University of Minnesota
Carolyn R. Miller, North Carolina State University
Russell Rutter, Illinois State University
Gary R. Stephens, New York Institute of Technology
William Van Pelt, University of Wisconsin at Milwaukee
Thomas L. Warren, Oklahoma State University
Robert Zwart, Rochester Community College

The Houghton Mifflin staff guided me through the exhausting, yet exhilarating process of revising draft after draft after draft.

At Honeywell Information Systems, Robert C. Hesser, Vice President of Marketing for the Large Computer Products Division, supported my efforts through his pride in me and through his gracious permission for me to cite several exemplary Honeywell documents.

Most of all, I must thank John F. Emerson, who put up with the three years of late-night and weekend work that this book required of me. His sincere and loving support of me and his deep belief in this book were a constant source of comfort.

F.B.E.

PART I

The Writing Process

CHAPTER 1

Writing about Technical Subjects

CHAPTER PLAN

The technological and scientific advances of recent decades have outstripped those of any other period in history. Vital to that progress is the efficient communication of technical information via reports, manuals, articles, proposals, and plans. By presenting their work in this manner, scientists and engineers share their findings with their peers. Often scientists and engineers also share technical information with people who have not had specialized training: doctors write articles about preventive medicine for general-interest magazines, data processing professionals recommend the acquisition of computer hardware to financial officers, construction firms submit bids on street repairs to city council members, and so on.

To equip you to participate effectively in this diverse communication environment, this chapter will

- define technical writing.
- introduce the technical writing process.

A DEFINITION OF TECHNICAL WRITING

Defined simply, technical writing is written communication that deals with engineering and scientific topics. The intent of technical writing is *to inform*—to communicate the writer's skills, observations, experience, theories, or knowledge to an audience. Technical writing covers a wide range of areas, including space exploration, the reclamation of waste water, genetic engineering, agricultural improvements, and disease control. Technical writing also covers more commonplace topics: housing construction, auto maintenance, medical techniques, machine design, and nutrition.

Through technical writing, engineers and scientists play important roles both in everyday experiences and in the continuation of human progress. Despite the importance of technical writing, however, few of those who are required to write technical documents have had formal training in technical communication. In fact, if you had asked today's technical professionals, when they were in school, how important they thought writing courses would be to their careers, many of them would have said, "Not very."

If you ask them *now* how important technical writing is to their careers, most will tell you that they spend at least 30% of their time on the job writing. Through their experiences in corporations, laboratories, government agencies, research institutes, and small firms, they have learned that it is not always the people with the best ideas who succeed, but the people who best communicate their ideas. They wish they knew efficient ways to organize and present their work—techniques that would improve the overall quality of their communication.

This textbook is designed to make you an effective communicator of technical information. In this introductory chapter, we look at the distinguishing characteristics of technical writing (the factors that make technical writing a reliable source of information). We also trace the basic steps in the technical writing process, by which the writer moves from the idea stage to the final document. These two topics will provide you with a solid foundation on which to build your technical writing skills.

DISTINGUISHING CHARACTERISTICS OF TECHNICAL WRITING

In addition to the specialized subject matter of the scientific and engineering worlds, five other qualities are often characteristic of technical writing: measurability, liability, objectivity, precision, and compliance. Although we will discuss these characteristics separately, in practice they work together and often are inseparable.

MEASURABILITY

As the medium of communication in such highly quantifiable disciplines as physics, biology, computer science, mechanical and electrical engineering, architecture, and chemistry, technical writing itself is frequently measurable. That is, the accuracy and effectiveness of a document can be evaluated. Many technical documents describe specific actions and ideas that *can be tested*. For instance, researchers describe their experiments with great care so that colleagues can achieve the same results (and hence verify them). Design engineers painstakingly explain the specifications for a new machine so that others can build a prototype. Theorists outline their observations and hypotheses to support or refute current thinking. In other words, the technical information that is communicated enables readers to achieve the desired results.

To further illustrate this notion of measurability, imagine that you have written a set of instructions for building a small addition to a home, perhaps a lean-to greenhouse. If the reader who follows your instructions exactly builds a satisfactory greenhouse, your technical writing was successful. And to a certain extent, we can quantify this success: the degree of conformity between the greenhouse you had in mind and the one your reader actually built determines the degree of your success. On the other hand, if the reader who diligently follows your instructions is unable to construct the greenhouse, your writing has failed. Here the variance between the greenhouse you envisioned and the unsatisfactory result determines the degree of failure.

LIABILITY

In our greenhouse example, readers expect results: they want to build a greenhouse. All they have to go on is the information you provide; they depend on you for accuracy and completeness. The writer of a technical document has an inherent obligation, or liability, to deliver the expected results. With the exception of legal writing, no other form of writing entails this degree of responsibility to fulfill the needs of its readers.

Consider the far lesser degree of liability found in fiction writing. Here the author may want readers to pity a particular character, and the readers themselves may expect to feel pity. If the author is unsuccessful in eliciting this response, the writing fails, but the damage is limited to mere disappointment.

When technical writing fails to deliver the expected results, however, it *can* inflict damage. If your greenhouse instructions do not stipulate ample drainage paths between rows of plants, excess humidity may harm the plants, mildew may form, water may seep into the house, and the builders who relied on you may be faced with costly repairs. In this case, unsuccessful writing could cause serious problems for both the readers and the writer (who could be held legally liable for the oversight).

An actual example is more dramatic. A writer working for a major supplier of natural gas developed a manual on the computerized operation of a compressor station. When an operator used the manual to control the station from a computer terminal at the home office, he followed the instruction to open the suction valve of the engine, whereupon natural gas rushed into the chamber. Unfortunately, the next step—to open the discharge value within 60 seconds—was omitted from the instructions. The effect of this oversight was a rapid increase in pressure within the engine chamber. Luckily, the chamber was equipped with pressure-detection devices that automatically triggered a "blowout" valve (one that vents the natural gas into the atmosphere above the station) when the pressure reached a dangerous level. Had the equipment not been fitted with such a device, the entire compressor station would have blown sky high. People could have been injured or killed, and many thousands of dollars' worth of damage would have been done. Now that's liability!

OBJECTIVITY

In addition to the traits of measurability and liability, technical writing exhibits objectivity. Whenever possible, technical writing is based on facts, reports, and inferences. A *fact* is verifiable information. For instance, a ton is 2000 pounds, water is two parts hydrogen and one part oxygen, and a right angle is 90°. *Reports* are legitimately witnessed actions or circumstances: Alexander Graham Bell invented the telephone; the Yankees won the World Series more than any other team in the 1950s; five tornadoes were sighted during yesterday's storm, but only one touched down. *Inferences* are conclusions drawn logically from facts and reports. An inference can best be demonstrated by listing the facts and reports that led us to draw it.

Fact: The driver of the red car had a blood/alcohol level that exceeded the legal limit.

Report: The driver of the red car did not yield the right of way to the blue car.

Inference: The elevated blood/alcohol level may have impaired the ability of the driver of the red car to drive safely.

In other words, the presentation of technical information resembles the scientific method by which the data are derived. Technical writing should be as free as possible of judgments and opinions. When information is presented objectively, readers can evaluate it fairly and can often agree on the results. (Agreement is another way of gauging the success of technical writing.)

The agreement of readers is essential to many types of technical documents, most notably the proposal. For instance, the Land Use Plan and Environmental Impact Statement are two documents that the Bureau of Land Management prepares to propose land management policies to Congress. Using these documents as the foundation on which their discussions are based, members of Congress generally hold public hearings in the local regions affected by the plans. The proposals would surely be rejected if they failed to reflect all the issues involved (such as the views of environmentalists, industry leaders, and those who promote tourism).

Before proceeding with our definition, let's look at a brief example of technical writing. The following excerpt is taken from a successful proposal in which an engineering firm seeks the opportunity to design an extensive sewage system for Pirtleville, Arizona. Here the writer provides important background information on which the firm bases its action plan.

Background Statement

The City of Douglas water system supplies 97 percent of Pirtleville's residents. Many residents report low water pressure, and some report problems with drinking water quality. Our study indicates that these conditions do, in fact, exist and that a variety of factors contribute to these effects.[2] The greatest single cause is substandard water supply lines, which are tapped into City of Douglas water mains and apparently are too small (3- to 4-inch diameter) to ensure adequate pressure. However, build-up of chemical deposits from the high mineral content of water in this area also reduces flow through the lines. Other factors include damage by road graders to shallow water lines, corrosion of the pipe, and seam separation. Individually or collectively, these conditions when combined with sewage-saturated soil, can contaminate the water lines and mains with hazardous bacteria.*

1. Objective presentation of facts and reports
2. Liability: will outline scope of, and reasons for, current conditions
3. Measurable observations listed (reader could replicate findings)

We all recognize the *specialized* nature of the information presented in this technical writing sample. We can also grant that the information presented is *measurable*. It would be possible for readers to replicate these findings—to determine the size of the pipes, analyze mineral deposits, or observe cracks caused by road graders—and ultimately agree with the writer's statements.

The writer of the proposal is responsible for understanding both the scope and the causes of the problems addressed in the proposal. The cata-

*Reprinted with permission from Morris, Clester, Abegglen & Associates, Inc.

logue of reasons why current conditions exist reflects the engineering firm's recognition of this *liability*. Finally, we can consider this analysis *objective* even though the engineering firm is bidding on the project, because the writer restricts the information she or he presents to facts, reports, and inferences. No highly emotive or subjective language is used; the style is straightforward and formal.

PRECISION

Communication of the skills and knowledge associated with engineering and scientific disciplines must be precise. Writers of technical documents continually strive to be as accurate as possible in vocabulary and emphasis. Such precise writing demands a great deal of the writer, not the least of which is patience. Also required are careful planning, in-depth analysis, and an overall sensitivity to the needs of readers.

Let's look at some factors that contribute to the precision of our excerpt from the engineering proposal. For instance, accurate *vocabulary* is used to describe the situation: "corrosion," "contaminate," "sewage-saturated soil," "hazardous bacteria." Each of these words or phrases is highly descriptive and hence preferable to longer, less concise discussion.

Emphasis is appropriately placed on the reasons underlying residents' complaints rather than on the complaints themselves, because the purpose of building the planned system is to eliminate those causes. This emphasis is reflected in the order in which information is presented (complaint first, reasons second) and in the amount of time and detail devoted to the reasons. Precise sentence structure contributes to emphasis as well. For instance, the two primary complaints are almost equally important, so the writer treats them in a single, compound sentence. But note how the water-*pressure* problem of *many* residents takes precedence, within that sentence, over the water-*quality* problem of *some* residents. Likewise, sentence structure helps us determine the importance of the various causes. The most important causes of the problems, substandard lines and the build-up of chemical deposits, are both accorded a whole sentence. Lesser factors that contribute to the problems are consolidated in a list.

The excerpt is analytical in that the problems are not merely presented; observation of them is attributed to residents of Pirtleville and corroborated by the engineering firm's study. If the writer had stated simply, "Water pressure is low and quality is poor," the reader might have thought, "Who says?" Citing sources contributes to the engineering firm's credibility—an important plus when one is bidding for a job.

Finally, the level of vocabulary, though precise, is not above the heads of people without hydrology degrees (the council members of Cochise County).

In our discussion of the writing process in this and the next few chapters, we will concentrate on achieving precision. We will explain how to assess the needs of an audience and how to balance these demands with the purpose of the document, the nature of the subject matter, and the role of the writer. We will also look at several methods of planning, research, and analysis in the prewriting stages of developing a document. We will examine strategies of effective organization and exposition during the initial writing phases. And we will investigate various tactics for tightening vocabulary and sentence structure during revision.

COMPLIANCE

Compliance with specifications and standards is a characteristic of technical writing that helps round out our understanding of this discipline. Many companies, government agencies, universities, and research institutes require technical documents to be prepared according to established guidelines. The military, for instance, has an elaborate system of "mil specs," or military specifications, appropriate for a wide range of writing purposes. Manufacturers have specific guidelines for documenting engineering designs, forecasting inventory requirements, and reporting sales figures. Government agencies empowered to grant research funds require applicants to meet particularly explicit specifications.

Although it is impossible within the scope of this textbook to cover the many guidelines in existence, it is important for each writer to learn both the general requirements that apply in the industry he or she works in and the particular demands of given tasks. It is up to the engineer to learn the format and standards customarily used in giving product specifications, for instance, and for the biologist to learn the publications standards of key scholarly journals in the field.

This textbook will explore basic strategies for writing several different kinds of documents, including the

- abstract
- brochure
- feasibility report
- letter
- manual
- memorandum
- progress report
- proposal
- research report
- trip report
- visual presentation

Established formats save both readers and writers time and effort. Specifications act as a mold to shape the contents into familiar form. Standardized presentations enable readers to concentrate on the content of a document and relieve writers of the task of devising new methods of presentation. Established formats do not eliminate creativity, however; even the most exacting formats leave ample room to present information in a new light.

The engineering proposal we quoted in this chapter is one of many that Cochise County received. Members of the council had created a Request for Proposal (RFP) that outlined the area requiring service, what they believed to be the problem, the solution they envisioned, and the specifications each applicant was to follow in submitting a proposal. These specifications included sections on the bidder's history and qualifications, technical approach, management and staffing, and experience with similar projects. Those proposals that did not comply with the specifications were returned with a request for revision. The council members simply would not read proposals that failed to follow the guidelines they stipulated.

Now that we have outlined the primary characteristics of technical writing, a summary of our expanded definition is in order. *Technical writing* is that form of written communication that deals with specialized subjects. Its content and effectiveness often are measurable, the writer has a responsibility to fulfill readers' expectations, the tone and style are generally objective, the language is precise, and technical documents often conform to an established set of specifications.

THE TECHNICAL WRITING PROCESS

Although each technical writing situation is unique, the overall writing process is similar to that used in many other writing tasks. It can be divided into four distinct phases:

- situational analysis
- gathering information
- planning and arranging
- writing and revising

The remainder of this section of the textbook is dedicated to exploring the technical writing process and to each of the steps you must take within these phases of development. "Conducting a Situational Analysis," the next chapter, concentrates on the many factors affecting the writing situation, such as audience characteristics, needs, and purpose; writer roles and purposes; and constraints on time, budget, format, and resources. Chapter 3, "Gathering Information," focuses on a variety of research sources and techniques that ensure the completeness and accuracy of technical information. Chapter 4, "Planning and Arranging," discusses ways of focusing your topic, determining what key points to make, and ordering information to support your findings logically.

1. Situational analysis

a. Audience
 –Characteristics
 –Needs
 –Expectations
 –Abilities
b. Purpose
 –Writer's
 –Organization's
 –Audience's
c. Role
 –Individual
 –Team
d. Content
 –Scope
 –Complexity
e. Constraints
 –Format specifications
 –Schedule
 –Budget
 –Availability of information
 –Human resources
 –Production capabilities

2. Gathering information

a. Determining what data you need
b. Secondary research
 –Library materials
 –Government documents
c. Primary research
 –Surveys and interviews
 –Experimentation
d. Summarizing data

3. Planning and arranging

a. Focusing topic
b. Listing information
c. Arranging information
d. Developing outline

4. Writing and revising

a. Developing rough draft
b. Polishing rough draft
c. Review cycle
d. Incorporating changes
e. Casting in final form
f. Approval
g. Production
h. Distribution
i. Follow-up

Figure 1-1 Always be aware of these phases and supporting steps in the technical writing process.

Chapter 5, "Writing and Revising," guides you through the development of a rough draft. You will learn to polish your draft, via several important revision techniques, into an acceptable review document—one you can confidently submit to your peers for their comments. Next you will learn to cast your document in final form by incorporating the suggestions for revision that such a review yields. Only then should the document be submitted to your supervisor for approval, production, and distribution.

The chart in Figure 1-1 lists the four phases in the technical writing process. Each phase comprises several steps, some of which you may need to repeat before you can move on to the next phase. Whether you are an engineer or a scientist writing in your field or a professional technical writer, it is important to follow these steps. Observing them will serve you well whether you are writing as an individual or as a member of a team and whether or not your audience is highly trained in the subject area.

The length of the list of supporting steps dedicated to situational analysis demonstrates the importance of preparation in the technical writing process. The more carefully you lay the groundwork for a document, the less reworking you will have to do and the more effective your results will be.

The initial step in the information-gathering phase really continues the situational analysis: you must determine how much and what type of research is called for. Having done so, you can proceed to the lab or the library. As you collect data, take notes that summarize the facts and reports you encounter. Soon you will discern patterns of information that will help you move effectively into the next phase of the process, planning and arranging.

This third phase is the time to limit the topic to a manageable size and tailor it to meeting the requirements you established in phase one. You should identify the key issues and points you are likely to make, the most appropriate tone, and the type of document that will work best. You may, for instance, find that an oral presentation will work better than a written report, or that an informal memo is all that is required. Having made these decisions, you can develop a detailed outline of the document—one that, through organization and emphasis, best addresses the subject matter, purpose, and audience you have selected.

A solid outline puts you in an excellent position to begin writing your rough draft. Through a series of revisions you can polish this draft into one you can subject to review. A review document is circulated to peers (and sometimes to legal counsel) to ensure accuracy, consistency with organizational guidelines and policies, and quality of presentation. Most people who write technical documents channel their work through a review cycle before crafting the final document. Of course, it is up to the writer to determine which suggested changes to incorporate. Substantive changes should always be made; stylistic changes are generally implemented at the writer's discretion.

In the cases of proposals, reports, manuals, and books, the final document may go through a production process after having been approved. The writer may even work with the printer to ensure the quality of the resulting document.

CHAPTER SUMMARY

Technical writing plays an important role in the communication of technological and scientific discoveries. Despite the importance of good communication skills, scientists and engineers generally place too little emphasis on them as students, only to regret this oversight later.

Good technical writing requires extensive planning and analysis, careful research, and skillful drafting and revision. When these efforts succeed, the result is specialized information presented in a measurable, responsible, objective, precise, and useful manner—information that fulfills the readers' expectations.

EXERCISES

1. Locate and analyze three sources of information on any one of the following topics. Consider the degree to which the information is testable (measurable). Is the writing objective? If not, circle words or phrases that are subjective. Does the writer fulfill your expectations? If not, why not?

acid rain	protein synthesis
carbon dating	radiation treatment of cancer
compaction	recombinant DNA
data base management	strip mining
Ohm's law	volcanic cones

2. Contact three people who hold jobs in your field or area of interest. Ask them what kind of writing they do on the job. How much of their time is spent writing? How much formal training did they receive in writing while they were in school? In retrospect, would any of these people have preferred more training? Why or why not?

WRITING ASSIGNMENT

Imagine that the Product Safety Commission has selected you to evaluate a product of your own choosing (real or fictitious). Either accept or reject the product, and support your opinion with information relevant to the commission. Be sure your "document" lives up to the definition of technical writing introduced in this chapter.

CHAPTER

2

Conducting a Situational Analysis

CHAPTER PLAN

Rarely, if ever, do you write a technical document simply because you have an urge to do so. Rather, an elaborate set of circumstances surrounds each writing situation. You need to consider all the factors that make up the *situation,* or context, that surrounds your writing task and accommodate them in preparing your document. In fact, the more time, thought, and effort you invest in analyzing the demands of a writing project, the more likely you are to communicate successfully.

This chapter will

- acquaint you with the elements that affect technical writing situations.
- enable you to recognize how these elements influence your own writing projects.
- enable you to accommodate these elements in your own documents.

THE IMPORTANCE OF A SITUATIONAL ANALYSIS

According to the overview presented in Chapter 1, the first phase of the technical writing process is conducting a situational analysis—an investigation of all factors affecting the project. These elements include

- audience
- role
- purpose
- subject matter
- constraints

The *audience* consists of the reader(s) of your document; you must understand their characteristics, expectations, abilities, and needs. Your *role* is the perspective you take toward the information and the audience. Are you an instructor informing your readers or a student demonstrating your mastery of a subject to your professor? Perhaps you are a design engineer explaining the rationale behind your plans to a budget committee that will fund additional work on the project. The *purpose* of the document includes both your purpose as writer and that of your readers. Why are you presenting the information and why are they going to read it? The *subject matter* must be analyzed in terms of its scope and complexity. What are the most important aspects of the topic—both to you and to your audience? *Constraints* may be many and varied; they include schedule, budget, resources, and format. When you consider all these factors *before* you write, you prepare yourself to produce work that is clearly focused and even memorable!

Of course, we all have an intuitive sense of many of these factors. Suppose, for instance that your biology lab instructor announces at the end of your Wednesday session, "I want all lab reports by the end of the week." You know several important things immediately:

1. The person you must please with this report is your lab instructor.
2. The report's subject matter is limited to your activities since you submitted your last lab report.
3. The format should adhere to the specifications distributed by the instructor at the beginning of the semester.
4. Last, but certainly not least, you have a deadline—let's say 4:00 P.M. Friday.

The low groans you hear around the room indicate that each student in the class is conducting this same brief situational analysis (and no one is very pleased with the request). The explicit, easily recognizable factors you considered in this evaluation were audience, subject matter, format, and schedule. It is likely that all the students in your class considered these areas and agree with your findings.

However, other factors are at work that are specific to your needs or reflect your own individual understanding of the situation. These implicit factors might include pressure, conflict, and insecurity.

You exert *pressure* on yourself to make this report better than the others you have submitted in this course. You have a total of five lab reports to do this semester (this is your fourth), and 50% of your biology grade is based on them; the other 50% of the grade is based on classroom examinations. You are doing quite well in the classroom exams, but you have only a C average on the three lab reports you have developed thus far. You know you have to pull up your grade on this and the next lab report, or your overall grade will suffer.

Conflict arises because you know you have a math test tomorrow that requires you to study tonight. Hence you won't get to the lab report until Thursday night, which is really cutting it close if you want to do a good job.

You have a sense of *insecurity* about this lab course, because the three C's you have received are lower than you feel you deserve and lower than the grades you normally receive in other classes. You are convinced that the instructor has something against you and, no matter what you do, will view your reports negatively.

Does this situation sound familiar? If not in a lab course, you may encounter it in a graduate seminar. If not there, you may encounter it in your construction engineering firm as you write a bid proposal, or you may encounter it as you prepare a paper for your research institute's weekend symposium on the therapeutic effects of low-level electromagnetic fields. Most people *are* unsure of themselves as writers. Most people are intimidated by the thought of writing and procrastinate in the most imaginative ways to avoid the activity.

But it doesn't have to be this way. The key to a more stable, secure writing environment—the key to feeling better about yourself as a writer—is understanding all the variables that affect the message you must communicate. Preparation can help you avoid the ill effects of pressure, conflict, and insecurity. *And preparation is nothing more than a solid situational analysis.*

Let's explore this brief, hypothetical situation further to learn more about the value of preparation. The analysis you conducted when you heard about the deadline for the lab report is a step in the right direction, but it leaves you far from your goal of writing a superlative lab report. To accomplish that, you need a more detailed analysis and a better understanding of the factors at work. For instance, it is insufficient to say to yourself, "My audience is my lab instructor." You have to understand who that person is and what he or she requires of you.

If you were to analyze your audience very carefully, you might discover some revealing character traits to help you present your report more effectively. For instance, you might notice that your instructor uses a great many graphic aids when presenting information in class and that this person frequently prods students who ask questions in roundabout ways to "Get to the point; I don't have all day!" You might also notice that this instructor usually presents the main point of a lab exercise at the beginning of class and then backs it up with supporting detail as the session progresses. Rarely do illustrative examples or anecdotes intervene.

Such a profile of your audience tells you that your lab instructor is interested in getting to the heart of the matter right away and knows the value of effective visual aids. Hence, to create a report that will appeal to this reader, you should

- keep your comments succinct and direct.
- use graphic aids to support your narrative.
- summarize your conclusions at the beginning of the report.
- support your findings with details as the report develops.
- order your points according to the importance you believe the lab instructor will place on them.

These accommodations that you make to meet the needs of your audience will improve your lab report considerably. However, there are other steps you should take as well. For example, you told yourself that the report should cover everything you have done in the lab since the last report. To be effective, you must be more specific. Generate a set of questions you believe a comprehensive and noteworthy lab report should answer. Here is an example:

1. Precisely which experiments did you conduct?
2. What was the objective of each?
3. What materials, methods, and equipment did you use?
4. What were the date and time of each experiment? How much time elapsed during the running of each experiment?
5. Did anyone assist you? If so, what were your respective roles?
6. What results did you observe? Did you observe any unexpected results? Were there any results you expected but did not achieve?
7. What *logical* conclusions can you draw from your observations?
8. Are there any unanswered questions or unresolved problems regarding your experiments?

9. Can you make any recommendations for action or further study?
10. In what way are the findings you report in this document related to activities conducted in previous report periods?

This example gives you an idea of how you might handle explicit factors involved in the writing project, but what about the implicit factors? First let's look at the pressure you exert on yourself to write your best report. The objective is admirable, but you are doing little more than making yourself nervous. What you need instead is confidence—the confidence born of knowing you have prepared well and can write well. This chapter will teach you how to prepare for your writing projects; this textbook will help you develop your writing skills to the point where you are confident of your ability.

One suggestion for improving your confidence is that you objectively review the comments the instructor provided on your three previous reports (on which you got C's). Identify the errors or omissions this person pinpoints, especially any that recur. Next consider what specific things you can do to eliminate these problems. Criticism hurts, but an objective analysis and plan for corrective action will help you eliminate the shortcomings your instructor cited. If you are leaving out required information, be certain to accommodate such data in the next report. If your instructor claims your conclusions are unsupported, work inductively rather than deductively to provide as many proofs for your assertions as possible.

A second implicit factor affecting the writing situation is your schedule conflict. Certainly you must study for your math test, but that doesn't mean you must write all of your lab report—start to finish—in a single block of time. The technical writing process can be divided into several easily managed phases of activity. You can start the planning stage right after class. You have a 45-minute break before your statistics class, and your math textbook is back at your apartment. Don't waste time having a cup of coffee at the Student Union; go to the library and jot down some notes about your audience, your lab activities, the important points to make, the best order of discussion, and so on. Having done that, you will begin thinking about the lab report (perhaps subconsciously) during your next class and even while you study for your math test. You might wake up tomorrow morning with two or three ideas for the report, and at lunch you'll find yourself discussing them with your friends.

Essentially, the longer you put off getting started on the report, the more difficult it will be to start. But having done some planning *before* you write, you will sit down Thursday night with all the materials you need and be in a frame of mind to get the work done. Armed with an analysis of the shortcomings of your previous reports and having got a head start on the planning and writing activities, you should be feeling far more secure. The report will be easier to write and will be better than your other documents for this lab.

With this extended example of a real writing situation in mind, let us now focus in detail on the factors affecting typical writing situations.

MEETING THE NEEDS OF YOUR AUDIENCE

Technical documents often address specific audiences. The more closely your writing corresponds to the interests, abilities, and needs of that audience, the more successful your message will be. Therefore, before you begin writing, you should discover as much about your intended audience as possible. The answers to some very basic questions will equip you to interest your reader in what you have to say, establish an appropriate tone in your document, select pertinent examples, choose a manageable level of detail, order information logically, and move your audience toward agreement with your statements (the true measure of a technical document's success).

The first step in a situational analysis is to write down answers to the following questions about your audience.

1. *Who will be reading the document?* For instance, will they be business executives, nurses, your boss, co-workers, a budget committee of laypersons, hobbyists, students, children, homemakers, your subordinates, or maintenance personnel?
2. *What is the audience's level of expertise?* Are your readers as experienced as you, or are they unskilled trainees? Perhaps they are highly educated theorists or accomplished technicians.
3. *Why do they need the information in your document?* Is your document simply providing information for their consideration, or will they be acting on the points you raise? They might be allocating funds, building a machine, rejecting a proposal, or implementing a program, for instance.
4. *Which points are most important to them, and why?* Are your readers most interested in the cost of implementing your recommendations? Are your methods of greatest concern to them? Do they need to learn how to operate the machinery you are describing, or are they going to be fixing it instead?

These sample questions indicate the kind of probing you must do each time you begin to plan a writing project. Other audience characteristics you should consider include

age	hobbies	opportunities
concerns	income	political affiliation
conflicts	interests	responsibilities
education	location	social status
gender	occupation	

The information you derive from this analysis helps you tailor your documents—right down to such elements as vocabulary, sentence structure, graphic aids, and format. For a good example of the impact of audience characteristics on the presentation of information, let's examine two descriptions of a type of electrical circuit called "Current Mode Logic" (CML). The first is addressed to a top-level executive who would allocate funds to pay for the computer that uses this circuitry. The second description is earmarked for the data processing professional responsible for evaluating the efficiency of the computer's design.

Document 1

[1] A major efficiency booster is the circuitry used by DPS 88. Requiring 40% less power than conventional types, DPS 88 circuitry also is five times faster and requires considerably less space. It can improve processing speed to better serve the rapidly changing demands of the large organization. It also helps minimize heat build-up in the system cabinetry. Less heat translates into better circuit preservation and greater reliability.[2]	1. Benefits and corroborating details 2. An answer to a concern of the reader
[3] System efficiency is further enhanced by the innovative packaging used for DPS 88 circuits and logic. Approximately 20% of the circuitry in these "micropacks" is devoted to ensuring integrity and diagnosing malfunctions. In the event that a micropack should falter, it can be snapped out and replaced easily. The system remains available to those who need it.[4]	3. A benefit and corroborating details 4. An answer to a concern of the reader
[5] Add to this safeguard, outstanding compactness—the micropack is built on a tile no larger than the palm of a hand, yet it holds as much circuitry as two conventional circuit boards. Within these compact components, signals travel shorter distances for additional improvements in processing speed.*	5. A benefit and corroborating details

*DPS 88 brochure (Waltham, MA: Honeywell Information Systems, Inc., 1984).

Document 2

The DPS 88 system uses Current Mode Logic (CML)
circuitry in the Central Processing Unit (CPU), Central
Interface Unit (CIU), Input/Output Processor (IOP), and
the System Support Unit (SSU).[1] CML is a circuit technol- 2
ogy utilizing current-mode steering techniques to achieve
switching speeds faster than Transistor-to-Transistor
Logic (TTL) circuit technology. Current mode is defined
as the technique of steering currents without altering their
values, while voltage mode actually turns the current on
and off, creating delays. CML circuits are characterized by
high switching speeds (less than one nanosecond) and low
power requirements, which result in less heat and higher
reliability and density of logic gates per chip than TTL.[3]

A DPS 88 system employs high-technology integrated 4
micropackaging for the CML circuitry. A micropackage is made up of densely packaged CML circuit chips mounted on multiple film layers of interconnecting circuits fabricated on a 3-inch-square ceramic substrate only .08-inch thick. A single micropackage will average about 72 chips and 5000 logic gates—approximately the same amount of logic found on two conventional 12-inch-square boards using TTL. Eight micropackages are mounted on a logic board and four pairs of such boards can be mounted back-to-back to form a CML module.*

1. Product description
2. Definition
3. Benefits and corroborating details
4. Product description

Note the relatively nontechnical vocabulary and simple sentence structure of the first example; the writer discusses only one idea per sentence so as not to confuse the nontechnical reader. (Although top executives are well-educated professionals, they usually are not "computer-literate.") The writer is careful to concentrate on the benefits of the design rather than on the design itself. It is as though the reader is asking, "What's in it for me?" and the writer is answering this query.

Both excerpts contain the same information. Yet consider how different the version addressed to the data processing professional is. This second example concentrates on the product design—the *how* behind the results. In this case, the writer seems to be responding to the demand "Tell me what the DPS 88 does and how it does it."

The writer of these two documents decided what kind of presentation was appropriate for each of these two audiences and then developed descriptions to match their disparate needs. Even so, the writer wanted to convince

*DPS 88 product brief (Waltham, MA: Honeywell Information Systems, Inc., 1984).

both audiences of the efficiency of the design and the overall integrity of the computer. To accomplish this objective, he or she might have developed the following answers to our set of basic questions.

Document 1	**Document 2**
1. *Who will be reading the document?*	
Top-level business executives of very large organizations.	Data processing (DP) professionals of large and sophisticated DP departments.
2. *What is the audience's level of expertise?*	
This person may be exposed to personal computing technology but has little understanding of mainframe engineering. However, this reader does recognize that the organization depends on a well-functioning system.	Highly trained expert who must keep informed to do a good job. Attends symposia and conventions, subscribes to major trade publications, and understands the computer market.
3. *Why do they need the information in the document?*	
These readers are responsible for approving the purchase of a multimillion-dollar computer system to meet the current and forecasted needs of the organization. They will use this document as a foundation for discussion with their data processing advisors.	These readers are responsible for recommending purchase of a system that will serve the extensive data processing needs of the organization. They will use this document as the basis for requesting additional information from a sales or technical representative from Honeywell.
4. *What points are most important to them, and why?*	
Because these readers are responsible for the financial health and stability of their organization, they are concerned with the return on investment these machines afford them in terms of increased productivity, accessibility of information, and dependability. They can't allow themselves to err in this area.	Because these readers are responsible for supporting an entire organization's processing needs, they are most interested in ease of maintenance, reliability, power, and engineering design. They need to believe in the integrity of the system.

The writer might supplement these answers to the four basic questions about his or her audience with the following profiles:

Document 1

Age: 50 and older
Concerns: Keeping business competitive, getting best value for the money
Conflicts: Pressure from peers to contract services from an industry giant
Demands: Little or no time to read product documentation
Education: College graduate with degree in business, management, marketing, finance, or law
Gender: Vast majority are male
Hobbies: Tennis, golf, boating, travel
Income: $100,000+ per year
Interests: Community projects, fund raising, art patronage
Location: Large industrial sites located throughout the United States and Canada
Occupation: Vice president, general manager level or above
Opportunities: To increase business within the marketplace
Politics: Conservative, likely to be Republican
Responsibilities: Control and plan the direction of the business
Social status: Wealthy, highly regarded member of business and local community

Document 2

Age: 40 and older
Concerns: Providing reliable computing to organization's users
Conflicts: Convincing senior management that more computer power is required.
Demands: Little or no time to read product documentation
Education: College graduate, possibly graduate work, in electrical engineering or computer science
Gender: Vast majority are male
Hobbies: Personal computer programming, music, reading
Income: $60,000 per year
Interests: Science, math, computers, volunteer work (such as Scouts)
Location: Same
Occupation: Director of management information services, manager of data processing
Opportunities: To increase DP domain within the organization
Politics: Conservative or liberal, likely to be Independent
Responsibilities: Keep the computer system up and running at all times
Social status: Middle-class, respected member of local community

Let's return to the product descriptions we quoted to examine how the writer used this information about the two audiences. Aside from the differences in sentence structure that we have mentioned, other key differences are apparent. First consider vocabulary. Through the use of nontechnical language in the first sample, the writer accommodates the executive's level of responsibility and his or her insistence on a dependable system: "ensuring integrity," "safeguard," "remains available," and "greater reliability." Readers feel they are getting the best value for their money when they see such design descriptions as "innovative packaging," "better circuit preservation," and "improvements in processing speed." In none of these cases, however, are the underlying details outlined. Rather, the writer takes a benefits-oriented approach to the product description.

Now look at the vocabulary *and level of detail* selected for the data processing professional. This audience is also told about "improvements in processing speeds," but here the description is accorded four complex sentences and couched in technical language: "steering techniques," "switching speeds," "altering values," "voltage mode," and "nanosecond." The writer assumes the reader understands these terms, and she or he relies on that understanding to make subsequent points. Similarly, the technical reader is told about "innovative packaging," which gets a paragraph of its own. The data processing professional learns the exact size of the package, how it is constructed, the number of circuits it carries, and its capacity compared to traditional designs. (The executive learned only that the package is on a tile that can fit in the palm of a hand and has as much capacity as two conventional circuit boards.) And rather than explicitly stating a set of benefits in the second version, the writer allows the data processing professional to infer these benefits from the details provided.

Audience characteristics also account for other differences between these presentations. Consider the structure of each excerpt. The executive's version is an integrated discussion of speed, reliability of design, and efficiency. These points are driven home through repetition and the interweaving of ideas from paragraph to paragraph. The first paragraph begins with "a major efficiency booster" and ends with "better circuit preservation and greater reliability." The second paragraph picks up this thread, beginning with "system efficiency is further enhanced" and ending with "the system remains available." The third paragraph continues the story with a clear transition ("Add to this safeguard, outstanding compactness") and ends with "additional improvements in processing speed." This very tight arrangement of information leads the reader to accept the writer's main point: that the DPS 88 can "better serve the rapidly changing demands of the large organization."

The data processing professional's version has a more compartmentalized structure than that used for the executive. The first paragraph describes current mode logic and its effects on the performance of the machine. The second paragraph concentrates on the micropackage and the eight-part module into which it fits. The data in these two paragraphs are related, but far fewer interconnections are made than in the first excerpt.

ESTABLISHING AN APPROPRIATE ROLE

Although understanding the audience is crucial to successful writing, it is not sufficient. Another factor you must consider is the writer's role. For our purposes, let us define the term *role* as those aspects of a writer's personality and experience that are brought to bear on a given document; that is, your role as a writer is a subset of your total self. As a writer of technical documents, you may play the role of scientist, analyst, researcher, critic, engineer, designer, or advocate.

Effective writing requires you to select a specific, appropriate role and to apply it consistently throughout a document. The appropriateness of the role can be judged only in terms of the audience's expectations of you. If it requires an analyst, it should get one; if it requires a friend, that role is more fitting. A writer's role affects his or her perspective on the subject matter of the document. Like the audience, this role influences vocabulary, tone, level of detail, choice of graphics, order of information, and points made. For this reason, the audience and the writer's role must be compatible.

When role and audience do not match, the result can be a lack of understanding on the reader's part. Recently a pamphlet was published by the Secretary of State's office in Arizona to inform voters about propositions to be voted on in the upcoming election. The issues included the right of teachers (state employees) to strike, bond offerings for city park maintenance, and the regulation of prices charged by ambulance services within the county. By law, the pamphlet was required to present the following items:

text of each proposition

analysis of each issue by the Legislative Council

pro and con arguments (properly filed with the Secretary of State's office by concerned individuals or organizations)

The audience for this document consisted of voters of all ages from various socioeconomic backgrounds. Their occupations and interests were varied, but they were all citizens of Arizona and had a right to vote. The members of this audience expected the pamphlet to give them enough information so that they could draw their own conclusions on election day. Although readers knew that the pro and con arguments from specific individuals and groups would represent special interests, they believed the information provided by the Legislative Council would be unbiased and analytical. The council members, then, were expected to play the role of "objective lawmakers."

Unfortunately, some of the remarks of the Legislative Council were inconsistent with the audience's expectations. Their comments included the following emotionally charged statement (italics were not in the original text): "If this proposition were to be passed, it would prevent reputable ambulance services from *being run out of business by cutthroat competitors* offering poorer services."

The council members could have fulfilled their role as objective lawmakers by making the point in less biased language: "The passage of this proposition would help protect reputable ambulance services against unfair price competition from agencies offering poorer services."

It is very likely that the judgmental wording of this "analysis" undermined the credibility of the legislators with some voters. Many became confused over this proposition; others were manipulated by the high-powered rhetoric. Of course they wanted to squelch "cutthroat competitors." (Wouldn't you?) But did they want price regulation imposed on ambulance services, the real issue at hand?

THE ROLE OF THE INDIVIDUAL

Though your role will vary from document to document, it is important to select roles consistent with the requirement that technical writing be objective. It is usually important that the role you choose remain the same throughout the project. The following are some of the roles that are appropriate in various technical writing situations.

advocate
analyst
buyer
chairperson
clinician
competitor
consultant
consumer
coordinator
critic
delegate
designer
developer
diagnostician
doctor
employee
engineer
experimenter
expert
laborer
manager
manufacturer
marketing rep.
member
observer
operator
opposer
planner
programmer
project leader
proposer
recruiter
repairperson
reporter
researcher
reviewer
scientist
skeptic
student
supervisor
taxpayer
teacher
technician
trainer
voter

Because roles can best be understood in terms of the expectations we have of those who fill them, let us consider some of the characteristics we associate with ten of these roles.

- *Advocate:* Exhibits a commitment to his or her cause that is both apparent and credible; presents a set of "bullet-proof" arguments in support of the stand taken.

- *Consultant:* Has readers' best interests at heart, considers a range of variables as objectively as possible, and recommends action on the basis of comparative options and his or her experience.
- *Designer:* Is well-acquainted with features and design elements of subject, understands the rationale behind each element, and takes pride in his or her work.
- *Experimenter:* Maintains a scientific attitude toward each experiment; concentrates on objectives, methods, and results to substantiate conclusions empirically.
- *Planner:* Anticipates future directions of growth and development in order to recommend investments, opportunities, and activities.
- *Project leader:* Remains "on top of" all areas within his or her jurisdiction to ensure the success of the project; provides supervision, incentives, and feedback to other members of the team.
- *Reporter:* Presents timely summaries of observations and events; expected to be accurate and comprehensive.
- *Student:* Demonstrates the acquisition of knowledge and skills (and expects to be evaluated on his or her success), strives for improvements, thrives on positive reinforcement.
- *Teacher:* Provides information and skills that his or her audience is compelled to learn, evaluates the progress of the learner, encourages learner's growth and development.
- *Technician:* Concerned with the proper installation and functioning of equipment, familiar with design strengths and weaknesses, generally a good improvisor.

For each writing situation you encounter, select a role that fits the expectations you believe your audience will have of you. These expectations will emerge during your analysis of the audience. Look back for a moment at the audience analysis of the top-level executive who would read the product description given in the first excerpt in this chapter. On the basis of her or his interpretation of this audience (concerns: keeping the business competitive, getting the most for the money; responsibilities: plan and control the direction of the business), the writer assumes the role of a marketing representative. In executing this role, the writer successfully portrays the computer system as able to fulfill the executive's requirements.

On the other hand, consider the audience analysis of the data processing professional. This reader is concerned with providing reliable computing equipment to the organization and is responsible for keeping the system running at all times. The same writer selected a different role in approaching this second audience—that of the design engineer. By presenting the product description from this vantage point, the writer convincingly establishes the computer system as reliable because of its state-of-the-art technology.

Likewise, the information you decide to include, the language and tone you adopt, and the order of presentation you select are all affected by the role you assume in each writing situation. In most cases, your role should be consistent throughout your document. If you start out as a ski enthusiast who is describing the best runs in Aspen, Colorado, or Park City, Utah, don't suddenly switch to the role of meteorologist as you describe the amount of snowfall each run receives. Maintain the ski enthusiast's perspective.

There *are* times when more than one role is appropriate and can be implemented in conjunction with another or in different parts of a document. For instance, consider a petroleum engineer who has just returned from the tar sands of Canada. The majority of his or her trip report will be written from the perspective of an observer: descriptions of terrain, operating procedures, personnel qualifications, capacities, problems, and yields. However, if the writer is expected to make recommendations for improvements, he or she might legitimately switch to the role of planner or advocate.

Even in such cases, the role should be readily identifiable and consistent within that part of the document to which it is applied. Haphazard shifting of the writer's role confuses the audience and works against successful communication. To help yourself identify and stick to an appropriate role, write a one- or two-sentence description of it as part of your situational analysis.

THE ROLE OF THE TEAM MEMBER

In practice, many technical documents are written by a group of people rather than by an individual. Each person working on the project must understand and carry out his or her role in the production of the final document. One engineer may be responsible only for outlining the problems at a construction site; another may be charged with planning the proposed work schedule; still another may write the methods-and-materials section of the proposal; and the project leader may itemize the firm's qualifications and combine the individual parts into the complete proposal.

Group activity of this type requires coordination and cooperation. Each writer keeps the other team members apprised of his or her progress so that work can be synchronized. As new data emerge, the first writer to encounter them informs the other members of the team. And each writer must stay on schedule because others depend on each portion of the document being written at the right time.

Ideally, discussions at the outset of group writing activity should yield agreement among team members about the objectives of the project, the main points to be made, the characteristics and expectations of the audience, and the role each writer should assume. Remember that consistency is important: even if different writing roles are appropriate at different points in the project, everyone on the team should agree on these various perspectives, understand each writer's intent, and determine whether the several roles will prove complementary in the final product.

The two documents we have been studying in this chapter are part of an entire product announcement kit that includes the brochure (from which the first excerpt is taken), a product brief (from which the second sample is taken), a very detailed description of each component within the system, and specification sheets that explain how much electric power and space these machines require. The writer of the brochure assumes the role of marketing representative, the author of the product brief takes the design engineer's viewpoint, the writer who provides detailed descriptions acts as an electrical engineer, and the specification sheet was written from the perspective of a site planner.

These documents are distributed in an attractive cover that graphically complements the executive brochure. The product brief uses color and graphics that work well with the kit; the detailed description exhibits the same kind of paper and lettering as the brochure; and the specification sheet uses a large graphic consistent with the other pieces. Each person responsible for a part of the kit (three in all) worked together diligently and agreed on these details.

Even more important than the visual coordination of the announcement kit is consistency of content. From piece to piece, the various readers come away with a similar understanding (although tailored to their needs). The executive reads the brochure and gleans an appreciation of the efficiency of the computer system. The data processing professional reads the product brief and learns how this efficiency can be achieved. The operator reads the detailed description and understands the rationale for each design element. The technician reads the specification sheet to understand how his or her organization's physical facilities can best accommodate the new computer system.

IDENTIFYING PURPOSES AT WORK IN THE WRITING SITUATION

In Chapter 1 of this textbook we learned that the underlying purpose of technical writing is to communicate skills, information, and experience. The need to communicate is usually prompted by other, more immediate purposes as well, such as passage of a land-use plan, allocation of budget dollars, implementation of a nutrition program, or dissemination of safety procedures. By determining the more immediate purpose of a document, you can focus discussion and better fulfill your obligation (liability) to your reader. Like the consideration of audience and role, investigation of purpose helps you select appropriate vocabulary, sentence structure, order of information, examples, graphic aids, level of detail, tone, and style.

Recall the greenhouse example used in Chapter 1. If the readers want to *construct* a greenhouse rather than merely to learn about greenhouse construction, the document should be focused on this objective. Such readers do not require a detailed history of greenhouses; they require discussion of materials, tools, dimensions, order of steps, and any possible difficulties. By considering which points are pertinent to both your purpose and that of your readers, you save time and energy. As you write your document, continually evaluate the worth of the information you are providing in terms of your purpose. If it does not support your overall objective, leave it out (provided that you are not distorting the issue by telling only one side of the story).

The following are a few of the many purposes you may have in technical writing situations.

affirm	design	pattern
analyze	detect	perform
apply	deter	plan
argue	determine	posit
arrange	disprove	promote
calculate	dispute	propose
challenge	evaluate	prove
clarify	examine	question
classify	experiment	reason
compare	explain	recognize
conclude	extract	recommend
construct	forecast	re-establish
contrast	guide	retract
control	hypothesize	revaluate
convince	index	schedule
corroborate	infer	summarize
deduce	instruct	survey
define	interest	synthesize
demand	investigate	test
demonstrate	measure	theorize
describe	model	validate

In some cases your readers will share your purpose, but just as often they will have their own purposes for reading your document. In order to answer their needs as well as your own, *develop a complete statement of purpose that includes both sets of objectives as part of your situational analysis.* Here are several samples of such statements:

I will *compare* the price and performance features of the 1987 Pontiac Trans AM to that of the Dodge Charger so that my readers can *choose* the car that best suits their needs.

I will *describe* the methods and materials I used in my experiment so that my instructor can *evaluate* my performance in recent lab sessions.

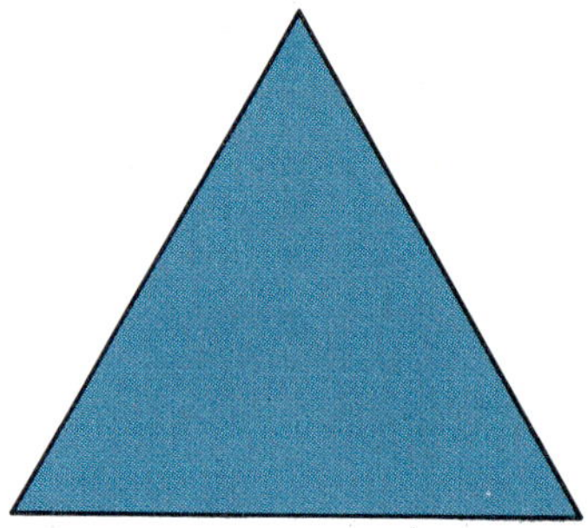

Figure 2-1 The Aristotelian triangle demonstrates the interdependence of audience, purpose, and role.

I will *recreate* the circumstances surrounding the accident in order to *demonstrate* my innocence to the insurance company.

The writer of the two product descriptions we quoted earlier in this chapter might have developed the following purpose statements:

For the executive, I will *list* reliability features to *convince* my reader of the DPS 88's ability to support the needs of large and growing organizations.

For the data processing professional, I will *define* the means by which processing speed is improved so that my reader will *appreciate* the engineering efficiency of these new systems.

Using these two purpose statements, let us examine the effect of varying a document's purpose on the resulting message. In the first sample, the writer accomplishes the purpose by detailing benefits such as speed, diagnostic circuitry, reduced heat build-up, and compactness. On the other hand, when addressing the data processing professional, the writer describes the inner workings of the system so that readers can appreciate the technology involved. To that end, emphasis is placed on description and explanation of the product, rather than on the benefits it has to offer.

As you can guess, purpose must interact effectively with both the audience and the role the writer assumes. An equilateral triangle symbolizing the interdependence and equal importance of these factors appears in Figure 2-1. Aristotle was the first person of note to discuss this relationship; hence this triangle bears his name.

The equilateral triangle draws its strength and stability from the fact that all its sides and all the angles between them are equal. No element receives more emphasis than another. Envision this symbol as you analyze your audience, role, and purpose in each writing situation you encounter. The stronger the interrelationship, the more successful your document will be.

EVALUATING YOUR SUBJECT MATTER

In most writing situations you will know what your subject matter is at the outset. However, time spent thinking about your topic *before* you understand your audience, role, and purpose can be time ill-spent. Say you have to write a lab report and the subject matter will be the experiments you have conducted. Or you know you will be writing about sewage problems for an engineering proposal to win a reconstruction project. Or perhaps the overall subject will be acid rain, recombinant DNA, the use of gabions in conduit construction, fourth-generation computer languages, or chimpanzees.

Knowing what you now know about audience, role, and purpose, you can appreciate how ineffective your output would be if you wrote about any of these subjects *without* giving serious consideration to these other factors. Your chances of coming up with the most effective presentation would be very poor—and, in any case, too much is at stake to leave the outcome to chance.

Don't misunderstand; subject matter is very important to the overall technical writing situation. Yet it is no more important than the three elements we have already explored. Without audience, role, and purpose supporting each other, you have no foundation on which to discuss your subject. The modified Aristotelian triangle shown in Figure 2-2 illustrates the relationship between subject matter and audience, role, and purpose.

Subject matter is delicately balanced on these other interrelated factors. The level of detail, the order of information, the examples provided—all of these aspects of subject matter are shaped by the support system of audience, role, and purpose. We noticed this phenomenon earlier, when we examined the detailed definition of Current Mode Logic circuitry in the excerpt addressed to data processing professionals. Top-level executives, though learning about the very same subject, do not demand such explanations; they want only to know the effect of CML circuits.

Consider one other example of how subject matter is affected by audience, role, and purpose. It is the end of the academic year and you have just returned to your parents' home to live and work for the summer. You pay for most of your school expenses through a scholarship, a work–study position, and summer employment. Your parents help you out only if you have an emergency and are not in a position to provide much additional assistance. Unfortunately, your grades for this past semester suffered for a

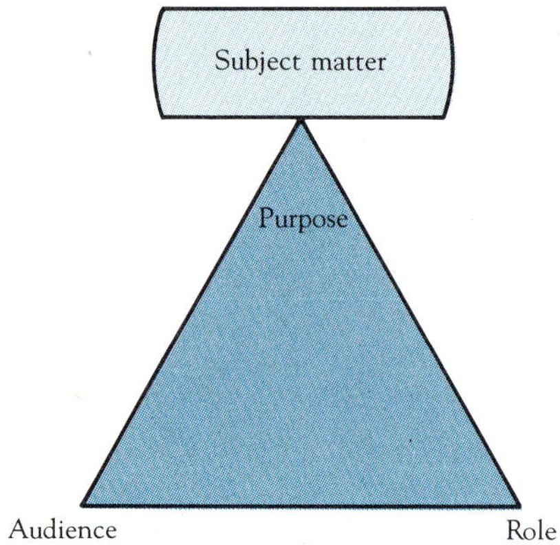

Figure 2-2 Considerations of audience, role, and purpose support your subject matter.

number of reasons (known only to you), and your scholarship is in jeopardy. Tuition is also going up again, and if you must pay the difference yourself, you won't be able to return to the same school you have been attending for the past three years. (You will be a senior next semester.)

Imagine that you must go to your parents to discuss the problem and to ask for help (even though you know it would be a burden). In this case your audience is your parents, your role is son or daughter, and your purpose is to gain financial help from them. Assuming that you would not be deceitful in any of these dealings, consider the differences that would probably exist if you were discussing this same subject with each of the other two combinations of audience, role, and purpose cited below.

Audience	Role	Purpose
bank loan officer	applicant	borrow money
your best friend	friend	enlist sympathy

Would you provide identical information in each case? Probably not. In appealing to your parents, you would probably emphasize the reasons for your lower grades, apologize (if necessary), point out options available to both you and them, and express your appreciation for their moral support through the years. As an understanding son or daughter, you might try to convince them to cosign a loan with you, rather than asking them to pay your next year's tuition out of their savings. Your discussions with a loan officer would be more likely to focus on your current financial condition and the impending tuition hike. You would point to your excellent grade point average (last semester need not be featured), your responsible nature (evidenced by your work–study and summer employment), and your maturity (evidenced by your not wanting to burden your parents with additional expenses). Finally, your discussions with your best friend would probably focus on the reasons behind your lower grades. You might dwell on the person you used to date (who broke up with you two months ago and for whom you continue to pine), the noisy roommates you have, the unfair biology lab instructor who never gave your reports an even chance, and so on.

This kind of versatility is not deceitful or manipulative. You are simply analyzing the overall communication situation and accommodating as many factors as possible. The more flexible you remain as you approach each writing situation, the more successful you can be.

The next chapter introduces various techniques for uncovering information about your subject, and in Chapter 4 we will explore methods of zeroing in on the subject and making it manageable.

RECOGNIZING CONSTRAINTS

As we will use the term, a *constraint* is a limitation or prescribed method that affects a writing situation.

These elements include

- format specifications
- schedule
- budget
- availability of information
- human resources and abilities
- production and distribution capabilities

Although none of these factors usually presents unsurmountable obstacles to the writer, you should take them all into account. The better aware you are of these limitations, the less limiting they will be!

FORMAT SPECIFICATIONS

The required format of a document can have almost as much impact on the nature of your message as your audience, role, purpose, and subject matter. A description of the effects of a low-level electromagnetic field on brain waves might work well in a scholarly journal, for instance, but it would have to be entirely recast to be printed in the science section of the Sunday newspaper. Likewise, the technical writing style used in a manual differs from that used in a product brief, proposal, report, or letter. The format in which information is to appear affects style, tone, level of detail, vocabulary, examples, and graphics.

Not all technical writing is produced in a prescribed format, but many companies, government agencies, universities, and research institutes establish their own guidelines for preparing documents. As a scientist or engineer associated with one of these organizations, you will have to see that your written work conforms to specifications.

When you are required to write in a certain format, consider all the headings you must provide, the order of information within those categories, the kind of graphic aids you can incorporate, and the emphasis given to each section of the document. By pondering these variables at the outset, you can begin to earmark material for the proper section as you uncover information or draw conclusions, thereby saving yourself time and effort.

To help you balance such factors as audience, role, purpose, and format, this textbook calls for a written situational analysis with each assignment. Preparing such a statement has several benefits: it helps you focus your work; it enables instructors to evaluate your understanding of the writing situation; and it allows you the flexibility of addressing audiences other than your college instructor. The following is a sample situational analysis that accompanied a student's definition of organic mulches (an assignment given in Chapter 7).

AUDIENCE	Home gardeners. Men and women who need to enrich their garden soil. They have at least a high school education, know basic gardening tools, and are highly motivated to apply this information.
ROLE	Nursery employee. Expectations are that I will know a great deal about Arizona soil types and desert fauna.
PURPOSE	To familiarize readers with the function of organic mulches so that they can evaluate their own gardens' needs.
FORMAT	Brief article that would appear in the Sunday "Sun Living" section of the *Arizona Republic.*

Each situational analysis will be a synthesis of the more detailed analyses you will have made of audience, role, and purpose. Although creating this statement may seem artificial to you at first, you soon will recognize how much it helps you meet the requirements of each writing assignment. In somewhat altered form, such statements sometimes actually appear in technical documents. Here is an excerpt of one taken from a service manual for Volkswagens.

For the VW owner with mechanical skills and for independent garages, this manual gives all the specifications that are available to VW dealers at the time of printing. In addition, a VW owner who has no intention of working on his or her car or truck will find that reading and owning this manual will make it possible to discuss repairs intelligently with a professional mechanic.

The aim throughout has been simplicity, clarity, and completeness with step-by-step procedures and accurate specifications. Every human effort has been made to ensure the highest degree of accuracy possible. When the vast array of data presented in this manual is taken into account, however, no claim to infallibility can be made.*

The writer identifies the audience, the chief characteristics of its members, what they can hope to derive from the manual, and what he intends to provide. He recognizes his own liability in the situation and even includes a disclaimer of sorts at the end of his comments.

SCHEDULE

By this point in your academic career, you are well acquainted with the phenomenon of deadlines. Few writing situations are without imposed time constraints; you must recognize their existence and not let them make you nervous. Early in this chapter we explored a hypothetical situation wherein you had a lab report due in two days but thought you couldn't work on it until the evening before it was due. We noted that, because the technical writing process unfolds in several stages, it was not necessary to do everything in one block of time. This point is even more important in complex writing situations.

As part of your situational analysis, *list the many activities that must be planned before you write.* These might include

- situational analysis
- clarification of requirements
- research
- compilation of data
- initial writing
- review cycle
- revision cycle
- approval process
- production
- distribution

Using a timetable or a simple calendar, estimate how much time each activity will require. Can any activities take place simultaneously? Is the

*Robert Bentley, *Volkswagen Rabbit/Jetta Diesel Service Manual* (Cambridge, MA: Robert Bentley, Inc., 1977–1984), pp. 5–4 and 5–5.

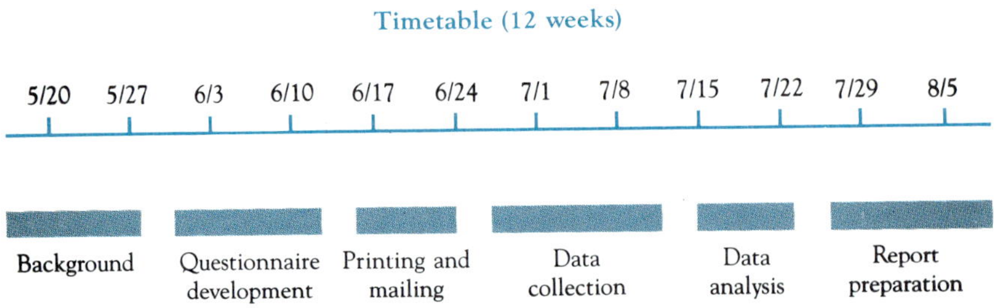

Figure 2-3 Use a timetable to plot activities and the time each will require.

sequence of activities flexible? If you are working with a team, assign activities to each project member as fairly as possible. Plan the required time accurately by drawing on previous experience or on the advice of people who have done similar work. Sometimes you will find that the person who requested the document (perhaps your boss) did not realize how much was involved in pulling the project together and therefore gave you an unrealistic deadline. The sooner you make this problem known, the more likely you will be able to extend the time allotted. If you leave writing for the last minute, your pleas will probably fall on deaf ears.

Generally, creating a schedule during the situational analysis will help you relax, organize, and plan your work. Then you must observe the schedule. Unless you are disciplined enough to meet the interim deadlines you set, you will find yourself "pulling all-nighters" or working nights and weekends at the office.

The timetable given in Figure 2-3 shows how a schedule of activities for preparing a survey report might look. The writer simply listed the phases of activity he planned and then associated periods of time with them by using a bar to indicate each interval.

The assignments in this textbook, of course, offer practice in meeting deadlines. The key to doing so is worth repeating: The more planning and analysis you invest during the preparation stages of the technical writing process, the less time it will take you to complete your work.

OTHER CONSTRAINTS TO CONSIDER

You may encounter several other constraints, including budget, availability of information, resources or abilities, and production and distribution capabilities. For instance, there may be times when you will be responsible for deciding how to spend the production dollars allocated for the preparation of a technical document. Bid proposals generally are printed professionally and bound in a cover that carries your firm's logo; brochures for products often justify the use of expensive photography and glossy paper; manuals may be printed by an outside vendor, and so on. In this case, during the situational analysis you should consider several means of producing the final documents at various prices. These prices should be drawn either from competitive bids submitted for the project or from previous experience. Factors you should consider when evaluating expenses include the importance of the document, the purpose of the document, the needs of the audience, and the return on investment.

It makes little sense to spend $35,000 on a manual for a personal computer product, for example, that will yield only $50,000 in annual revenues. It *does* pay to spend an extra $800 to use quality binding and paper on a bid proposal that may generate $500,000 worth of construction work for your firm. On the other hand, even though your audience may think bound proposals have a more professional look, they may not allow this factor to influence their decisions in the least. Such are the dilemmas you will confront when you face budget constraints.

Similarly, it may be necessary to "bring your project in" below a specific dollar amount—an amount that may or may not be adequate. Here's where creativity is particularly useful. The first area to consider when you are trying to conserve funds is the medium itself. Perhaps a pamphlet on flimsy stock would be just as effective as a report on costly paper. Or maybe a looseleaf notebook containing letter-quality computer print-out copy would work as well as a bound, typeset manual. It might be possible to save money by using computer-generated graphics rather than illustrations drawn by an artist. Or perhaps size and typeface could be altered to reduce the size of the document with no sacrifice in quality.

Another constraint to consider during your situational analysis is the availability of information. Very often only a single person (a "guru") knows the facts about a forthcoming design. Just as often, key reports are available only to those with certain security clearances. And most troublesome, some information may be filed in a government document depository 50 miles away. All such factors affect your writing. You will have to work into your schedule such activities as getting a copy of the confidential document through acceptable means (usually convincing someone with the proper clearance to secure it for you), interviewing the individual with the ideas locked in his or her head, or driving to the university library to examine the government document. In this last case, you will probably have to order the document and wait (anywhere from 10 days to 10 weeks) for it to arrive. We'll learn more about gathering information, even under adverse conditions, in the next chapter.

Your situational analysis may also have to take into account the human resources required to conduct research and write the document. Much technical writing is done by teams, so it is important to have enough team members. Sometimes all that is required is telling a few people about your project and getting them to "sign up" to work on it. At other times you will have to lobby actively for additional resources to get the job done well and on time. The lack of sufficient writing resources is a key problem in business and industry. Many employers do not view writing as a time-consuming part of your job; rather, they expect you to continue your computer programming, design work, or research *and* somehow find a way to churn out documentation, analyses, and reports.

Finally, you may need to consider the production and distribution resources required to do the job. Will you have to prepare your document early enough to get it to an outside vendor for typesetting, layout, and printing? If so, allow ample time in your schedule for the production and review process. Are these services available in-house, perhaps through the organization's communications department? Who will receive the document—and how? Will it be distributed via internal mail or through the U.S. Post Office? Perhaps the document will be hand-delivered to readers at trade shows, seminars, and demonstrations. You may be responsible for ensuring that this kind of activity is arranged. Put yourself rigorously through the paces during your situational analysis in order to anticipate all the things that might go wrong and all the things that must be done to guarantee smooth sailing.

CHAPTER SUMMARY

A situational analysis is a survey of all explicit and implicit factors affecting the development of a technical document. By planning carefully at this preparatory stage, writers can save a great deal of time and effort later in the technical writing process. Conducting the situational analysis includes

- developing a thorough understanding of the audience's characteristics, needs, and abilities.
- selecting an appropriate writer's role, which will then be applied consistently throughout preparation of the document.
- clearly articulating both the writer's and the audience's purpose.
- determining the most pertinent materials to include in the subject matter.
- preparing to accommodate such constraints as format, schedule, budget, resources, and production demands.

The writing assignments in this textbook, like writing projects in your future careers, require you to perform at least a four-part situational analysis. Evaluating the writing situation in such detail will help you develop essential planning and focusing skills as well as an appreciation of the many factors that must be controlled in the technical writing process.

EXERCISES

1. Find two different discussions of a topic in your field that are addressed to two different audiences. Judging from the level of detail, order of information, vocabulary, style, and tone, describe the characteristics the writers assume each audience exhibits. Do you believe the writers meet the needs of their respective readers? Why or why not?

2. Make an extensive list of roles for yourself. Choose three roles from this list and, for each, formulate a possible treatment of one of the following topics. Consider tone, order, focus, and word choice.

bicycles	space exploration
foreign policy	taxation
health foods	video games
pollution	weight watching

3. Select one of the topics from Exercise 2. Write a brief description of this topic that addresses the needs of one of the following audiences. After you have completed this description, select another audience from the list. How would the description change if you were to write another document for the new audience? (You don't have to write it.)

athletes	financiers
members of the clergy	hobbyists
doctors	managers
engineers	researchers

4. Review a report or other written assignment that you recently developed for a class in your major course of study. Produce a situational analysis that describes your audience, role, purpose (and that of your audience), and format and that accurately details their effect on the document.

WRITING ASSIGNMENT

For the following writing assignment, begin by developing a complete situational analysis. Attach it to your final draft so that your instructor can evaluate the factors affecting your work.

Select an article from a popular journal in your field (such as *Computer Decisions, Psychology Today, Modern Health, IEEE Spectrum*) and write a 200-word letter to the editor that does one of the following:

agrees with the author's conclusions
refutes the author's conclusions
suggests an application of the points raised in the article
relates the points raised in the article to another, seemingly dissimilar situation

CHAPTER 3

Gathering Information

CHAPTER PLAN

Although you may be familiar with the subject matter of a writing project, the preparation of most technical documents requires some research. You may simply have to speak with a few people to get the information you require, but more often your needs will lead you to the library, lab, or field. Knowing where to find the information you need, how to use it, and how to discover new sources of information will help ensure the accuracy and completeness of your work. The more precise and comprehensive the research effort, the better the foundation you create for your document.

This chapter will

- distinguish between primary and secondary research.
- explore research strategies that are useful in the technical writing process.
- survey various resources of technical information.
- present a research problem to illustrate the tactics that will serve your future needs.

THE ROLE OF RESEARCH IN THE TECHNICAL WRITING PROCESS

Research, the act of discovering or gathering information, is the second phase of the technical writing process. Most writing projects require some degree of research, whether you are confirming information you already have or are learning new material. What research you conduct for any given project depends on your analysis of the writing situation (the first phase of the technical writing process), because purpose, audience, subject matter, and role must be served by your efforts.

TYPES OF RESEARCH

The degree of research required can vary from a quite elementary to a highly sophisticated level. In the simplest case, merely reading an introductory textbook on a subject will provide you with sufficient information. Under complex conditions, you might undertake a long-term laboratory project to determine the suitability of treating certain forms of cancer with interferon. In the first situation, you are compiling facts and figures discovered by experts in the subject. This compilation process is called *secondary* research; you are using previously written or published materials. In the case of determining the suitability of interferon for cancer treatment, although some data may already exist in medical journals, much of the information you need will have to be discovered through laboratory experiments and case studies. This discovery process is called *primary* research; you are using material that has not been codified, written, or published before.

Because each technical discipline has its own range of secondary resources and its own methods and standards of accuracy for the conduct of primary research, it is beyond the scope of this textbook to do more than survey the types of materials available to you for research. It is up to you—through courses in your major field of study, discussions with colleagues, and observation—to determine the requirements of your own technical discipline.

The many kinds of resources for secondary research include

abstracts	case studies	manuals
almanacs	dissertations	pamphlets
articles	essays	periodicals
bibliographies	handbooks	proceedings
biographies	histories	proposals
books	indexes	reports
brochures	journals	theses

Resources for primary research include

case histories	meetings
field experiments	questionnaires
interviews	personal experience
laboratory experiments	statistical surveys

Often, writing projects require you to combine primary and secondary research methods to answer the questions you pose. For instance, as a laboratory scientist, you may test a prevailing theory that has been challenged. Or, as a compiler of information, you may conduct a survey to demonstrate the points you discovered during secondary research.

THE RESEARCH STRATEGY

Most research begins with work at the general level and then narrows to focus on specific areas of concern. To conduct research well, it is best to devise a research strategy at the outset—one that defines your objectives, identifies potential sources, allows you time to work and evaluate the information you uncover, and looks ahead to the synthesis of materials. One useful strategy is to follow ten basic steps:

1. stating the research problem
2. determining what you know
3. determining what you don't know
4. gathering general information
5. taking notes
6. evaluating general information
7. gathering specific information
8. taking notes
9. evaluating specific information
10. synthesizing your findings

1. *Stating the research problem.* Use a concise, three-part statement to describe your research objectives, the subject matter to be considered, and the purpose of the document you intend to write. With the aid of such a statement, you can concentrate on specific issues more quickly than if your research efforts were undirected. Once again, planning helps save time and effort.

Let's look at the three elements of a good statement of the research problem: "The purpose of this study is to determine the physiological effects of video display terminals on their operators, in order to recommend appropriate protective measures." Here the research objective is to "determine"; in other circumstances, it might be to "prove," "test," or "define." The subject matter is video display terminals, and the purpose of the document is to provide recommendations to its readers.

2. *Determining what you know.* Dedicate five or ten minutes to listing as much information as you possibly can about your subject matter by free association: this is called *brainstorming*. Do not worry about order; just keep jotting down one point at a time until you can no longer add to your list. Now examine the list and begin grouping points into similar categories of information. For the foregoing example, you might have categories such as *terminal design, operator habits, undocumented complaints, media reports of effects,* and *the work environment.* All these areas will be important to your subject matter, even though you may know far more about the work environment than about the design of terminals.

This exercise will help you discover how much you already know about the subject. It should help build your confidence and make you eager to get started on the research project.

3. *Determining what you don't know.* You must determine in which areas of the subject you require additional information in order to write a document that will accomplish its purpose. In terms of our example, you may feel you need to define unfamiliar terms and to know considerably more about electromagnetic fields and computer terminal designs. You may want to have documented proof of cases in which employees were adversely affected by working with computer terminals, and you may wish to identify any experts in the field. List these questions and arrange them in order of their importance to your research requirements.

4. *Gathering general information.* In many cases, you will need valuable background information before you can focus your research: key terms in the field, an overview or history of the subject, important contributors to the subject, trends, accepted methods, basic theories, and important developments and events. In the case of the problem statement we have been using, you might require an introduction to electromagnetic field theories.

5. *Taking notes.* You must summarize the information you encounter and keep track of the resources you use. Complete instructions for constructing a bibliography are provided later in this textbook. For now, be sure to develop an index card for each source you use.

For books, cite the following information (see Figure 3-1):

Author's last name, First name. Full title of book, edition number. City of publication: Publisher, copyright date, page numbers.

For articles, cite the following information (see Figure 3-1):

Author's last name, First name. Full title of article. Name of publication (Issue month or number), page numbers.

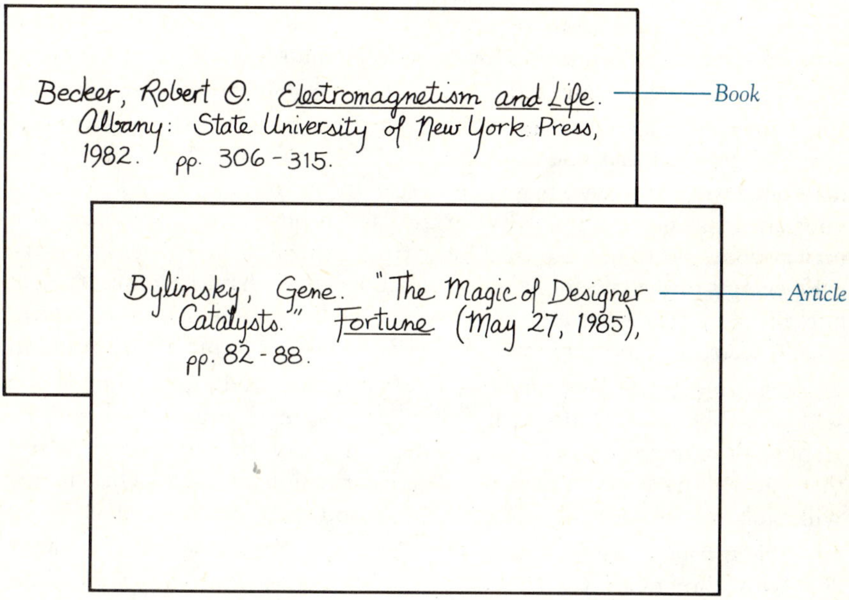

Figure 3-1 Citation cards for books and articles supply the information you and your readers will need to find the documents cited.

On separate index cards, key notes to the bibliographic citation by using the author's last name or an appropriate code. Keep track of page numbers on these notecards in case you decide to quote or paraphrase material.

Taking notes should not be a process of simply jotting down points of interest as you read. Rather, it should reflect your *understanding* of your source's entire content and main points. Here are some simple steps to help ensure that you record the kind of information you will need when you start writing.

First read an *entire* logical unit of the source—a complete article, paper, chapter, or report. Then return to the source and reread it, this time underlining or highlighting the key points you come across. (To prevent damage to library materials, either make photocopies or, on a piece of paper, write down what you would otherwise underline.) Review this material and eliminate any unnecessary words or phrases. At the same time, make sure that the points you have highlighted reflect the main ideas of sentences, paragraphs, and sections.

Then and only then are you ready to *take notes*. On notecards or notebook paper, rewrite the underlined statements in your own words as briefly as possible. Only when it is essential to a point you expect to make in your report should you jot down a direct quotation. Follow the order in which the points are made in the original document. At this time, you may wish to cross-reference statements to other notecards you have developed or to make evaluative comments about these key points.

6. *Evaluating general information.* Sift through your notecards to identify common points, conflicts, important issues, and basic constructs. Consider what information that you need to know is still missing. At this point you may need to continue your general research, or you may be able to focus your attention more specifically. In the case of our research problem, you may now be able to narrow your focus from general theories of electromagnetic energy to the specific ways in which its effects have been observed in humans. And from that point, you may be able to home in on the ways in which electromagnetic emissions from video display terminals affect users.

7. *Gathering specific information.* With a more focused topic in mind, you will probably be working with periodical literature, government reports, laboratory experiments, and proceedings of symposia and conferences so that you will have the most up-to-date information available. This chapter will explain how to identify these sources and use them efficiently.

8. *Taking notes.* Again you will be taking notes from your source materials. Now, however, you are likely to be more inclined to copy direct quotations that will prove useful to your document. Be certain to identify the person making the remarks if he or she is not the author of the source materials.

9. *Evaluating specific information.* Now you should be striving for completeness and accuracy as well as looking for the relationships among sources and for inferences you can form. If you do not have enough information to begin writing the document in earnest, identify the remaining gaps and fill them.

10. *Synthesizing your research.* Summarize your notes in a brief statement of the key points, issues, and concerns you have encountered. Such a statement will serve as the foundation for your document as you approach the third phase of the technical writing process: planning and arranging.

With this overview of the research strategy in mind, let us turn to a closer examination of the resources and techniques available to you.

USING THE LIBRARY

Although research on technical subjects may lead you down many different paths, the best starting point for most projects is the school library. Information found in the library will serve as a springboard for detailed analysis of virtually any subject, from the functions of the abomasum to the formation of zygospores.

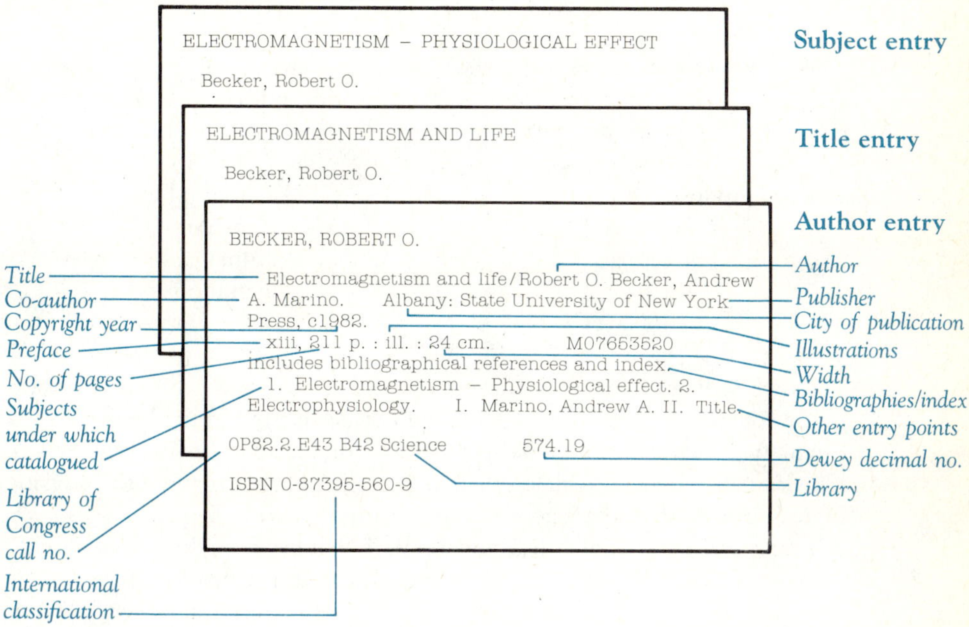

Figure 3-2 The card catalogue categorizes materials by author, title, and subject.

LIST OF LIBRARY HOLDINGS

An increasing number of libraries are placing their list of holdings onto computer data bases for easy access and maintenance. Some have placed more recent acquisitions on the computer while retaining older materials in the card catalogue. Microfiche is also being used in many libraries; still others rely solely on cabinet after cabinet of cards to list their holdings.

Whatever the physical source of the list, information can be retrieved alphabetically according to three categories: author, title, and subject. Although the order of information on each entry may vary, these cross-referenced entries frequently provide a wide range of items, including

author	full title
content description	co-authors
edition number	publisher
place of publication	number of pages
whether illustrated	whether includes bibliography
whether indexed	subjects under which catalogued
call numbers	ISBN number

See Figure 3-2 for an explanation of how this material is presented.

THE REFERENCE ROOM

The *reference room* acts as a threshold in almost any research project, especially one relying on current rather than historical information. In the reference room you can find vast collections of information, including

guides to literature	bibliographies
indexes	abstracts
directories	almanacs
atlases	handbooks
dictionaries	dissertations

As we will see later in this chapter, these sources can provide you with many sources of information that have not been listed in the library's card catalogue, such as articles. It is possible, for instance, to use the *Science Citation Index* to locate articles written during the last year about the physiological effects of electromagnetic fields. Or, with the help of *Dissertation Abstracts,* you can sample the graduate work that has been done in this field.

Likewise, texts located in the reference room can provide a wealth of overview information that is not likely to be covered in other research sources. A handbook might help explain measurement systems used in bioelectricity testing. A dictionary might provide definitions of such key terms as *electromagnetic field, conductivity, Hall effect, Boze–Einstein condensation,* and *plasma.*

THE STACKS

Most libraries store the majority of their books and bound periodicals on compact shelves called *stacks.* You will spend a good deal of research time here, tracking down books, reports, and journals important to your project. Generally, categories of related documents are grouped in the stacks by *call number.* Be certain to copy down this number accurately for each of the documents you need, or you will waste valuable time futilely trying to locate your source materials. Consult the library map to find the stacks you want.

PERIODICALS READING ROOM

Almost all libraries provide specially designated areas for consulting current periodicals and journals. Daily newspapers from around the country, magazines of popular or highly technical interest, scholarly bulletins, catalogues, and even telephone directories can be found in these rooms. The vast majority of information here, is not catalogued in the main library classification system. The periodicals listing is usually located both in the reference area and in the periodicals room.

Tips on important articles appear in a variety of places, including the field-specific indexes available in the reference room, bibliographies and notes within texts you have found, the *Science Citation Index,* and (for nontechnical subjects) the *Reader's Guide to Periodical Literature.* Both the reference librarian and the clerk in the periodicals room should be able to help you locate the sources you require. Reference librarians are generally quite familiar with the documents in their charge and may be able to introduce you to some very important resources you did not know about.

GOVERNMENT DOCUMENT HOLDINGS

Some libraries have been designated as repositories of government documents. These include proceedings of Congressional sessions, legislative measures, a wide range of reports from government agencies, an unbelievable number of documents from the Government Printing Office, research reports from government-subsidized laboratories, and information made available by all military branches.

If you are fortunate enough to attend a school whose library is a government documents repository, you have at your fingertips the output of the largest publisher in the world—the United States. Unfortunately, *none* of the information available in a government documents collection is catalogued in the main library's classification system. To learn more about the classification of government documents, speak with the librarian at the repository.

One timely source of information is called the *Monthly Catalog of U.S. Government Publications.* This massive index is relatively easy to use and will steer you to recent government work on your research topic. Other clues to government documents you might need are located in such indexes as *Science Citation Index, Biological Abstracts,* and *Government Reports Announcements.* The reference librarian can suggest the best indexes to use to find a combination of government, academic, and private-sector documentation on your research subject.

SPECIAL COLLECTIONS

University libraries frequently have special collections of information that might be of interest. For instance, the University of Arizona has one of the largest collections of information on arid land management, and Arizona State University has a tremendous collection of materials on solar energy.

Most libraries devote an area to a special collection and take great pride in it. If special collections exist at your school and are appropriate to your field, by all means take advantage of them.

SUPPORT SERVICES

Libraries generally provide a variety of support services, including a reserve room for books in high demand and an interlibrary loan program for obtaining materials not available in the library.

Materials that professors place on reserve are available to all students in a given course. The interlibrary loan program, on the other hand, is a network of libraries all across the country that lend books to other libraries. This reciprocal arrangement allows libraries to make fuller use of available information without having to increase their individual holdings.

COMPUTER SEARCHES

A computer search is a survey of local, state, national, and/or corporate data bases for a specific topic or keyword. This service may be subsidized, but you are likely to have to pay a fee based on the amount of computer time the search requires. Depending on the importance of a research project, the cost may be well worth the investment. With the aid of a computer, you can generate a list of *all* known documents in a given subject area. A task that might normally take years to complete can be accomplished in hours.

It is possible to narrow a computer search in order to limit the expense. For instance, if you want only the most current information in a field, you might focus an inquiry session on only those documents produced in the past two years in a given subject area. Of course, narrowing the topic effectively limits your search as well. Once a comprehensive list of documents has been generated, it is possible to focus only on those documents held in your library.

Most university libraries subscribe to national computer search services such as DIALOG, a data-base access and retrieval system available through Lockheed Online Services. This service taps a wide range of data bases, including

CIS/INDEX of Congressional Information Services, Inc.
GPO Monthly Catalog of the U.S. Government Printing Office
AGRICOLA from the National Agricultural Library

NTIS from the National Technical Information Service of the U.S. Department of Commerce, which is concerned with government-sponsored research, development, and engineering

ENERGYLINE from the Environmental Information Center, Inc.

To learn more about computer searches, talk to the librarian in charge of this service on your campus.

SECONDARY RESEARCH MATERIALS

Now that we have examined the facilities of a typical university library, let's look at the kinds of secondary research materials available. In addition to exploring such familiar areas as books and journals, we will survey some lesser known but highly valuable resources. For research projects requiring recent information, it is especially important that you become acquainted with these resources. And you will turn to these materials throughout your career to learn of emerging technologies, theories, and developments. These sources are arranged within this chapter according to the type of information they provide: general or specific.

MATERIALS PROVIDING GENERAL INFORMATION

If you are a first- or second-time researcher, you are probably not familiar with the full range of research materials available to you in your field. The following paragraphs describe several different kinds of overview documents that are designed to provide newcomers with a timely, comprehensive view of their disciplines.

Bibliographies and Directories

To become acquainted with research resources, consult a general *bibliography* within your broad subject area, such as electrical engineering or agriculture. These lists detail the important journals, indexes, government documents, societies and organizations, and various other sources of information. Bibliographies in your subject area are filed in the subject card catalogue of the library in the following manner:

CHEMISTRY—BIBLIOGRAPHIES
MATHEMATICS—BIBLIOGRAPHIES

In addition to the citation, some bibliographies provide brief descriptions (annotations) of the resources cited. These bibliographies can be especially helpful as you try to determine the value of a resource to your research problem.

Some important annotated bibliographies in specific fields include *Literature of Agricultural Research, The Use of Biological Literature, A Guide to the Literature of Chemistry, Civil Engineering Reference Book,* and *A Guide to Computer Literature: An Introductory Survey of the Sources of Information*. If you do not uncover a similar document for your major field of study, a librarian may be able to refer you to related bibliographies or slightly more general treatments.

The listings in most bibliographies are not annotated, and they may appear with or without library call numbers. Books, articles, and reports may be integrated in a single, extensive listing, such as

"Birth defects and miscarriage clusters stir up more fears over VDTs," *Microwave News,* Nov. 1981. pp. 1–3.

Ergonomic Aspects of Visual Display Terminals, ed. E. Grandjean and E. Vigliani. London: Taylor and Francis, 1980.

Eliot Marshall, "FDA sees no radiation risk in VDT screens," *Science* v. 212, no. 4499. June 5, 1981. pp. 1120–1121.

M. M. Weiss and R. C. Peterson, "Electromagnetic radiation emitted from video computer terminals," *American Industrial Hygienists Assn. Journal,* v. 40, April 1979. p. 300.

The search for a bibliography will undoubtedly provide you with a long list of possible resource materials. Note the complete citations so that you can look up individual works later.

Directories list the names and addresses of persons or organizations in a particular field. This information is especially valuable when it becomes necessary to attend meetings in your field, to receive proceedings from previous meetings, or to establish contacts with your peers. Directories include *American Medical Directory* and *Scientific and Technical Societies of the United States.*

Handbooks, Encyclopedias, Dictionaries, and Biographies

To become acquainted with the basic premises, theories, terms, and prominent figures in a field, consult key reference materials such as handbooks, encyclopedias, dictionaries, and biographies. *Handbooks* provide formulas, definitions, equations, and mathematical and statistical data that may prove valuable in your research. These texts are often shelved in the reference room rather than in the stacks. Most scientific and engineering fields have at least one handbook; some of the most popular are *Gray's Manual of Botany, Handbook of Chemistry, Standard Handbook for Electrical Engineers, Structural Engineering Handbook,* and *American Institute of Physics Handbook.*

Almost every field has its own *encyclopedia* that provides a concise description of key topics. In chemistry, for instance, there are four: *The Encyclopedia of Chemistry, Encyclopedia of Chemical Technology, Encyclopedia of Chemical Elements,* and the *Encyclopedia of Chemical Reactions.* Check the subject card catalogue for an encyclopedia in your field or speak with the reference librarian.

Many disciplines also have field-specific *dictionaries* that explain key terms and concepts. If you are new to a subject area, consult a dictionary to be sure you understand the words, phrases, and acronyms that you encounter. Dictionaries range in subject from agriculture, such as *A Dictionary of Agricultural and Allied Terminology,* to zoology (*A Dictionary of Zoology*), and they include everything imaginable in between.

A final general document is the biographical listing. *Biographies* are important because they provide the names of key contributors in a field and succinct descriptions of the contributions these people made. Some of the most useful biographies are *American Men and Women of Science, Asimov's Biographical Encyclopedia of Science and Technology, Dictionary of Scientific Biography, Who's Who in Engineering,* and *Who's Who in Science.*

Field Guides and Manuals

For practical, applied information, field guides and manuals can prove quite helpful. *Field guides* are illustrated roadmaps to the sights, smells, sounds, and natural phenomena in a given science. Generally illustrated with photos or drawings to facilitate quick identification of the item described, such texts can enhance your ability to recognize these elements. Donald Menzel's *Field Guide to the Stars and Planets,* for instance, is designed for the novice astronomer and includes charts, maps, and astronomical tables as well. Paleontologists can consult *Fossils in America: Their Nature, Origin, Identification, Classification and Guide to Collecting Sites* to determine where remains can be found. Another good example is *A Field Guide to Western Reptiles and Amphibians: Field Marks of All Species in Western North America,* which describes and illustrates 207 species of reptiles and amphibians found in the western portion of the continent (including Alaska).

Manuals are procedure-oriented documents that outline the specific steps involved in important activities within a given discipline. Typical of this type of reference text is *Engineering Manual: A Practical Reference of Data and Methods in Architectural, Chemical, Civil, Electrical, Mechanical, and Nuclear Engineering.* This book presents formulas, methods, and tables commonly used in the engineering fields given in the title.

Almanacs, Atlases, and Yearbooks

Many scientific and engineering fields have almanacs specific to their disciplines. An *almanac* is an annual publication that provides current and retrospective information as well as projected developments in the field. Its frequency of publication makes the almanac a quick reference for recent information. The National Industrial Conference Board, for instance, produces the *Economic Almanac,* which includes the latest pricing, banking, finance, income, resource, manufacturing, and agricultural information. Another example is *The Almanac of American Politics,* which provides information by state, district, senator, and representative.

Atlases provide demographic data on a timely basis (they are usually revised yearly or biennially). *Books in Print* lists dozens of atlases on such diverse themes as surgery, politics, history, and the solar system. Typical of this genre are the *Atlas of the World's Resources,* in which both agricultural and mineral resources are discussed and plotted, and *Hammond's Nature Atlas of America,* which illustrates—via mapping—the distribution of minerals, flowers, and wildlife in this country.

Yearbooks are generally addenda to other books, such as indexes and encyclopedias, that serve to keep these main bodies of work up to date. The U.S. Department of Agriculture, for instance, puts out the *Yearbook of Agriculture* to inform our agricultural community of recent developments.

MATERIALS PROVIDING SPECIFIC INFORMATION

After surveying the materials described in the previous section, you should have a general understanding of the terms, theories, and developments in your field. At this point it is possible to focus a general topic to a more specific level for more productive research. With a more focused topic, you can return to the subject catalogue to find books related to your research. Rely on the brief descriptions in the catalogue to weed out extraneous materials. Concentrate on resources that are designated as important in annotated bibliographies or encyclopedias.

To be sure you have the most up-to-date information, however, you should work with other resources in addition to books (which generally take several years to produce). Consult the library's periodicals holdings file to find field-specific journals of interest. To find particular current information, consult collections of abstracts or indexes in your field. These will cite recent articles, reports, proceedings, government documents, and graduate papers on a wide range of topics. Ask your librarian to show you the best resources of this type.

Abstracts and Indexes

In most scientific and engineering disciplines, collections of abstracts are published annually. An *abstract* is a terse description of a longer document, such as a report or journal article. The key information provided usually includes objectives, methods, results, conclusions, and recommendations. This brief description can help you determine whether the resource in question is worth tracking down for your research project.

To make finding the cited document easier, the abstract usually contains the full name of the article, report, government document, or paper; the author's name; the publication in which the item appeared; the publication date; and the page number(s). Armed with this information, check the listings of periodicals, government documents, and reports to determine whether the library has the document in question.

The most frequently used collections of abstracts in scientific and engineering fields are *Applied Science and Technology, Biological Abstracts, Chemical Abstracts, Mathematical Reviews, Psychological Abstracts,* and *Science Abstracts.*

Indexes are field-specific listings—rather than descriptions—of articles, reports, and research. These texts are issued frequently, in some cases weekly. Use the citation to locate resources that interest you. Some well-known indexes are *Bibliography and Index of Geology, Biological and Agricultural Index, Directory of Published Proceedings, Engineering Index, Index Medicus,* and the *Science Citation Index.*

The *Science Citation Index* is a unique kind of index; it tracks the mention by an author of another individual. Knowing how often others are referring to her or him can help you determine how important a figure is and how influential her or his work and ideas are.

The *Science Citation Index* is arranged in four different volumes each year (with quarterly updates) to give you four ways to access information: by author of the document, by subject area, by citation, and by organization (agency, corporation, laboratory, or the like).

Reports and Meeting Papers

Reports—the findings of researchers, corporations, and government agencies—can be valuable sources of information for your project. To find reports you are interested in, consult the *Government Reports Announcements,* the *Index to U.S. Government Periodicals,* the *Monthly Catalog of U.S. Government Publications,* and such indexes of government agencies as *Scientific and Technical Aerospace Reports.* The librarian in the government holdings room will help you learn the best methods to use in finding documents listed in these collections.

Although many papers presented at meetings, conferences, and symposia are accessible through the abstracts and indexes mentioned earlier in this chapter, additional documents may be listed in such resources as the *Directory of Published Proceedings*. Most libraries do not carry proceedings from corporate sources; they focus on government and university research instead. But it is often possible to get a copy of proceedings by contacting the chairperson of the conference or the appropriate agency, research laboratory, or corporation to request a copy.

Dissertations and Theses

Often a university library makes available to researchers the *dissertations* (extended treatments of a subject, submitted in fulfillment of requirements for a doctorate) and *theses* (extended arguments submitted in fulfillment of requirements for a Master of Arts or Master of Science degree) of its graduate students. Likewise, faculty research and scholarly reports may be held for reference. These documents can be very important in secondary research, because they may synthesize recent research in a specialty field and they may provide you with extensive bibliographies.

Ask the librarian whether your library houses such documents (sometimes they are on microfilm). For access to dissertations not held at your school, consult *Dissertation Abstracts,* which provides an overview of the dissertation's treatment of a subject and a dissertation number by which the document can be ordered. Again, allow plenty of time for receiving your copy of the dissertation if you are depending on it for your research.

SAMPLE RESEARCH PROBLEM

Up to this point, we have talked in general terms about library facilities and types of secondary research materials. Now let us turn to a real research problem to see how these resources are applied.

STATING THE RESEARCH PROBLEM

In February 1985, a subscriber to a popular science magazine came across an article describing research that raised serious questions about the potential health hazards of low-level electromagnetic fields (EMFs).

One of the emitters of low-level electromagnetic radiation cited several times in the article was the video display terminal (VDT). The article suggested a possible connection between this device and problem pregnancies and other health problems. The reader, who often used a VDT in her work, read with great interest and concern. She was so upset by the potential for danger that she approached the chairperson of the Employee Health and Safety Council at her company. This individual, whom we will call Sharon, promised the employee that she would research the problem and determine whether corrective actions were required. As soon as she read the article for herself, Sharon shared the employee's concern. Realizing that she had an obligation both to employees (who could be at risk) and to the company (which could be at fault), Sharon made this research a top-priority project. She began by stating the research problem: "The purpose of this study is to determine the physiological effects of video display terminals on their operators, in order to recommend appropriate protective measures."

DETERMINING WHAT YOU KNOW

Sharon then tried to determine what she already knew about the topic. She knew all of the points raised in the magazine article, but she did not know whether they were valid. Certain other information, however, she was sure of:

1. The VDT operators in her company complained of headaches, eye-strain, stress, and back pain.
2. VDTs emit low-level electromagnetic fields.
3. VDTs are designed to meet federal design standards for health and safety.
4. Operators in her department used VDTs an average of 6 hours a day, 5 days a week.
5. Several experts were cited in the magazine article, including W. Ross Adey, Andrew Bassett, Suzanne Bawin, Robert Becker, Carl Blackman, Arthur Guy, Charlotte LeGates, A. Liboff, Samuel Milham, Jerry Phillips, A. S. Presman, Louis Slesin, Nicholas Steneck, and Richard Tell.

DETERMINING WHAT YOU DON'T KNOW

Sharon went on to list some questions she wanted her research to answer:

1. What are electromagnetic fields (EMFs)?
2. How are EMFs measured?
3. What are the effects of EMFs on humans?
4. What kinds of tests have been conducted on the physiological effects of EMFs on animals? on humans?

5. Do tests indicate that dangerous levels of EMFs are emitted from VDTs?

6. Are there arguments against the points the article made? in support of them?

The order of the questions listed reflects the order in which Sharon believed she needed the answers to support the development of her report. She knew, therefore, that certain general concerns had to be addressed first. Once she had a sense of what electromagnetic fields were, she would be ready to apply her general understanding to an investigation of visual display terminals.

GATHERING GENERAL INFORMATION

Sharon began her research at the card catalogue, where she looked up ELECTROMAGNETISM. That entry led her to several introductory texts, including *Introduction to Electromagnetic Fields and Waves* by C. A. Holt (1963), *Modern Electromagnetic Fields* by P. Silvester (1968), and a more current text, R. L. Ferrari's *An Introduction to Electromagnetic Fields* (1975). She also found two books by experts cited in the magazine article. From the article, Sharon recognized an outspoken American researcher and a Russian scientist whose theories on the functions of electromagnetic fields within biological systems had been the basis of much research. For each of these sources, Sharon wrote down the complete name of the author, the book's title and its call number, and she noted that the American researcher's text included an index and a bibliography.

Sharon looked farther through the listings under ELECTROMAGNETISM. She soon found the subcategory ELECTROMAGNETISM—PHYSIOLOGICAL EFFECTS, under which were not only the two texts she had already uncovered but also several other intriguing titles. Sharon copied down the call numbers for each of these listings and headed for the stacks.

After locating these texts, Sharon leafed through them, noting the chapter titles in the table of contents, the bibliographies, and the topics covered in the indexes. Each of these texts, it seemed, would supply her with valuable information. One anthology, for instance, carried an article by Adey and one by Liboff, two of the scientists cited in the article. The IEEE symposium carried papers by Phillips, Tell, Adey, Guy, and Bawin. Sharon was now well on her way to assembling the materials she needed; she had collected works by 8 of the 15 scientists mentioned in the article.

Back at the office with her texts, Sharon began her initial reading. By reading salient chapters in each of the books, Sharon developed a basic understanding of electromagnetism.

With this foundation, Sharon concentrated on a text that detailed theories about intrinsic electromagnetic energy—that is, the amounts of electromagnetic energy resident in living organisms and the roles this energy plays in the functioning of the organism. It seems, Sharon learned, that incoming sensory stimuli and motor activity register a negative charge, whereas decreased activity (such as sleep) is measured as a positive charge. The author inferred relationships between the EMFs of organisms and their central nervous systems. He found that changes in electromagnetic fields could be correlated with changes in the living organisms. Sharon noted that although many questions were raised through these experiments, measurements and controls were never sufficiently precise to allow for replication of results by other experimenters, and, as the author himself admitted, none of the tests were conclusive.

TAKING NOTES

After Sharon read through each of the works she had collected, she reread them for the purpose of taking useful notes. On photocopied pages, she used a highlighting pen to mark the main point of each paragraph. Within that highlighting, she underlined key words and phrases. Then she reviewed the highlighted material to select only those statements that would be the most useful. She also made sure she had not overlooked any essential points that the source made. Finally, on notecards keyed to the bibliographic citation for each source, Sharon briefly stated these highlighted points in objective language of her own. She then entered comments and cross-references, as appropriate. Figure 3-3 shows a notecard Sharon created from a chapter in one of her sources.

EVALUATING GENERAL INFORMATION

Sharon organized her notes and began to evaluate what she had learned. She now had a basic understanding of electromagnetic fields and believed she could follow accounts of specific experiments. From her reading, she also felt confident that she would be able to apply this general information to any data she might uncover about video display terminals. Without further delay, Sharon returned to the library to find more specific information.

GATHERING SPECIFIC INFORMATION, TAKING NOTES, AND EVALUATING SPECIFIC INFORMATION

Now Sharon needed to find current, specific information on both EMF testing and VDTs. She brought with her the list of scientists cited in the

Becker, ELECTROMAGNETISM, Ch. 2 "Physiological Function of Intrinsic Electromagnetic Energy," pp. 25-59.

- By 1961, research indicated that DC potentials did play an important functional role in brain activity:
  - Incoming sensory stimuli & motor activity = Neg. DC potential
  - Decreased activity (e.g., sleep) = Pos. DC potential
- Experiments w/ salamanders indicate that some component of central nervous system generated and transmitted direct currents that produced measurable voltages on surfaces of animals, inc. humans.
- Surface potentials vary in definite pattern w/ subject's level of consciousness.

Figure 3-3 Notes should include only key points and the source.

magazine article as well as names and citations she encountered in her preliminary reading. She first went to the reference room of the library and consulted *Science Citation Index,* where she found lists of reports, articles, papers, proceedings, and organizational meetings on the subject.

After checking the periodicals holdings list, Sharon located many of her sources in the reading room. Others of her sources were bound volumes of the periodical and were shelved in the stacks by call number. Sharon took meticulous notes from each of these sources on electromagnetism. From all of her reading up to this point, some important patterns were beginning to emerge:

1. *No* researcher felt absolutely certain about the functional role of EMFs within biological systems.
2. No one could conclusively prove how intrinsic EMFs were generated within organisms.
3. No one could measure very much more than applied EMF forces on the surfaces of organisms (rather than within organisms).
4. Almost no experiments had been done with human subjects.
5. The few experiments done on humans had not been carefully controlled or replicated.

Before she could feel absolutely sure about these observations, Sharon wanted to consult a report by the Navy that was cited by many of the experimenters. She supplied the report number to the clerk in the Government Documents collections but was disappointed to learn that the library was not a repository for the reports of the Naval Aerospace Medical Research Laboratory. The librarian helped Sharon fill out the order form that was required to get a microfiche copy of the report from the National Technical Information Service. It would be two to three weeks before the report was received.

Not being able to wait two to three weeks for the Navy's report to come in before continuing her quest, Sharon decided to begin focusing her research on the more specific topic of the effects of EMFs emitted by video display terminals on their human operators. Back to the card catalog—this time to look up the subject VIDEO DISPLAY TERMINALS. One book (*Video Displays, Work, and Vision*) was listed, and it was only about eyestrain. She could remember reading articles in newspapers about problems with VDTs, and she remembered a talk show host exploring the subject with a labor union leader. Surely there must be something more on the topic in a library as extensive as the one she was using! Sharon reflected for a moment and realized that the abbreviation CRT (cathode ray tube) was used interchangeably with VDT. Perhaps if she looked under this listing, she might get more material.

Her insight paid off when she came across a recent, extensive bibliography dedicated to the subject of CRTs. This list of articles, papers, test results, reports, and books spanned over 60 pages. Sharon read each entry and recorded the location of those that, on the basis of their titles, she thought would provide a representative sample of views on the safety or hazardousness of VDTs. Among these were popular news articles citing user complaints that surfaced during labor union disputes and stories about clusters of problem pregnancies (in *Computerworld* and *Popular Computing*), scientific journal articles (IEEE *Spectrum* and *American Industrial Hygienists Association Journal,* for example), government reports commissioned by the National Institute for Occupational Safety and Health (NIOSH), the proceedings of Congressional hearings, and the results of a National Academy of Science symposium held in Washington, D.C., during August 1981.

Sharon created her own list of 26 documents worth gathering. She then checked all of these articles against the periodicals holdings and wrote down their call numbers. Of the 26 works, 18 were journal articles, but only 12 were available in the library. Her first priority was to determine the availability of the government reports she thought she needed. Sharon ordered several with the help of the government documents librarian.

Before she left the library for the evening, Sharon had time to look up two particularly interesting articles, which were written by the researchers responsible for some of the government reports she had just ordered. She hoped these writers would allude to their findings and thereby give her a hint of the reports' contents before they actually arrived. Her hunch paid off.

William Murray's "A Radiation and Industrial Hygiene Survey of Video Display Terminal Operations" (*Human Factors,* 1981, 23(4), 413–420) described tests conducted on 136 terminals at 3 different locations. In addition to tests performed on the devices, samples of the air in the workroom were analyzed to determine whether workers were exposed to any airborne contaminants. The results, according to Murray, suggested that users of VDTs are not being exposed to hazardous levels of EMF radiation or chemicals. The research indicated that the predominant emitter of radiation in a terminal is the CRT face. The strength of the magnetic field was $0.49A^2/m^2$ at a distance of 30 cm; a level that Sharon had already learned from her reading on EMFs has no discernible biological significance.

An article by Michael J. Smith called "An Investigation of Health Complaints and Job Stress in Video Display Operations," appearing in the same issue of *Human Factors,* discussed the results of an extensive survey of VDT users and nonusers. The health complaints showing the greatest differences between the two groups were visual and muscular, and they were unrelated in any detectable way to radiation.

SYNTHESIZING YOUR FINDINGS

Sharon left the library that night feeling relatively sure that EMF radiation was not going to be a cause for concern and that the magazine article—the impetus for this research exercise—was not representative of opinion in the field. But then again, she hadn't read three key reports, nor had she studied any of the claims of hazardous effects.

The next day Sharon read the proceedings of the Subcommittee on Oversight and Investigations of the House Committee on Science and Technology. Representatives from scientific and medical research laboratories, government agencies, various labor groups, the military, and equipment manufacturers were called on to testify. At the end of the session, its chairperson concluded that "the preponderance of evidence suggests that VDTs are safe." The abstract of the Bureau of Radiological Health report entitled "An Evaluation of Radiation Emissions from Video Display Terminals" supported this view. Thirty-four VDTs made by 31 manufacturers were tested and it was found that none of them posed a health threat. However, the report suggested that work conditions such as lighting, chairs, and table height be designed to eliminate any physical discomforts that might be associated with VDT use.

The next time Sharon went to the library, she looked up several news articles that carried negative statements in their titles, such as: "Birth Defect and Miscarriage Clusters Stir Up More Fears Over VDTs" and "VDTs and the Fetus—Cause for Concern?" These kinds of articles were written from a sensational viewpoint, but a relatively high incidence of problem pregnancies were reported in a Toronto newspaper office, an Atlanta government contracting office, and a Dallas retail chain. Upon further investigation, Sharon was able to find more enlightening research into these problems. For instance, careful measurement of all 296 terminals at the newspaper office failed to reveal any radiation, insufficient ventilation, or hazardous chemicals as problems. Sharon found other "clusters" of miscarriages reported in companies where the workers were not using VDTs at all.

By now Sharon was confident she could report to the Health and Safety Council that employees had nothing to fear from their VDTs. The fields emitted by such devices are of such low intensity that they are very unlikely to have *any* biological effects.

Once the government reports she had ordered came in, Sharon was certain that she had evaluated the data accurately. She began to write up her conclusions in a formal report that would be available to employees at her company. During her research she had come across several suggestions for minimizing eyestrain and back pain, so she incorporated these into her report as a set of recommendations:

1. Operators should spend no longer than two hours at a time in front of a VDT. Schedules should allow for frequent breaks of five or ten minutes.
2. Anti-glare screens should be placed on top of the phosphorescent screens of the terminals.
3. Chairs should provide greater support for the lower back.
4. Workers should engage in a variety of activities to reduce stress and the ill effects of immobility.

CONDUCTING PRIMARY RESEARCH

Primary research, the process of creating and discovering information, can take a variety of forms. When your writing project necessitates primary research, you may find yourself interviewing, surveying, experimenting, taking notes at meetings, and recording personal experiences.

As we have noted, it is beyond the scope of this text to explore all of these procedures in detail. Taking notes at meetings and recording personal experiences are both self-explanatory. However, let's look briefly at the primary objectives and techniques of interviewing, surveying, and experimenting.

INTERVIEWING

The preparation of some technical documents will require you to conduct interviews with experts in the field, either because these people are the only ones familiar with the subject (in the case of a new-product design) or because their vantage point provides you with an important interpretation of data you have compiled. Interviews are usually conducted in person, because much information is conveyed through nonverbal means such as facial expressions, hand gestures, and posture. However, it is possible to conduct an interview over the phone or even by mail.

The quality of an interview depends on the quality of the questions you pose to your subject. If the interview is not conducted well, you may overlook important areas and have to go back to your source (who may not appreciate your taking more of his or her time). First, consider the information you want to extract from your subject(s) and design questions that are *open-ended* (not easily answered by "yes" or "no") and allow the expert to speak to the point. Prepare a set of written questions that emphasize explanation, relationships, and trends. During the interview, listen to the subject's responses carefully. Be certain to ask more specific questions when a response requires further explanation.

In conducting the interview, follow a basic sequence that includes an introductory portion, general questions, specific questions, and recapitulation. In the introductory phase, explain your purpose in a way that establishes rapport with the interviewee. Let this person know how you became interested in the project and what experience you have in the field. Having discussed your own background, move on to questions that bring out the expert's experience and basic credentials. These credentials will serve you well in evaluating the importance of this person's remarks vis-à-vis that of the remarks of other experts.

Move through your question set from general to specific, being careful to listen and take notes. Avoid asking long-winded questions; the interviewee should do most of the talking. While taking notes, be sure to pace the interview so that you can pay attention to the speaker as well as your note pad. Notice the speaker's animation, tone of voice, gestures, and facial expressions to help you recognize points of emphasis and importance. Establish eye contact in order to provide him or her with cues (head-nodding, "uh-hum," "oh, yes," or "it sounds like this point is very important to you . . .") that acknowledge what is being said or encourage further elaboration.

To make certain you understand key points, summarize what you believe the person is saying. For instance, after a long response that requires verification, you might inquire, "Do you mean that if I add the test compound to the disease specimen, the bacteria will be destroyed?" If you are correct, this type of question elicits a simple "yes." If you are mistaken, the interviewee will clarify the point for you.

Taking notes is important but should be done discretely. Jot down only key points, using words and brief phrases. If you wish to quote your speaker directly, stop him or her and ask permission to do so. Then ask the person to repeat slowly the statement you feel is so important. Tape recording is a good way to keep a complete record of the interview; however, few people feel comfortable enough in front of a microphone to speak naturally and candidly. You are far more likely to acquire substantive information if you do *not* use a tape recorder. Under no circumstances should you conceal a recorder; that is a violation of the subject's basic rights.

Toward the end of the interview, summarize the major points you have learned in the order you believe reflects their value to your research. Ask the interviewee to confirm the information or clarify any misunderstanding you may have. Likewise, ask whether there is anything he or she would like to add. Before you leave, determine whether you may call on this person again and suggest that he or she phone you if additional points come to mind after you have left. Be sure to send the person a courtesy copy of your document when it is completed.

SURVEYING

A survey is an examination of a specific condition, situation, issue, attitude, or product by means of a prepared set of questions presented to a fair sample of some larger population. A survey may be very important to a technical document. For example, had our researcher Sharon found that indeed there *was* cause to be concerned about health hazards of VDTs, she might have conducted a survey of all operators in her group to determine their exact complaints, work habits, and degree of exposure. She would then have expanded her sample to include other word processing groups within her corporation. Using the same set of questions, she could have drawn comparisons among several departments to determine the degree of consistency. With these results, she might have discovered an important trend or corroborated the article that stimulated her investigation of the issue. But to test her results accurately, she would have had to compare her findings with those for typing pools (where no VDTs are used) and for other kinds of departments that use VDTs in different applications, such as programming, modeling, financial analysis, and graphics.

The quality of the results of any survey depends on the statistical validity (size and representativeness) of the sample, the objectivity of the questions posed, and the accuracy of measurement and interpretation. These elements are the subject of specialized courses in statistics and are too complex to be discussed here. Although there are certain standard procedures in applying statistics to survey results, each technical field has some unique statistical methods and standards that you must learn and apply to the research you conduct.

EXPERIMENTING

In both your major course of study and your future career, you may be called on to conduct experiments to test or establish an hypothesis. Two major types of experiments exist: field experiments and laboratory experiments. These two types of experiments differ primarily in the environment in which they are conducted and the degree of control exercised during test procedures. Although the conditions involved in each are somewhat different, the objective of supporting or disproving a specific hypothesis remains the same.

Both field and laboratory experiments should be designed to include the following:

- objective—the aim of the experiment.
- hypothesis—the principle being tested or established.
- procedure—the steps taken to achieve the objective. These include the methods and materials used at each point. Procedures can vary drastically and will affect the validity of the experiment.
- results—the calculations or measurements taken during the procedure.
- conclusions—the inferences drawn logically and directly from the results.
- recommendations—the logical suggestions based on the conclusions.

A *field experiment* is conducted in the environment in which the condition normally exists—in the "field." You can carefully design the experiment in order to create the most realistic conditions possible and to control as many variables as possible. (In any type of experiment, the less control, the more the results are suspect.)

If you require greater control than is possible in a field experiment, you will want to perform a *laboratory experiment* (if it is logistically and financially feasible to do so). To minimize the expense of some laboratory testing, experiments are often conducted on a few isolated cases or on animals rather than humans. In both of these circumstances, results must be extrapolated to broader populations via statistically acceptable measures. Again, it is essential to learn the standards and practices specific to your area of interest.

CHAPTER SUMMARY

After conducting a situational analysis for each writing project you undertake, you will probably need to perform some research to accomplish the objective of the document. Using the ten-step research strategy we have outlined, you should state the problem to be investigated, evaluate both known and unknown data, perform some general background research, move on to more focused research, and synthesize your findings.

Libraries harbor a wealth of secondary research materials in their stacks, reference rooms, periodicals, special collections, and government documents. By getting acquainted with the facilities available to you and by familiarizing yourself with the fundamental research texts in your field, you will become an efficient investigator. Handbooks, dictionaries, and guides to the literature help establish a strong foundation upon which you can build more sophisticated research. Bibliographies, indexes, abstracts, and periodicals are focused, timely sources of information. Government documents, proceedings, and reports represent a pool of current, expert views of your discipline.

Primary research tactics include interviewing, surveying, and experimentation, all of which depend on measures and methods specific to your field. Interviews, the most personal means of conducting research, provide you with expert "testimony" regarding your subject. Conducting surveys and experiments enables you to discover new information and test hypotheses. Frequently, technical writing projects will require you to conduct both primary and secondary research. To do so, you must anticipate time requirements, assess the availability of resources, and accommodate budgetary constraints (in the case of experimentation, especially).

EXERCISES

1. To become acquainted with the key resources in your field, locate the following in your library: abstracts, bibliographies, dictionaries, field guides, handbooks, and indexes. List the call number of each, and write a one- or two-sentence description of how each is used and may be helpful to you in your future research.
2. Locate the *Science Citation Index* in the reference room at the library. Using the "citation" volume, locate the name of a prominent person in your field or one whose work has interested you. Jot down the articles and other documents in which that person is cited for the year. Then turn to the "source" volume (list of authors) and cross-reference the articles. In what ways are the ideas and works of this person being cited by others?

WRITING ASSIGNMENTS

1. Speak with a reference librarian at your school library to determine what computer search facilities are available. Describe the computer services that would be useful to you in your discipline and explain how one gains access to them.
2. Select a topic from a popular magazine in your field and construct a set of bibliographic materials that would enable you to test the validity of points raised in the article. Describe your research strategy.

C H A P T E R

4

Planning and Arranging

CHAPTER PLAN

Planning and arranging the elements of a document, the third phase of the technical writing process, depends heavily on both the information you compiled in phase two and the situational analysis you performed in phase one. Conscientious planning helps you balance crucial elements of the writing situation, such as audience needs, purpose, subject matter, the writer's role, and format. This phase of the process involves focusing the topic further, grouping related pieces of information into categories, formulating a thesis, and structuring the document to best support your purpose and the requirements of the audience.

This chapter will

- introduce ways of limiting your subject matter to your requirements.
- enable you to develop a thesis suitable to the writing situation.
- explore various methods of organizing a document.
- introduce standard formats in which to cast an outline.

THE PURPOSE OF A PLAN

Just as architects develop blueprints, engineers plot designs, and programmers create flowcharts, writers of technical documents plan the outcome of their work. Using an outline—even an informal one—enables you to evaluate the merits of your intended document much as these professionals analyze their future products before they are produced in final form. Plans of this sort save developers money, time, and effort.

The optimal plan for a technical document is determined more by audience needs and the writer's purpose than by the subject matter itself. The optimal plan continually provides a context for the information that will follow, a context that reflects the audience's level of expertise, concerns, responsibilities, expectations, and interests. In other words, as you build your writing plans, you should answer two basic questions that stem from the situational analysis:

What do the readers need to know?

What do they need to know first before they can understand the whole message?

By asking these questions continually, you can tailor the scope of the document to your audience's needs, come to understand the crucial relationships between ideas that emerge, emphasize key points, and order information to best support these points.

Your plan becomes not merely an inventory of all the information you have gathered during the research phase of the technical writing process but also a carefully constructed *model* for the document. As a model, an outline enables you to manipulate variables and evaluate the effects of change, and it will vary in length and complexity with the nature of the writing situation and the subject matter. By anticipating all the requirements of the writing situation in your plan, you will be able to modify your presentation more easily and effectively if changes are required.

After completing phase one of the technical writing process, you have a fair idea of the interests, needs, and characteristics of your audience for a document. During this situational analysis you constructed a research problem statement, which helped you gather information in keeping with the nature and scope of the topic. Then, in the research phase, you evaluated your data and began identifying inherent relationships among elements. With this much accomplished, you are ready to begin building a document plan.

ESTABLISHING A THESIS

In Chapter 3 of this book we examined the research problem confronting "Sharon," the supervisor of a word processing department who was concerned about the possible health risks involved with using video display terminals (VDTs). Through her secondary research activity, she was able to uncover general information about electromagnetic radiation and its effects on animals and humans. She then gathered more specific information on the emission of electromagnetic energy by VDTs and the health-related issues surrounding these devices. After evaluating this information, she felt confident that she could write a report demonstrating to her subordinates that they were not in danger. In other words, she inferred from her findings the following statement: "Video display terminals do not emit sufficient electromagnetic radiation to cause a health risk."

This type of statement, called a *thesis* or main point, serves as the primary focus of Sharon's document. It is this point that Sharon's report will have to *prove* to her word processing operators. This clear, one-sentence statement of Sharon's thesis reflects her subject matter (video display terminals and electromagnetic emissions), her purpose (to explore the potential risks of VDT use), the concerns of her audience (that these devices could be dangerous), and her findings (emissions are insufficient to cause health risk). She was able to formulate this thesis only after scrutinizing the writing situation and evaluating the information she had gathered.

To progress further in the technical writing process, you—like Sharon—should begin formulating a thesis at this stage. Armed with a single statement of your research findings, which also reflects the requirements of the writing situation, you will be able to direct your continuing efforts toward a well-focused goal.

LISTING INFORMATION

The main point of the document you must write may not always be apparent at this point; some further study of the information you have uncovered may be required. First, to determine the extent of your post-research understanding of your topic and your audience, try brainstorming once again (a technique you used earlier in the technical writing process to determine the extent of your preliminary knowledge of the subject). This listing exercise will intensify your understanding of your materials and help you establish your thesis. Here is a brainstorming method you might wish to employ at this time:

1. Review your research notes and situational analysis.
2. Leaving your notes aside, go off to a quiet area where there is a cleared

writing surface and where you will not be interrupted for at least five minutes.

3. Bring a good pen, several sheets of clean paper, and a timer.

4. Set the timer for three minutes.

5. Place pen to paper and begin writing down every point you remember from your research that is important to both your purpose and the needs of your audience. Let the information pour out onto the paper as quickly as possible. Use key words, phrases, or sentences.

   NOTE: Do not worry about order, relationships, or details. Do not lift your head or stop writing until the three minutes are up.

6. At the end of three minutes, stop writing.

Small teams can also use brainstorming tactics. Standing with a flipchart at the front of a meeting room, a transcriber can write down all the ideas that are voiced during an oral brainstorming session. Ideas should be recorded without comment; evaluation will take place later. At this point, no idea is a bad idea.

After you have generated a list, review it carefully in a reflective, less hurried atmosphere. Add to it any other ideas that come to mind. Go to your notes and reference materials to add facts and depth. Even at this stage, it is not important to worry about emphasis or order; you are simply putting down as much information as possible.

Let us look at a student's use of this method for a writing assignment in an introductory computer science class. The instructor had asked students to research the history of computers and to write a 1000-word report. No other direction had been given (that is, no description of purpose, audience, or occasion). The student conducted a situational analysis and came up with the following statement:

AUDIENCE Computer science professor who is well acquainted with the history of the discipline but is interested in determining the writer's research and analytical skills. His evaluation of the report will affect his perception of me.

ROLE Beginning student of computer science. I am committed to learning about the discipline but have had very little exposure to computers thus far in my academic career. I have used a computer at home for nontechnical applications such as budgeting and keeping my calendar.

PURPOSE Demonstrate that the use of computers dates back to ancient history rather than the beginning of World War II.

FORMAT Brief report.

This student conducted secondary research that included use of his textbook and, at the university library, use of both periodicals and books on the subject. After following the steps of the research strategy outlined in Chapter 3, he did not have an apparent thesis statement, but he knew that before he worked much longer on the project, he ought to know where he was headed. Hence he sat down to do some brainstorming to help crystallize his information. Here is an excerpt from his effort:

several different kinds of computers

manual, mechanical, early electronic

long history of using computers to improve job efficiency

various mechanisms: punched cards, vacuum tubes, sand in grooves, abacus

people with new ideas: Pascal, Hollerith, Jacquard

Harvard professors

gears

Charles Babbage

calculating functions

speed

much larger computers than modern day

available technology limited developments

Chinese abacus still in use today, but invented some 2500 years ago in Tigris–Euphrates valley

ENIAC: 5000 calculations per second, vacuum tubes, electricity turned flow on or off in one-millionth of a second

metalsmiths of Babbage's day did not have technological prowess to build his design

"Analytical Engine" by Babbage, to use punched cards, would have been programmable if it had been built

Mark I of Harvard, mechanical computer, translated analog to digital, used for weapons analysis, 760,000 working parts

U.S. Census in 1890 aided by Hollerith's tabulating machine

abacus—beads to left, space to right. Addition: move beads to right. Subtraction: move beads to left. Go to next row when one fills.

ENIAC was not programmable

ENIAC used vacuum tubes

UNIVAC was first-generation modern computer: invented in 1951, used punched cards, was programmable, four tons

mechanical switches took long time

vacuum tubes had to be replaced often, would burn out, one gone every 7 minutes
UNIVAC could add 237,000 five-digit numbers in one minute

Note that items are not in any particular order (though some ideas are grouped through the process of association) and that some redundant information appears. Note also that the student provides considerably more detail in some areas than in others. This imbalance can be corrected at the next focusing stage.

GROUPING RELATED PIECES OF INFORMATION

Once you have created as complete a list of elements as you can through both brainstorming and review, you can begin to identify relationships between items—relationships that are important to your audience and will support your purpose. Here is a small sampling of the many ways in which items on your list can be related:

agreement: substantiation, corroboration, proof
balance: symmetry, asymmetry
causality: action to reaction, cause to effect
chronology: history, time
class: general to specific, genus to species
comparison: similarity, analogy, parallelism
conflict: opposition, denial, disproof
contrast: dissimilarity, lack of parallelism, variation, difference, heterogeneity
degree: intensity, complexity, measure
equality: sameness, equivalency, homogeneity
function: purpose, objective, output
inquiry: problem to solution, question to answer
logic: facts and reports to inference, major premise and minor premise to conclusion
membership: individual to group, member to set
partition: whole to parts, part to subparts
physical connection: dimension, attachment
sequence: process, steps, stages, phases

Returning to the computer science student's list, you can readily see general-to-specific relationships, chronological relationships, and even whole-to-parts connections. For instance, "several kinds of computers" is a general statement substantiated by such other items in the list as "manual, mechanical, early electronic." Each of these elements, in turn, is the label for an even more specific set of items on the list. The terms "abacus" and "grooves in sand" apply to "manual"; "tabulating machine" and the "Analytical Engine" belong with "mechanical" computers; and "ENIAC" and "UNIVAC" are members of the "early electronic" set. Chronological relationships include the inventions of the abacus, the "Analytical Engine," the tabulating machine, Mark I, ENIAC, and UNIVAC. Whole-to-parts relationships include the vacuum tubes, punched cards, programming capabilities, and overall size of ENIAC.

At this stage of the planning and arranging phase of the technical writing process, you too should be able to identify relationships between items on your list. To do so, review the list and evaluate the status of each item by asking the following questions:

1. On the basis of the relationships we have listed, what categories can be created?
2. Given the purpose of my writing project and the needs of my audience, which of these categories are important and why?
3. Given these important categories, which items can be included in each?
4. Which items belong in more than one category and why?
5. Which items do not belong in any category?
6. Of the items unassigned to a category, which are important to the purpose of my writing project and the needs of my audience?
7. Which categories are redundant (and therefore unnecessary)?
8. Which items within categories are redundant (and therefore unnecessary)?
9. What relationships exist between categories that might lead to the formulation of an overall thesis?

Actually label the type of relationship at work and reorganize the elements of the list into the categories you identify. You may find that you have far more information on one topic than on another of equal importance to your audience and purpose. If that is the case, you should conduct further research to balance your treatment. At this point it is also a good idea to resolve any conflicts that are apparent. For instance, if one of your sources claims that a specific computer meets industry communications standards, whereas another source claims the machine does not, you must determine which is correct. To do so, you may have to do further research or rely on an expert's opinion. Likewise, now is a good time to eliminate redundant or unnecessary information from the list.

Our computer science student came up with the categories previously cited as well as several others when he asked the foregoing questions about the items on his list. He filled in some important gaps involving the individual computing devices and their inventors. However, one item did not belong in any particular category; rather, it applied to all of them. The student recognized that he had happened on a potential thesis statement in the following cryptic notation: "long history of using computers to improve job efficiency." He realized that he would be able to shape this statement into one that would provide an interesting perspective on the history of computers (his ultimate purpose) and would probably satisfy the needs of his audience (the instructor). The subject matter supported this statement, and it would be easy to use the remark as a focusing point for his report. After a bit of thinking, he revised the phrase into a full-blown thesis statement: "Although we tend to think of computers as modern inventions, they have been used throughout history to help people do their jobs quickly and efficiently."

From the categories that you discern, you ought to be able to formulate a thesis statement as well. You should create a sentence that is succinct, direct, and supportive of your subject matter, audience needs, and purpose.

ARRANGING INFORMATION TO SUIT THE WRITING SITUATION

Armed with a thesis statement and related categories of information supporting that statement, the writer of a technical document is ready to consider methods of arranging information to best meet the requirements of the writing situation. The order in which information is presented supports your thesis, creates a context for learning or understanding, and helps persuade readers to agree with the findings of your document.

We will look at five overall patterns (each of which has variations) that will serve you well as you develop technical documents. These are the causal, graded, deductive/inductive, sequential/chronological, and physical patterns of organization.

CAUSAL ARRANGEMENT

Four basic types of cause-and-effect relationships exist, each of which calls for a unique arrangement of material. Although the phrase *cause and effect* is used here, these principles apply equally well to problem-and-solution, action-and-reaction, and input-and-output formats. These types of relationships are

single cause, single effect
single cause, multiple effects
multiple causes, single effect
causal chain

When dealing with a single cause and a single effect, the writer has several options. For instance, if the cause and the effect are of equal value, devote equal time to each of these two components. However, if the cause is more important to the audience than the effect, place the cause first and spend more time discussing it. If the opposite is true, reverse the order and increase the time spent on the effect. This structure can also be used to focus on a primary cause and its primary effect or on a particular event and the main reason why it occurred. As an example, consider the definition of an intrauterine device. A physician might be more interested in the way in which the conception process is stymied (cause), whereas a woman who does not want to become pregnant might be more concerned with its interruption of the conception process. The arrangement of a single cause and a single effect can be represented in either of the following ways:

SINGLE CAUSE ⟶ SINGLE EFFECT

SINGLE EFFECT ⟵ SINGLE CAUSE

depending on the interests of the audience and the purpose of the document.

A relationship involving a single cause and multiple effects can be arranged with the cause presented first and the effects next, or the effects can be presented before the cause. In either case, the effects are usually mentioned in descending order of importance to your readers or in the order in which they occurred. The emphasis you wish to place on one effect can be further strengthened by discussion length and by the use of transitional words and phrases such as: *most important, the primary effect, the first results we observed,* and *the final results were.* This arrangement can be represented graphically in either of the following ways:

SINGLE CAUSE ⟶ 1. EFFECT, 2. EFFECT, 3. EFFECT

1. EFFECT, 2. EFFECT, 3. EFFECT ⟵ SINGLE CAUSE

An example of a technical document in which arrangement in terms of a single cause and multiple effects is appropriate is a position statement issued jointly by several U.S. computer manufacturers and addressed to an audience of design engineers for data communications products. The background portion of the paper used the following arrangement:

Single Cause

The French government proclaimed that it would use only those computer systems that complied with the communications standards devised by the International Standards Organization (ISO).

Multiple Effects

1. Computer vendors not meeting these standards were not able to conduct business in France.
2. Vendors meeting them suddenly were in a more competitive position than they had enjoyed previously.
3. Several computer vendors banded together to participate in ISO in order to exercise greater control over future communications standards.

A third cause-and-effect relationship involves multiple causes that yield a single effect. As we will learn later in this textbook, causes can be classified as contributory, necessary, sufficient, or insufficient, depending on the degree of impact they have on the effect. For now, just remember that more than one cause is at work under these circumstances. The causes might come first in order of importance or occurrence and be followed by the single effect, or the effect might be followed by a discussion of its causes. This arrangement is represented graphically in either of the following ways:

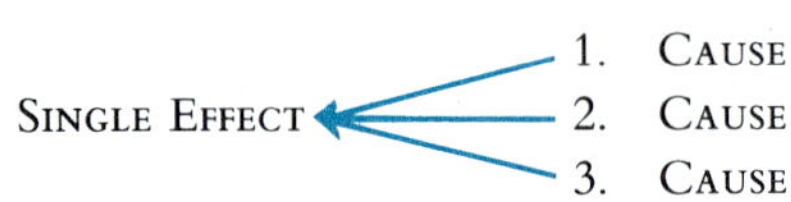

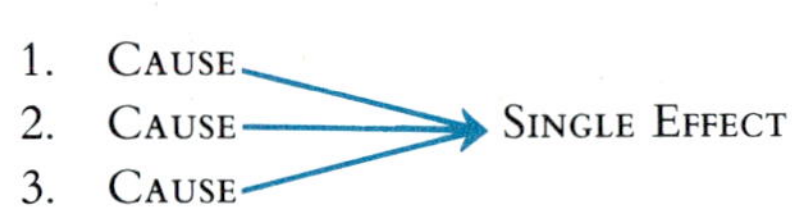

As an example of a document that describes the single effect of multiple causes, consider a physician's report on a patient's heart attack.

Multiple Causes

1. Coronary thrombosis
2. Elevated levels of triglycerides in blood
3. Excessive use of alcohol
4. Overweight
5. Insufficient exercise

Single Effect

Myocardial infarction

The final cause-and-effect relationship is called a causal chain or chain reaction and is characterized by one cause triggering an effect that in turn triggers another effect, and so on. In this case, sequential order of causes and effects works best. However, you can best meet the needs of your audience by arranging information in "straight-flow" order (start to finish) and sometimes in "reverse-flow" order (finish to start). This pattern exhibits one of the following forms:

INITIAL CAUSE → EFFECT [CAUSE] → EFFECT [CAUSE] → FINAL EFFECT

FINAL EFFECT ← CAUSE [EFFECT] ← CAUSE [EFFECT] ← INITIAL CAUSE

The power cycle of a diesel engine is a good example of a causal chain: one action prepares the way for the next. The following straight-flow arrangement of the intake stroke (one of four strokes in the cycle) demonstrates the order a writer might use to discuss the topic in a technical document.

INITIAL CAUSE		EFFECT [CAUSE]		EFFECT [CAUSE]		FINAL EFFECT
intake valve opens	→	piston goes down	→	air is sucked in	→	valve closes

Each of the four basic causal relationships can have variations as well. For instance, it is possible for a single cause to have multiple effects that in some cases cause other events. If you were to consider the topic of inflation, you would find multiple causes for the condition, which in turn has multiple effects. Such an arrangement, known as a tree or as a branching pattern, might look like this:

1. CAUSE ↘
2. CAUSE → EFFECT [CAUSE] → 1. EFFECT, 2. EFFECT, 3. EFFECT
3. CAUSE ↗

The order in which you discuss any of these items of information depends on the importance of the causes and effects to your audience and purpose. If your audience is most interested in high unemployment figures, for instance, you might start with the effects of inflation and work backward to the causes of inflation. On the other hand, if your audience is historically inclined, you might use a straight-flow arrangement beginning with the first cause that contributed to the condition.

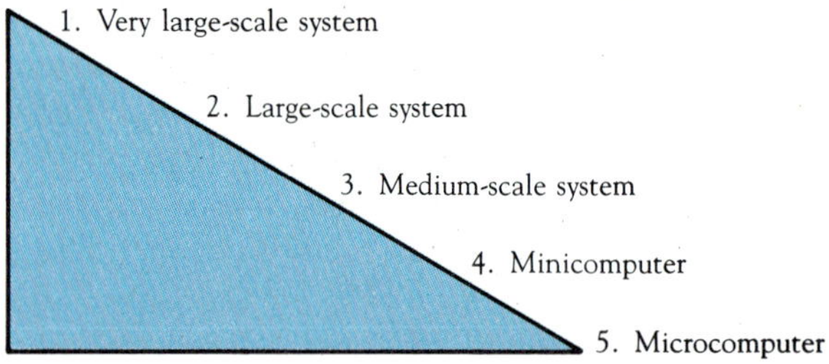

Figure 4-1 A descending arrangement of information helps readers appreciate the power of very large computers compared to that of other computer systems.

GRADED ARRANGEMENT

The graded arrangement of information is appropriate when an audience needs to understand degree or status as key aspects of the subject matter. Generally, if you were to use this pattern, you would notice that information on your brainstorming list falls into logical and/or quantifiable increments of quantities, degrees, qualities, or intensities. When dealing with concrete matters (such as car engines, computers, or airplanes), you might arrange your discussion according to size, performance figures, or speeds. When dealing with qualities (such as reliability or cost-effectiveness), you might arrange information in terms of "least to most" or "most to least." The four basic graded arrangements are descending, ascending, familiar to unfamiliar, and simple to complex.

Descending order of information is used effectively when the audience is most concerned with the biggest, best, fastest, most cost-efficient, or most powerful aspects of your topic. For instance, in a technical document defining types of computers, the writer might present information about the computers in descending order of their power. A very large-scale system might support thousands of simultaneous users, a large-scale system a thousand, a medium-scale computer several hundred, a minicomputer a dozen, and a microcomputer only one user at a time.

On a more abstract level, descending order would be effective in a proposal wherein you list recommendations from most vital to least important. This direct approach works well for audiences of busy executives, reviewers of technical documents, and budget committees. Each of these readers is usually pressed for time and needs to know the "bottom line" in a hurry.

The descending arrangement of points is represented graphically in Figure 4-1.

The converse of the descending arrangement is ascending order, or the "least-to-most" structure. This method is useful when information builds

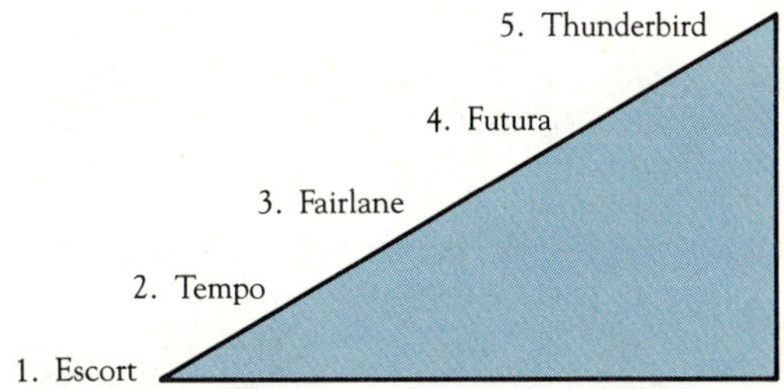

Figure 4-2 An ascending arrangement of information helps readers compare the benefits of small, fuel-efficient cars to increasingly larger, less economical models.

in degree or intensity and when that quality is important to your audience. For instance, with a budget-conscious audience in the market for a new car, you might begin your discussion with economy models that are devoid of accessories, small, and fuel-efficient. In progressive steps "upward," you would then describe increasingly more luxurious models that have more space and frills but also consume more fuel and cost more. Such an ascending arrangement of information is represented graphically in Figure 4-2.

When dealing with abstractions rather than concrete entities, writers can use either an ascending or a descending arrangement of information to reinforce such least-to-most or most-to-least attributes as being

accessible	expensive	promising
advanced	feasible	rare
affordable	important	reliable
common	innovative	remote
dynamic	justified	sophisticated
effective	likely	successful
efficient	modern	trustworthy
enjoyable	probable	valuable

Another type of graded arrangement presents a combination of familiar and unfamiliar information to an audience. This arrangement continually creates the appropriate context to ease readers from the comfort of the known to acceptance of the unfamiliar. This kind of arrangement is sometimes useful for drawing comparisons or analogies between similar circumstances, locations, or materials. Because the writer is equipping the reader to understand something new and different, the familiar information generally precedes the unfamiliar (though reverse flow is possible).

Consider an astronaut who is explaining the problems of weightlessness to earthbound readers. He or she might begin by mentioning weightless-

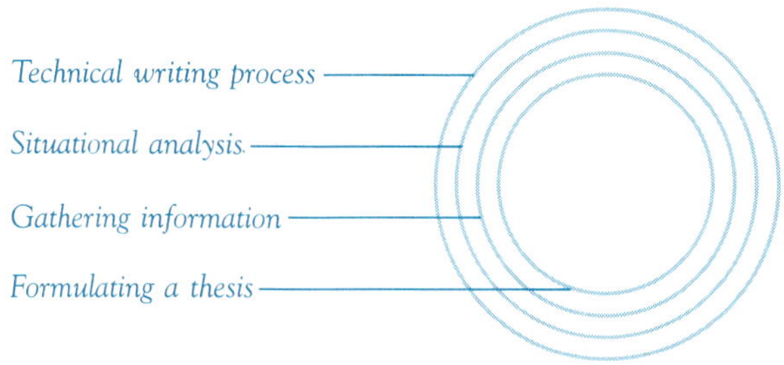

Figure 4-3 A deductive arrangement explains general information first, then proceeds to increasingly specific information.

ness but then turn quickly to a familiar situation: swimming under water. The audience's familiarity with the buoyancy and lack of control experienced in this situation serves as the foundation for the astronaut's discussion of similar sensations and problems that occur when one is trying to maneuver in space.

A final type of graded arrangement is suited to writing situations in which readers must progress from relatively simple to relatively complex levels of information. This structure advances your argument in increasingly more sophisticated steps to equip readers to learn new material gradually. The organization of this textbook and most others reflects this overall pattern. Students receive basic instruction that serves as the foundation for mastering more complicated material, which in turn serves as the basis for learning still more sophisticated information.

DEDUCTIVE/INDUCTIVE ARRANGEMENT

Deductive arrangement presents information in a general-to-specific order. For instance, our discussion of the technical writing process has been deductive thus far: we started with a general understanding of the process (Chapter 1), moved to a closer look at the whole picture (the situational analysis of Chapter 2), progressed through several layers of research—all the while narrowing our focus—to the very specific activity of defining a thesis statement. Each successive activity has been increasingly specific and directed; nonetheless, each activity depends on completion of the broader activity preceding it. A deductive arrangement can be represented as a series of concentric circles, where the movement of ideas progresses from the outside inward (see Figure 4-3).

Inductive arrangement, on the other hand, presents specific cases first and draws from them a general statement or conclusion; it is the piecing together of particular *bits* of information to construct a general theory or statement. If, for example, you are conducting a series of laboratory tests on the reactions of various *Plantago ovata* strains to several different herbicides, you might discuss each experiment with each combination of herbicide and plant strain and conclude your report with any generalizations that the results support about toxicity, reactions, survival rates, or recommended methods of treatment.

SEQUENTIAL/CHRONOLOGICAL ARRANGEMENT

A sequential or chronological arrangement is best used for presenting information that falls into a time-ordered pattern. Audiences who want to know *how* to do something (such as set up an experiment or clean a handgun) expect to follow a step-by-step (sequential) arrangement of information. Audiences who are concerned with the *history* of a subject (such as the evolution of *Homo sapiens* or computer science) expect to follow a date-by-date (chronological) arrangement of the information. In both cases it is possible to use either straight flow or reverse flow. However, in the case of procedural instructions, it rarely is a good idea to work backwards.

If we return to our computer science student's brainstorming list and begin arranging elements according to the needs of the audience and the writer's purpose, we see that a chronological pattern makes very good sense. The student's report might begin with a brief glimpse of the use of "computers" in the marketplaces of the Tigris–Euphrates valley, where grooves were formed in sand and small stones manipulated to register transactions. He might progress to the abacus (the next invention), move on to "La Pascaline," cite the "Analytical Engine," and introduce the tabulating machine, Mark I, ENIAC, and UNIVAC. He might then survey several important modern-day machines and bring his report to a close by looking at the ways in which today's most powerful computers register transactions.

PHYSICAL ARRANGEMENT

Naturally, all of the arrangements we have been discussing are physical in that bits of information are placed in some physical relationship to one another on a page. Here, however, the word *physical* applies to an arrangement of information about concrete items—mechanisms, buildings, locations—wherein discussion of physical attributes is important to your audience. When this is the case, information can be arranged spatially (according to the structural relationship of parts) or functionally (according to the operating relationships or interdependencies of parts). As with other patterns, it is possible to use straight flow or reverse flow, depending on the emphasis you are trying to achieve.

Imagine for a moment that you are the design engineer for a new, more powerful lawnmower. In your engineering specifications report, you might present a design drawing and then discuss each of the parts of the device as they are physically connected to one another (spatial arrangement). You might start with the rotary blades at the base and move upward through the engine and gears to the handle. Or you might start with the comfortable grip of the handle and move downward toward the powerful rotary blades. Another alternative is to trace the functional relationships between parts in your discussion, demonstrating the inner workings of the engine and the ways in which the blades and wheels are driven.

CREATING AN OUTLINE

Whichever arrangement you choose—causal, graded, deductive/inductive, sequential/chronological, or physical—it must help you strike a balance between the scope of your subject matter and the requirements of your audience. At this stage of the technical writing process, you should return to the brainstorming list and impose the most appropriate arrangement on the categories that you have identified. For some writers, the simple act of arranging the listed items into the appropriate pattern will be an adequate foundation for writing the document. In fact, at this point you already have an *informal* outline. However, many writers work more effectively from a *formal* outline, which is especially useful for longer documents.

Two formal outline styles exist: the Roman style, which you have probably used in other writing courses, and the decimal style, which you may have noticed being used in scientific and engineering disciplines.

THE ROMAN OUTLINE STYLE

The *Roman outline* incorporates Roman numerals, capital letters, Arabic numerals, lower-case letters, and various punctuation conventions. Its objective is to demonstrate categories of information and general-to-specific relationships within categories. You must decide whether to use full sentences or phrases for each level of subordination. The most useful style is the full-sentence outline. From this kind of outline, it is a relatively simple task to launch the actual document.

Each level of subordination is indented from its more superior element, and ideas are developed in parallel. That is, if an idea has one subelement, it should have at least two. If an idea cannot be broken into two subelements, it is more efficient to remain at the higher level of treatment.

Let us look at a small portion of the computer science student's brainstorming list, now arranged chronologically and recast as a formal full-sentence, Roman outline.

THESIS Although we tend to think of computers as modern inventions, they have been used throughout history to help people do their jobs quickly and efficiently.

I. Manual computers were used by ancient cultures.
   A. One type of manual computer involved sets of regular-sized stones.
      1. Egyptians used these primitive computers in their marketplaces to calculate prices and register transactions.
      2. The stones were moved along established grooves.
   B. A second early computer is the abacus.
      1. It was invented in the Tigris–Euphrates valley 2500 years ago.
      2. The abacus is still used widely in China.
      3. Its operation is relatively straightforward.
         a. Inoperative, its beads are to the left with a space to the right.
         b. For addition, the beads are moved to the right.
         c. For subtraction, the beads are moved to the left.
         d. When one row fills, the person calculating moves to the next higher (addition) or lower (subtraction) row.
II. Mechanical computers were first invented in the mid-1600s.
   A. "La Pascaline" is the first documented mechanical computer.
      1. It was invented by the famous mathematician Blaise Pascal in 1642.
      2. He designed a box that, at first, could perform only addition and subtraction.
         a. Functions were performed via a series of gears.
            (1) The rightmost gear represented units, the next represented tens, and so on.
            (2) The long tooth on the units gear moved the tens gear one notch for each revolution.

(3) For addition, gears were moved to the left; they were moved to the right for subtraction.

b. The device was enhanced several years later to perform multiplication and division.

B. The "Analytical Engine" was invented by Charles Babbage in the mid-19th century.

1. He borrowed Joseph Jacquard's notion of using punched cards to indicate patterns.

a. Jacquard used punched cards to indicate weaving patterns.

b. Babbage imagined the same kinds of cards could be used to indicate mathematical "patterns."

2. Babbage's ideas were sound; however, the technology of his day did not provide him with metalsmiths who could create the sensitive gears required.

3. His machine was never completed.

4. His machine's design was incredibly innovative for his day.

a. It would have been programmable.

b. It would have had a limited form of memory.

THE DECIMAL OUTLINE STYLE

The *decimal outline* differs from the Roman in that it uses numbers and subordinate numbers rather than a combination of alphanumeric characters. Many scientists and engineers find this style easier to use; the interrelationships between parts are always apparent, which facilitates cohesive writing. The first point of discussion is designated 1. Points subordinate to it are numbered 1.1., 1.2., 1.3., and so on. If any of these secondary points have tertiary points, they are designated 1.1.1., 1.1.2., and so on. As more points are placed at the same level of subordination, only the last numeral changes to reflect their addition. The numbers to the left act as referents to the point being discussed and help the writer keep track of the level of subordination. Most veteran writers admit that this style becomes unwieldy when more than four levels of subordination are used.

What follows is a continuation of our computer science student's outline. To demonstrate the decimal outline, we focus on the third category of his list, "early electronic computers." In practice, of course, you would *not* mix Roman and decimal styles.

3. Early electronic computers were able to overcome the slowness of mechanical switches by using vacuum tubes.

- 3.1. The first electronic computer was called ENIAC (Electronic Numerical Integrator and Calculator).
  - 3.1.1. It was invented by Drs. Presper Eckert and John Mauchly of the University of Pennsylvania in 1947.
  - 3.1.2. Its vacuum tubes were connected as switches to control one another.
    - 3.1.2.1. Electricity in the tubes was able to turn the flow of current on and off in one-millionth of a second.
    - 3.1.2.2. A tube would burn out every 7 minutes.
  - 3.1.3. ENIAC was designed to predict weather, calculate weapons trajectories, and the like.
  - 3.1.4. It weighed 30 tons, occupied 1500 square feet, and stood 9 feet high.
  - 3.1.5. This machine could do 20 laborhours of work in only 30 seconds.
    - 3.1.5.1. However, it was not programmable.
    - 3.1.5.2. When functions were changed, plugs and cables had to be rearranged.
- 3.2. UNIVAC (Universal Automatic Computer) was the first true electronic computer.
  - 3.2.1. It was also invented by Eckert and Mauchly, only 4 years later.
  - 3.2.2. UNIVAC was programmable by means of punched cards.
    - 3.2.2.1. It could add 237,000 five-digit numbers in 1 minute.
    - 3.2.2.2. It could check for calculation errors.
    - 3.2.2.3. It could handle input other than numbers.
  - 3.2.3. UNIVAC was considerably smaller than ENIAC.
    - 3.2.3.1. It weighed only 4 tons.
    - 3.2.3.2. It occupied only 575 square feet.

As you can see, either style of outline will help you develop a well-organized, carefully planned document. With relative ease, you can move from the detailed outline or even from your brainstorming list to actually writing your document. The more detailed the outline you develop, the easier the writing job you face. With practice, you will become able to develop an outline quickly; you will anticipate its shape and many entries even as you perform some of the prewriting tasks of the technical writing process.

CHAPTER SUMMARY

Planning and arranging information to meet the requirements of the writing situation will ensure a better document. Using brainstorming techniques and a review of your research, you should list the vital information you must include in your document. Then, on the basis of the situational analysis you prepared in phase one of the technical writing process, evaluate the list and organize its elements into categories that make sense for your audience and purpose. A variety of relationships can exist between elements, including agreement, comparison, conflict, equality, functionality, membership, and partition. Within these categories, you should arrange information according to the needs of your audience and your purpose. Five basic arrangements exist: causal, graded, deductive/inductive, sequential/chronological, and physical. Armed with an appropriate arrangement of elements, you are ready to develop a formal outline, using either the Roman or the decimal style. This outline will be the framework on which you develop your written document in the final phase of the technical writing process.

EXERCISES

1. Reread Chapter 1 of this textbook and create a detailed, full-sentence decimal outline of it. (This outline will come in handy when you study for your final examination.)

2. Suppose you are a recruiter for your university and are writing a letter to a potential candidate for admission. Arrange the following random points in descending order of importance to this reader. Brainstorm to create a list of points you might make about each of these elements.

social life	quality of faculty
part-time job	tuition costs
grading standards	campus accommodations
availability of courses	location of college

3. Suppose you are a teaching assistant in a beginning biology class and must introduce simple life forms to your students. Arrange the following topics in simple-to-complex order and brainstorm to create a list of points you might make about each of these elements to these new students.

cell	protoplasm	protozoan
amoeba	paramecium	nucleus

WRITING ASSIGNMENT

Select a project in your major field that you are in the process of working on or that you have recently completed. Imagine that your supervisor has requested a written report on your progress. In preparation for writing this report, develop the following materials: situational analysis, list of reference materials, brainstorming list, and formal outline. On the formal outline, identify the arrangement you are using (causal, graded, deductive/inductive, sequential/chronological, or physical). Explain why the arrangement you selected is the most appropriate to the writing situation you have defined.

CHAPTER

5

Writing and Revising

CHAPTER PLAN

Having reached the last phase of the technical writing process, you are in a position to develop an effective technical document—one that commands attention, gives your readers the information they need in order to follow your ideas, and accomplishes your purpose. Armed with the outline you developed in the last chapter, you can graduate from planner to writer with confidence.

This chapter will

- briefly review basic techniques for developing paragraphs.
- introduce various matters to consider as you revise.
- discuss the importance to technical documents of a review cycle.
- help you create a final document worthy of distribution to your audience.
- provide editing criteria applicable to a variety of technical documents.

CREATING A ROUGH DRAFT

Technical writing addresses people who wish to be informed and who generally want facts without frills. Accordingly, you must present information precisely, striking a balance between conciseness and clarity that enables you to accomplish your purpose. Writing precisely takes skill, discipline, judgment, and revision. No one can write well without being able to revise well. Hence a good deal of this chapter is about specific techniques you can use to fine-tune your message into truly successful communication.

Precise writing begins with your thesis, the primary point you wish to present to your readers, and is carried through to the level of the sentence and even the individual word. Let us begin our examination of the final phase of the technical writing process with the thesis statement. Although the thesis is not always a proposition or argument in technical writing, a specific message usually takes precedence over all others in the document.

In the case of a status report, for instance, the focal point might be simply that all projects are on schedule and are progressing according to plan. In the case of a proposal, your main point might be that the contract should be awarded to your firm because it is best equipped to handle the complex job. A user's manual might convey the primary message that "using the IBM PC XT can help you increase productivity," whereas a mechanism description might point out the many ways in which the Honeywell DPS 88 computer system is fault-tolerant. In each of these cases, support for the thesis and elaboration of it constitute the body of the technical document.

DEVELOPING TOPIC SENTENCES FOR PARAGRAPHS

The outline you already have developed is the ideal platform from which to launch your writing activity. As you were developing the outline, you considered your primary message and arranged supporting information in ways that would meet the needs of your audience. From the outline you can develop skeletal paragraphs to flesh out as you continue writing.

To begin shaping paragraphs from your outline, examine the largest units of information in the list. These are usually found at the primary level (1 or 2 in decimal outlines, I or II in Roman outlines) or at the next lower level of subordination. These sentences, words, or phrases will serve as the foundation for topic sentences in support of the thesis of your document.

Topic sentences have at least two purposes: they corroborate the thesis or amplify it, and they promise the reader that the paragraph will discuss the topic. You must live up to this agreement. If you fail to discuss a topic you have announced, your readers will be disappointed or—worse yet—

confused. These sentences should be as simple and direct as possible and should strongly support your document's thesis.

Let us examine how a writer develops topic sentences from an outline by looking at the following excerpt from an outline.* Here the writer intends to write an introductory report on a specific human-made catalyst that helps petroleum industries develop a low-grade form of gasoline from natural gas. The audience is a group of businesspeople who tour the company plant; they will receive this document as a handout during the tour. These readers are neither skilled petroleum engineers nor chemists. They are interested in the investment potential of new technology or in ways in which a new catalyst might improve quality, productivity, or efficiency.

The overall arrangement of information in the outline is deductive. First the writer defines catalysts and discusses their biological and industrial uses. The outline then focuses on a specific industry: petroleum production. Focusing further still, the writer discusses the catalyst used widely in that industry. The focusing process culminates in a discussion of a specific strain of that catalyst, which is proving very useful to the company.

Note that this report is to have several sections (marked in the left margin) and that each entry is numbered with a Roman numeral and acts as the topic sentence for its section. In turn, the entry at the first level of subordination acts as the foundation for topic sentences that support the section's main point. All the sentences derived from the outline support at least a portion of the thesis.

Not every paragraph requires a topic sentence. In lengthy documents such as the one this outline will yield, each topic is more likely to require several paragraphs of amplification. Sentences created from the outline's lower levels of subordination serve as the basis for these paragraphs.

THESIS Modern industrialists use chemical catalysts to support innovative processes. Mobil Oil has made significant strides in the development of a synthetic catalyst that will help us turn natural gas into gasoline.

I. Living organisms and industrial processes depend on catalysts[1]
  A. Substances that speed or initiate chemical reactions
    1. Enzymes accomplish work in several ways:
      a. Divide molecule at specific point

1. Definition section. Section topic sentence: *Both living organisms and modern industries depend on catalysts to speed up or initiate processes.*

*Based on Gene Bylinsky, "The Magic of Designer Catalysts," *Fortune,* May 27, 1985.

b. Join 2 molecules in specific way
c. Transfer group of atoms from one molecule to another
d. Enzyme remains unchanged at end of process
2. Examples:
a. Human—digestion, muscle movement, tissue building, energy storage
b. Industry—milk into cheese, sugar and starch into alcohol
3. Most natural enzymes too fragile for widespread industrial use

B. Synthetic substances used in commercial processes[2]
1. Examples—mass-production of gasoline and fuel oil, fabrics, plastics
2. $1.5 billion/year business

C. Emergence of new research technology, called molecular engineering, widens scope of catalyst development[3]
1. Reflects understanding of relationship between molecular structure and function
2. Improves existing processes
a. Less wasteful
b. Less costly
c. Less time-consuming
3. Can aid in development of raw materials

II. U.S. petroleum industry especially dependent on synthetic catalysts[4]
A. Spurred by oil shortages
1. Need for higher yield from crude oil
2. Minimize dependency on foreign resources
3. Catalysts already exist to help turn methanol into high-octane gasoline
a. Costly
b. Energy-intensive process

2. Paragraph topic sentence: *Synthetic catalysts are used in a variety of commercial processes.*

3. Paragraph topic sentence: *A new research technology, molecular engineering, dramatically widens the scope of catalyst development.*

4. Overview of industry. Section topic sentence: *Because of recent oil shortages, the U.S. petroleum industry depends heavily on synthetic catalysts to yield more gasoline from crude oil.*

B. Aim is to reduce energy consumption of process and limit wasteful by-products[5]

   1. Molecular engineers study natural enzymes to learn method

      a. *Methanolococcus capsulatus* bacterium converts methane to methanol in single step

         (1) Methane principal ingredient of natural gas

         (2) Natural process involves breaking and making chemical bonds

   2. Biomimetics, a subset of molecular engineering, may give us clue

      a. Practitioners have learned enzymes are able to eliminate unstable molecules and stop reaction

      b. By mimicking natural enzyme, new synthetic catalyst may be able to be stable

5. Paragraph topic sentence: *At this time, considerable research is aimed at reducing the energy consumed by the synthetic catalyst and at limiting undesired by-products.*

III. Zeolites now most widely used catalyst in petroleum industry[6]

A. Act as catalytic sieve

   1. Composed of silicon dioxide and small amounts of aluminum

   2. Well-defined molecular geometry of holes, channels, and caves

      a. Provide large area for catalytic action

      b. Let in some molecules but not others

B. Mobil scientist realized their usefulness as catalysts[7]

   1. Aluminum atoms in channels act on specific molecules—joining or separating

   2. Break down the hydrocarbons in crude oil into high-octane gasoline

   3. Mobil experimenters fine-tuned dimensions of channels in synthetic zeolites

6. Focus narrows. Section topic sentence: *Zeolites are now the most widely used catalyst in the petroleum industry.*

7. Paragraph topic sentence: *Mobil Oil was one of the first companies to recognize the value of zeolites in the petroleum refinement process.*

C. Now 90% of refineries in non-Communist world use synthetic zeolites
   1. Yield increased by up to 40%
   2. Zeolite production itself has become $250 million/year business

IV. ZSM-5 can help produce gasoline molecules from methanol[8]
   A. Most versatile of many synthetic zeolites produced by Mobil
      1. Used in 10 different processes
      2. Upgrades petroleum and chemical products
      3. Makes synthetic fuels
   B. Will be used to turn methanol into high-octane gasoline in New Zealand
      1. Joint venture between Mobil, Bechtel, and New Zealand Synthetic Fuels
      2. Will turn smaller methanol molecules into larger gasoline molecules
   C. Ability of ZSM-5 to make gasoline wasn't designed[9]
      1. Just an outgrowth of development
      2. Process can be stopped along the way to create ethylene and propylene as well

8. Focus narrows further. Section topic sentence: *ZSM-5, the most versatile synthetic zeolite, is the most likely to help us produce gasoline from methanol.*

9. Paragraph topic sentence: *This ability to convert methanol into gasoline was not actually designed into ZSM-5 but was a synergistic outgrowth of experimentation.*

The writer has assessed the importance of each packet of information in the outline and converted the more significant items into topic sentences. At this point it is not important to create "perfect" sentences. Your goal is to begin writing and thereby sow the seeds from which fuller, more elegant sentences and paragraphs can grow.

DEVELOPING SUPPORT FOR TOPIC SENTENCES

The next logical step is to expand these topic sentences into preliminary supporting sentences that develop the idea they introduce into a complete paragraph. A *paragraph* is a distinct portion of a document dealing with a particular, single idea. Although it may be necessary to use more than one paragraph to cover a specific topic, an individual paragraph should not cover more than one topic. And when two or more paragraphs are required to discuss an idea, each paragraph becomes a subset of that topic. This happens at least twice in our sample outline: in the definition of enzymes and in the emphasis on reducing energy consumption and wasteful by-products.

The supporting sentences you create should be as closely linked to the topic sentence as your topic sentences are to the thesis. No set number of sentences can be prescribed, but a paragraph never should be so long that readers lose sight of its original point. The more closely linked each sentence is to the topic sentence, however, the easier it will be for readers to follow your reasoning, even in a relatively long paragraph.

Working from the outline in conjunction with the topic sentences you develop, you can create supporting sentences in several ways:

1. by answering the implied question "Why is this true?" and supplying your answers as the support.
2. by recognizing the whole-to-parts relationship between bits of information and making those connections clear to readers.
3. by moving from a general statement in the topic sentence to successively more specific points in the body of the paragraph.
4. by recognizing a chronological or sequential pattern (working forward or backward from the topic sentence).

In fact, the arrangement patterns we cited in Chapter 4, when we discussed the structure of your outline, also apply to individual paragraphs. Your sentences can reflect a causal chain, a comparison and contrast between bits of information, a movement from simple to complex information, and so on. Thinking about these relationships may help you generate the sentences you require if you are ever faced with "writer's block." But more often the piecing together of supporting sentences is a simple matter of putting pen to paper and letting the ideas flow.

To demonstrate the relationship between the topic sentence and its supporting sentences, let us return to the example involving synthetic catalysts. The writing is still rather disjointed at this point and somewhat repetitive; nonetheless, the writer makes clear both the focus of the research and its objectives.

TOPIC SENTENCE At this time, considerable research is aimed at reducing the energy consumed by the synthetic catalyst and at limiting undesired by-products.

Supporting sentences:

1. Molecular engineers are concentrating their efforts on the study of *Methanolococcus capsulatus.*

2. This natural catalyst is a bacterium "convertor" of methane and may provide a clue to increased processing efficiency for synthetic catalysts.

3. Practitioners of biomimetics (a subset of molecular engineering) have recently discovered that this bacterium has the ability to eliminate unstable molecules during its natural processing of methane.

4. It then halts the process at the point that yields gasoline.

5. There are researchers who expect to mimic this natural behavior in the synthetic catalysts they are developing.

6. If they are successful in mimicking the natural tendency of the bacterium to break and make chemical bonds without waste, they will also be able to reduce the energy consumed in the process.

Subject each of the topic sentences you have created to this process of developing supporting sentences. At the same time, incorporate any elements that format specifications call for so that you wind up with a rough draft of your document. In the case of our example, the format calls for little more than the use of headings.

REVISING THE ROUGH DRAFT

Certainly the resulting skeleton is not yet suitable for presentation to either your reviewers or your audience. It is sketchy, the ideas are not very clearly connected, repetition occurs, and some thoughts are not fully expressed. You should revise these paragraphs to ensure appropriate sentence structures, precise word choice, logical flow of ideas, completeness of exposition, and accurate mechanics.

SELECTING APPROPRIATE SENTENCE STRUCTURES

Few writers give much thought to the sentence structures they use; they just write down whatever they have to say without trying to match the medium to the message. To be a successful writer in a technical environment, however, you should consider both *what* you are saying and *how* you are saying it. The message should have the benefit of the most appropriate sentence structure possible.

Each of the four basic sentence structures has its own strengths and effects. Once you become acquainted with the impact of each type of sentence, you will be able to select the structure that best complements your message. The *simple sentence* is appropriate when your reader needs to concentrate on a single, unelaborated point. The *compound sentence* (really two or more simple sentences combined) is used to demonstrate a dynamic equality between two or more ideas. Use the *complex sentence,* the most common structure, to demonstrate a dependent relationship of one idea to another. Finally, the *compound-complex sentence* is useful for balancing two complex ideas.

The chart shown in Table 5-1 summarizes the positive and negative effects of each of these structures. The negative effects occur only when a writer depends solely on a single structure or misuses the type of sentence.

With this chart in mind, let us return briefly to the topic sentence and supporting sentences of our example. The writer does not seem to have given much thought to matching the structure to the message. The readers are relatively sophisticated, so the tendency to use longer sentences is appropriate. But do the two ideas in the compound sentence (supporting sentence 2) really deserve a balanced treatment? A complex sentence would be more appropriate to both the information and the writing situation.

Original: This natural catalyst is a bacterium "convertor" of methane and may provide a clue to increased processing efficiency for synthetic catalysts.

Revised: This bacterium, a natural converter of methane, may provide a clue to increased processing efficiency for synthetic catalysts.

Note in the revision that the notion "*Methanolococcus capsulatus* is a bacterium" is now subordinate to the idea that it "may provide a clue." It was inappropriate to place equal emphasis on these ideas. Similarly, it is both awkward and confusing to place the single idea "It then halts the process at the point that yields gasoline" in a simple sentence where it receives undue emphasis within the paragraph.

Reviewing your rough draft sentence by sentence helps you select the most efficient structures to use and improves the quality of the document. To determine whether a simple sentence is best in each situation, ask the following questions:

- Does the reader need to concentrate on a single, unelaborated idea?
- Should the idea be the focal point of the paragraph?
- Is an arresting beginning required for the paragraph in order to catch the reader's eye?
- Will rhythmic alternation of sentence patterns help drive home the information?

Sentence structure	Positive effects	Negative effects
Simple—Has a single main idea (independent clause). Example: *The boat capsized.*	1. *Simplicity*: Reader can concentrate on a single unit of thought. 2. *Emphasis*: Ideas are set apart from those in surrounding, longer sentences. 3. *Brevity*: As start of paragraph, compels reader to find out more. 4. *Rhythm*: Balanced with longer sentences, lures audience. 5. *Isolation*: Breaks complex subject into manageable units.	1. *Over-simplification*: Can alienate reader. 2. *Undue emphasis*: Contents automatically highlighted, even when wrong to do so. 3. *Fragmentation*: String of undifferentiated ideas.
Compound—Has two or more main ideas. Example: *The boat capsized and the passengers got wet.*	1. *Correlation*: Creates dynamic, inherent relationship between two ideas: equality, sequence, or juxtaposition. 2. *Rhythm*: Balance between two ideas of equal importance.	1. *Inappropriate correlation*: Makes inaccurate relationship of equality, sequence, or juxtaposition. Misleads reader. 2. *Fragmentation*: String of undifferentiated ideas.
Complex—Has one main idea and one or more lesser ideas. Example: *When the boat capsized, the passengers got wet.*	1. *Subordination*: Indicates clear-cut relationship between two or more ideas. 2. *Emphasis*: Idea in main clause receives greater attention. 3. *Flexibility*: Variety of conjunctions to express wide range of relationships. 4. *Economy*: Eliminates redundancy of separate sentences.	1. *Inappropriate subordination*: Depicts inaccurate relationship between two or more ideas. 2. *Inappropriate emphasis*: Idea in main clause does not deserve greater attention. 3. *Inappropriate relationship*: Inaccurate indicator of relationship selected.
Compound-complex—Has two or more main ideas and one or more lesser ideas. Example: *When the waves came up, the boat capsized and the passengers got wet.*	*Compactness*: Combines many ideas into a single packet of information.	1. *Density*: Too many ideas combined. 2. *Lack of clarity*: Relationships between ideas become jumbled. 3. *Lack of emphasis*: Importance of any one idea is lost in the mass.

Table 5-1 Each of the four types of sentence structures offers certain advantages. And the overuse or misuse of each leads to characteristic flaws.

- Is the subject sufficiently complex to be broken down into smaller, more manageable units?

If you are able to answer any of these questions affirmatively, the simple sentence is appropriate.

If you rely too much on the simple sentence, as Table 5-1 suggests, you can create negative effects instead. At the end of an entire paragraph of simple sentences, readers may forget the majority of points or, worse yet, infer meanings you do not intend. Often relationship indicators (found in other sentence types) are vital to comprehension.

To determine whether a compound sentence is appropriate, you need ask only one question:

- Should a dynamic relationship of equality, sequence, or tension between two or more ideas be stressed?

If you answer "yes," you should use a compound sentence because it is a powerful tool in establishing strong associations between separate items of information. Used inappropriately, however, the compound sentence misleads readers into placing equal value on unequal ideas.

The complex sentence also conveys at least two units of information. But in this case only one idea makes sense independently of the rest of the sentence. The other unit communicates an incomplete thought (even though it has both a subject and a verb). The incomplete thought is "subordinate" to the other. This process of subordination indicates relationships between ideas to your readers. As the complex sentence combines notions, it also economizes on words and sentences. Using a subordinate conjunction—a special type of word indicating a certain relationship—this sentence binds independent and dependent portions of the structure. For example, here are two related ideas (causality):

1. Zeolites make extra gasoline yield possible.
2. Some countries depend on extra gasoline yield in order to survive oil shortages.

Combining these into a single, complex sentence demonstrating that relationship, we get "Some countries survive oil shortages because of the extra gasoline yield that zeolites make possible."

To determine the effectiveness of a complex sentence in a given writing situation, you should ask these questions:

- Will information be clarified if ideas are combined?
- Do natural or logical relationships exist between two or more ideas that are important to readers?
- Will the message be more concise if such relationships are pinpointed?

Be sure the information warrants the close bond forged by the complex sentence, and *place the more important information in the independent portion of the sentence.*

The compound-complex sentence is appropriate only when two or more ideas of equal importance are modified by another, subordinate item of information. This type of structure can be complicated and difficult to follow. When you consider using it, ask yourself

- Are the two main ideas of this sentence equal?
- Are they equally supported by the subordinate information?

Unless you answer "yes" to *both* of these questions, you should select an appropriate alternative sentence pattern.

SELECTING APPROPRIATE LANGUAGE

After this brief refresher course in sentence structure, you can now concentrate on the words and phrases within your sentences. Almost all writers suffer from wordiness; they simply use too many words. Many writers also have trouble selecting the best word for the occasion. By eliminating extra "baggage" and by choosing your words carefully, you can make your documents far more effective.

To reduce wordiness in your rough draft, identify the following elements by circling them:

- repetition—the use of the same words or phrases several times in a single passage.
- prepositional phrases—used to indicate such relationships between a subject and an object as position, location, time, or condition. Common prepositions include *after, at, before, by, from, in, of, on, to,* and *with.*
- relative clauses—complete ideas that modify another part of the sentence and begin with the word *that, which, who,* and *whom.*
- expletives—words acting as fillers that add no meaning to the sentence. The most common expletives are *there is* or *there are.*

With the exception of expletives, all these elements have their proper place in technical documents. But when they are misused, the result is wordiness. Examine each use of these elements in your draft to determine whether it is appropriate.

Repetition is used effectively to create logical connections between sentences and paragraphs or to reinforce key ideas. In all other cases it should be eliminated.

Prepositional phrases are used to modify the subject or predicate (a verb or verb phrase and its object) of the sentence. Very often, they can be replaced by a more precise adjective (modifier of a noun) or adverb (modifier of a verb).

Relative clauses are used (1) to demonstrate subordination, (2) to modify a noun (in which case they generally appear after the noun), or (3) as a noun, operating in the sentence as either the subject or the object of a verb.

Expletives, primarily *there is* and *there are,* add absolutely no value to a sentence and usually require the inclusion of a relative clause as well.

The example that follows demonstrates how you can eliminate each of these extra elements without jeopardizing clarity or emphasis. First look at the rough draft after the writer has circled the elements that need to be scrutinized.

> At this time considerable research is aimed at re-
> ducing the energy consumed by the synthetic catalyst
> and at limiting undesired by-products. Molecular en-
> gineers are concentrating their efforts on the study of
> Methanolococcus capsulatus. This bacterium, a natural
> convertor of methane, may provide a clue to increased
> processing efficiency for synthetic catalysts. Practitioners
> of biomimetics (a subset of molecular engineering) re-
> cently have discovered that this bacterium has the abil-
> ity to eliminate unstable molecules during its natural

processing of methane. It then halts the process at the point that yields gasoline. There are researchers who expect to mimic this natural behavior in the synthetic catalysts they are developing. If they are successful in mimicking the natural tendency of the bacterium to break and make chemical bonds without waste, they will also be able to reduce the energy consumed in the process.

This analysis reveals that the passage is fraught with prepositional phrases, repetition, and unnecessary use of relative clauses, and an expletive has even crept in. While revising, the writer realized that other modifications could also be made—changes that enhance the meaning and make crucial relationships more apparent. Note especially that the topic sentence of the paragraph has been changed to emphasize the idea that previously occupied the second sentence of the paragraph. During the revision process, the writer realized that this is a more accurate way to convey the information.

Currently, molecular engineers are concentrating their research efforts on Methanolococcus capsulatus, in an attempt to limit both the energy consumed by synthetic catalysts and the undesired by-products they create in processing. Scientists have discovered that this bacterium, a natural convertor of methane to gasoline, eliminates unstable molecules during its catalytic action and halts processing once gasoline is produced. Biomimetics, a specialized subset of molecular engineering, is dedicated to creating synthetic catalysts that mimic this behavior. By producing artificial substances to break and make chemical bonds without waste, these researchers hope to develop a more direct conversion process that consumes less energy.

Because the writer is addressing a sophisticated audience of businesspersons, the complex sentence structure is retained. And repetition of key words—*catalyst, molecular engineering, synthetic, energy,* and *consume*—is still used to unify the passage. However, the word *natural,* which appeared four times in the original passage, appears only once, and *bacterium* now occurs once rather than three times. Relative clauses still abound, but now they occur only where no other structure is so appropriate. The expletive has been deleted.

We can all agree that this sample passage is much improved from its original and is light years ahead of the phrases and sentences in the outline. If such careful analysis were conducted throughout a rough draft, the result would undoubtedly be a succinct, informative document.

Before we go on to some other revision techniques that will help you polish your rough draft into an acceptable review copy, we want to emphasize the importance of eliminating verbiage. Revision is not easy, but most writers make it more difficult than it has to be. For some reason, many people are reluctant to drive the hard edge of a pencil across a word, sentence, or paragraph they have written. Pride of authorship is very real, but your goal is to construct a *document* you can be proud of, and that means one that addresses your readers' needs. In a technical writing environment, where most readers are in a hurry and want the facts (just the facts), you do them a tremendous service by eliminating extraneous material. If you remember nothing else from this chapter, remember this: no rough draft of a document is ever as good as you can make the final document. And one of the best ways to improve it is to pare it down to the essentials.

FINE-TUNING WORD CHOICE

To further improve word choice in your draft, identify the following elements by circling them:

- noun phrases—groups of words that function collectively as nouns (*all those participating* rather than *participants,* or *individuals being trained* rather than *trainees*).
- modifiers—words, phrases, or clauses that describe another part of the sentence ("the temperature *dropped extremely fast*" rather than "the temperature *plummeted*").
- passive constructions—grammatical constructions that emphasize the object of the action rather than the agent of the action. This structure usually exhibits the *-ed* form of a verb (called the past participle) plus a form of the very *to be.* ("the poison's effect *was slowed by* the antitoxin" rather than "the antitoxin slowed the poison's effect").

- jargon—technical terminology, generally understood only by those schooled in a specific discipline (*end-user computing, microscopy, biomimetics, debug, arpeggio, embolism, vascularization*).
- highly connotative words and phrases—emotionally charged, judgmental, or manipulative language rather than objective language (*fluff* rather than *extraneous material, despicable methods* rather than *suspect methods*).
- generalities—vague, nondescriptive language that can cause confusion (*thing, seemed, something*).
- clichés—trite, overworked expressions that dilute the message because of their tired nature (*viable alternative, Information Age, spectrum of needs, synergistic effect*).

With the exception of clichés, all these elements have their place in technical documents. But when misused, each represents a serious obstacle to precise writing. Noun phrases, for instance, should be used only when a single noun will not suffice. For a nontechnical audience, noun phrases may be better than technical jargon. And in rapidly advancing fields such as quantum physics, the noun phrase allows the scientist to describe new phenomena without coining new terms.

Modifiers can enhance the precision of your message by describing qualities, degree, characteristics, or conditions. Unfortunately, many writers who do not revise carefully create documents full of modifiers that add little value, such as *very, actually,* and *really*. The best solution to this problem is often a more precise verb or adjective.

We have somehow come to construe the passive voice as a sign of objectivity, when in fact it frequently omits information readers might require. This structure *seems* objective because either it uses no subject in the sentence or the emphasis falls on the object rather than the subject. Hence action is displaced from subject to object. It is correct to use the passive voice only when the subject is either obvious or irrelevant to your audience or when emphasis on the object is vital. The average student understands the note tacked on the closed classroom door:

CLASS IS CANCELED

Rarely does that reader care who canceled the class. However, if the statement were made to the dean of faculty, he or she might want to know both who canceled the class and why it was canceled. And if the statement were made to the teacher whose class had been canceled, he or she would *demand* an explanation.

Because the passive voice omits or minimizes the subject, writers of technical documents must use the structure discriminately. It should not be used to escape responsibility, nor should it be used to conceal information. In addition to often being inappropriate, the passive construction may add unnecessary words to a sentence. Compare "A recommendation *was made to* use heparin to dissolve the clot" to "Dr. Brewer recommended heparin to dissolve the clot." Using an active verb and dropping the prepositional phrase make the sentence less wordy and more informative.

The next element to consider in the revision process is jargon. It is often necessary to use technical terminology in a technical document, but keep in mind your audience's level of expertise. Technical terminology should not be used if it obscures information. Will the accountant understand the doctor who says, "Vascularization has created an embolism on your cornea"? Probably not. However, the data processing manager will understand the following description of a software product.

> This product runs as a utility and is supported by the operating system. The product may also be called from a COBOL, FORTRAN, or Assembly language program.

Too often, those steeped in the language of their discipline forget that outsiders perceive their vocabulary as a foreign tongue. Try to catch yourself when you are guilty of this writing error. If you restrict your use of jargon to field-specific documents addressed to experts, you will save your audience time and effort. And if inexperienced readers may be looking at the document, avoid jargon even at the expense of conciseness (sometimes it is necessary to use more words to describe the unfamiliar). When it is important for your audience to learn technical terminology, provide a definition of each new term at its first occurrence in your document.

The *connotation* of a word is an idea or attitude suggested by or associated with it. On the other hand, a word's *denotation* is its explicit meaning. Because the writer of technical documents should strive for objectivity, words with strong connotations—negative or positive—should be used with discretion. They may be appropriate, for instance, when you have an urgent message to convey, as in a speech before a nuclear disarmament committee:

> No more time can be spent irresponsibly dreaming up new war toys to mount in the sky pointed at our children. "Star wars" can be launched without human intervention, because the degree of computing reliability achieved in even the most advanced computers is only 99.4%. Is anyone willing to bet such high stakes against the malfunction of just one of these machines when, with more than 300 in the air, it is virtually a sure thing that there will be a disaster?

As a writer of technical documents, you should select the least biased, most accurate word. It should be appropriate to the writing situation in both denotation and connotation.

Generalities are vague words or phrases that provide little information. Technical documents are enhanced by concrete or specific language—words and phrases that can be verified or applied. Sometimes generalities are useful, as in a document that moves deductively from a broad overview to the particular case or in summary statements after the specifics have already been discussed. But here the details either follow or precede the discussion. Although eliminating generalities may add words to your message rather than shorten it, the result is a more precise message. Which description of a patient's eye problem is more informative?

The patient's eye seemed irritated.

The patient's eye was red, the lid was puffy, and yellow small-grained crystals adhered to the lashes.

Generality can usually be eliminated by paying attention to details and describing the situation rather than summing it up. When you find generalities in your draft, force yourself to itemize as many verifiable details as possible. Then select the most important elements from this list to include in the document.

One other obstacle to precise technical writing is the overworked expression, the cliché. You should avoid hollow phrases because, in falling back on them, you run the risk of boring or losing your readers. Seek accurate, lively ways to present your information.

The exercises at the end of this chapter will help you remove such conventions from your work. Meanwhile, let us see how concentration on word choice helps improve our draft on molecular engineering.

Currently, molecular engineers are concentrating

jargon

their research efforts on Methanolococcus capsulatus,

noun phrase

in an attempt to limit both the energy consumed by

highly connotative

synthetic catalysts and the undesired by-products they

create in processing. Scientists have discovered that this

jargon — noun phrase

bacterium, a natural convertor of methane to gasoline,

eliminates unstable molecules during its catalytic action

and halts processing once gasoline is produced. / passive voice

jargon Biomimetics, a specialized subset of molecular engineering, / noun phrase is dedicated / highly connotative to creating synthetic catalysts that

mimic this behavior. By producing artificial substances

to break and make chemical bonds without waste, these

researchers hope / highly connotative to develop a more direct conversion process that consumes less energy. / noun phrase

Currently, molecular engineers are concentrating their research efforts on a bacterium called Methanolococcus capsulatus, in an attempt to limit the high energy consumption and wasteful by-products of synthetic catalysts. They have discovered that this catalyst naturally eliminates any unstable molecules while it converts methane into gasoline. Using the technology of a specialized form of molecular engineering known as biomimetics, these researchers will try to develop synthetic catalysts that mimic this ability to break and make chemical bonds without waste. The anticipated result of such an accomplishment is a more direct methane-to-gasoline conversion process that consumes less energy.

During the revision process, it was possible to reduce the fifteen-word noun phrase that had ended the original first sentence to a mere nine words by consolidating elements. Substitution of the word *they* makes sense in the second sentence, because it is clear that molecular engineers are the ones who have made the discovery. Note that the word *catalyst* replaces *bacterium* and equates the two terms from the first sentence to the second. The noun phrase *natural convertor of gasoline* has been activated in the second part of the sentence, and the result is a clearer, more direct statement of the process.

The writer realized that jargon was used in several instances. Therefore, he provided a brief definition of *Methanolococcus capsulatus* when it was first introduced. The changes in the third sentence reflect the writer's recognition that introducing the word *technology* justifies using the word *biomimet-*

ics, which up to that point had brought no real value to a message directed at businesspeople. Likewise, the sentence connects this technology to its aim: developing catalysts that mimic natural behavior. The *conversion process* of the final sentence has been clarified through repetition to *methane-to-gasoline.*

The incremental steps we have followed thus far in our revision have been discussed in isolation. But writers actually make several kinds of changes at once, as you will after you learn the basics. The writer of our sample, for instance, has gotten ahead of us and has already made changes in logical flow and completeness. We will examine these and other changes as we progress to the next focal point of the revision process.

LOGIC AND COMPLETENESS

Prior to this point in the technical writing process, you have had ample opportunity to consider the completeness of your information. After intensive information gathering and planning, you should approach the writing and revising stages with confidence that your "data banks" are full. Nonetheless, during the revision phase it is especially important to be sure your information is logically and fully developed. We face problems in these two areas either because we overlook points when working from outline to draft or because we have become so familiar with the subject that we omit information we mistakenly assume our audience also knows.

Each paragraph of your rough draft should prepare the way for the next paragraph; it should create the context for understanding what follows. Likewise, sentences within each paragraph should unfold logically. And each point you bring up should be discussed thoroughly enough for your readers to appreciate it fully (though they may not agree with you). You address some of these concerns when you select an overall arrangement for your document (see Chapter 4) or a pattern within paragraphs. But here is another set of questions to ask as you revise your draft:

- Is each topic sentence directly related to the thesis statement?
- Is each sentence within the paragraph directly related to the topic sentence?
- Does each paragraph contain only a single point of discussion?
- Is that idea clear and is it discussed sufficiently, given the needs of the audience? If not, should additional information be inserted, or is a second (or third) paragraph of discussion warranted?
- Does each sentence belong precisely where it is? (Does it equip the reader for the information in the next sentence?)
- Are relationships between sentences clear?
- Are relationships between paragraphs clear?

Clear relationships between sentences and paragraphs often depend on transitional words and phrases that move the reader from point to point. The writing elements you employ to help your reader can be added to a sentence or paragraph as an introductory remark, or they can be an integral part of the message. The list that follows is a sampling of transitional devices you simply add to a sentence. (See any good writing handbook for many more, along with examples.)

1. To indicate causality: *because; since; as a result of; consequently; for; naturally; of course; therefore.*
2. To indicate comparison: *similarly; likewise; in the same vein; also; as well as; too.*
3. To indicate contrast: *by contrast; unlike; but; on the other hand; conversely; in opposition; juxtaposed.*
4. To indicate sequence: *and; first, second, third, . . . finally; now; next; then; in conclusion; subsequently; from now on; heretofore.*
5. To indicate simultaneity: *in conjunction; concurrently; simultaneously; at the same time.*
6. To indicate addition or continuation: *and; furthermore; in addition; again; also.*
7. To indicate negation, denial, or disagreement: *nonetheless; however; nevertheless; despite; although; contrary to.*

You can make subtler transitions by integrating certain indicators of relationship into sentences. The following methods are generally unobtrusive; use them to help readers follow the development of ideas.

1. Repeat a key word or idea. (In this case, repetition serves a purpose and should not be revised out of your document.)

Examples of processes in which living organisms use **enzymes** include digestion, muscle movement, tissue building, and energy storage. Cheese and beer processing are industrial **examples** of the use of **enzymes.**

2. Use a synonym of the key word.

Currently, molecular engineers are concentrating their research efforts on a **bacterium** called *Methanolococcus capsulatus,* in an attempt to limit the high energy consumption and wasteful by-products of synthetic catalysts. They have discovered that **this catalyst** naturally eliminates any unstable molecules while it converts methane into gasoline.

3. Use an antonym (opposite) of the key word.

Most **natural** enzymes are too **fragile** for widespread industrial use. **Synthetic** catalysts provide manufacturing and processing plants with a **resilient,** yet costly, alternative.

4. Use words that are commonly paired.

The natural **action** of *Methanolococcus capsulatus* on methane molecules causes the structure to expand into the larger gasoline molecule. This **reaction** can be replicated on a small scale in the laboratory but is not feasible for large-scale industrial use.

5. Use a pronoun (with a clear reference to the noun it replaces).

Zeolites are composed of silicon dioxide and small amounts of aluminum. **Their** well-defined molecular structure allows **them** to act as a catalytic sieve that lets in some molecules while blocking others.

Precise writing requires transitions between each and every sentence and between paragraphs. Whether these connections are additional or integral parts of a sentence, they help readers follow the discussion, they support the thesis by demonstrating relationships, and they ensure that all points fit together logically.

Let us look at the transitions between paragraphs and between sentences in the following excerpt.

Zeolites, which occur naturally in volcanic rocks and as deposits in claylike material, are the most versatile of all **catalysts** in wide commercial use. **They** are one of nature's gifts to chemists, because in many ways **they** are the ideal of what a **catalyst** should be. Composed of silicon dioxide and small amounts of aluminum, **zeolites** have a well-defined **molecular geometry** of holes, channels, and miniature caves. Because of that **geometry, zeolites** offer extremely large areas for **catalytic** action—several hundred square yards for every gram of **zeolite.** Depending on their size and shape, channels let in some **molecules** but not others.

Were **that** all, **zeolites** would have made interesting sieves to filter **molecules** of different sizes. In fact, **that** is what **they** were intended for when Union Carbide introduced **them** commercially in the late 1950s. Paul B. Weisz, a Mobil scientist, quickly figured out that the pores of many **zeolites** uncannily fitted the sizes and shapes of petroleum **molecules.** He began investigating **zeolites** as possible **catalysts,** on the theory that the aluminum atoms inside the channels would act **catalytically** on specific **molecules,** in some cases joining **them** together and in others separating **them.**

Weisz and other Mobil scientists found, for example, that **zeolites** break down, or **"crack,"** the hydrocarbons in crude oil into high-octane **gasoline. They** began a search for **synthetic zeolites,** using **molecular** engineering to fine-tune the dimensions of the channels to tolerances as small as the diameter of a single atom. **Their** first success was a **synthetic cracking catalyst,** which went on the market in 1962 and increased **gasoline** production by as much as **40%** with no change in the number of **cracking** units. Today 90% of the **catalytic crackers** in refineries in the non-Communist world use **zeolites,** which have become a $250-million-a-year business. Some countries survived the Arab oil embargo in large part because of the **extra gasoline yield** that **zeolites** made possible.*

Here the writer weaves a transitional pattern of repetition and reinforcement to guide the business audience through the passage. The word *zeolite* unobtrusively occurs eleven times, and pronouns are substituted for it in four other places. Synonyms for the term are used also; they include *catalyst* and *cracker.* Movement from the first paragraph to the second is accomplished via the relative pronoun *that,* which reminds readers of the sievelike nature of these catalysts while it provides a foundation for the research summary that follows. Readers move forward to the final paragraph through repetition of the Mobil scientist's name and his work.

After defining zeolites in the first paragraph, the writer presents a brief chronology of research events in the succeeding two paragraphs. Emphasis shifts from zeolites themselves to Dr. Weisz and his accomplishments.

This matter of emphasis is important to the revision process as well. You must be certain to give more time and more detailed coverage to the points that are most important to your readers. For instance, it is safe to assume that a business audience reading about zeolites will be interested in levels of efficiency, profits, and the overall commercial impact of technical accomplishments. On a separate sheet of paper, list the key ideas you wish your audience to grasp. Return to the draft and carefully weigh how much time and emphasis you gave each. Have you given minor points "equal time" or as much detail as you have accorded to key issues? Revise passages in which emphasis is weak or inappropriate.

ACCURATE MECHANICS

One last, very important, area of revision remains: mechanics. You must be certain that spelling, grammar, and punctuation are correct. Consult a dictionary whenever you are not sure of the right spelling; fewer errors alienate a reader more than misspellings. Similarly, you must check for grammatical accuracy. Here are some common errors:

*Gene Bylinsky, "The Magic of Designer Catalysts," *Fortune,* May 27, 1985.

- Subjects and verbs do not agree. (Singular subjects take singular verbs; plural subjects take plural verbs.)
- Nouns and pronouns do not agree in number and gender.
- Verb tense shifts. (If you are using the past tense, for instance, you must maintain it throughout the document.)
- Voice shifts without reason from passive to active and back again.
- Modifiers (adjectives, adverbs, and relative clauses) are not placed as close as possible to the part of the sentence they describe.
- Sentence fragments (incomplete sentences) occur.
- A comma is used when a conjunction, semicolon, or period is appropriate. (This error is called a comma splice.)
- Sentences "run on" when they should be broken into separate sentences.
- Metaphors are mixed. ("There's no use locking the barn door after the milk has been spilled.")
- There are dangling modifiers (places where, though there is no noun to be modified, a modifier occurs).
- Infinitives are split (a word is inappropriately inserted between *to* and the verb, such as "You are *to immediately revise* all sentences in which this problem occurs").

Consult a composition handbook to learn more about these grammatical problems and about proper punctuation. If you do not yet know when a semicolon (;), colon (:), or comma (,) is appropriate, you don't have to be ashamed (many of us have trouble remembering). But as in a court of law, ignorance is no excuse. Consult a handbook before you submit your draft.

Adhere scrupulously to format specifications concerning length, order, use of headings, indentation, pagination, type requirements, content, style, and dimensions. You may have a guide to publication standards in your organization, or a list of criteria may be provided in your classes. Take these guidelines to heart, because you will be held accountable for applying them. The following items are covered by one company's specifications for a document called a product brief (basically a mechanism description). This list should give you an idea of how exacting such standards can be.

fonts	stock	color
trimming	folding	drilling
image area	margins	masthead
logo	product line	title
rules and hairlines	artwork	copyright
file number	inside pages	listings
leading	product specs	footnotes
figures	captions	added spacing

notes	photography	framing
tables	ruling	special notices

It is not yet necessary to be concerned with the specifications for the final document, such as leading between typeset lines, image area, fonts, or masthead. (These elements become important when you work with a printer *after* your document is approved for production.) Rather, you should concentrate on the headings, captions, figures, tables, listings, and file numbers that ought to be included in your document as you prepare it for review.

Your last task at this stage is proofreading. Look over your document very carefully and correct any typographical errors, messy erasures, omitted words, skipped lines, and transposed characters. You now have a document worthy of submission to your reviewers.

THE REVIEW CYCLE

Through painstaking revision, you have brought your document from mere skeletal form to a review draft. In most technical writing situations, it is necessary to circulate this preliminary draft among colleagues, technical experts, and supervisors to ensure its accuracy and completeness. As a student, you rarely have this opportunity to have criticism applied to your work along the way. Instead you write a paper or report, hurriedly read it over once or twice, cross your fingers, and hand it in. The next time you see the document it bears a grade and some cryptic comments from your instructor.

One of the nicest things about technical writing in the workplace is the review cycle, a source of help with your document. Particular subjects are often assigned to specific reviewers (those who have management responsibility or expertise in the area). Because reading your document is a legitimate part of their job and because it is their responsibility to ensure your accuracy, these reviewers pay close attention to every word. If you have gathered enough information, planned its exposition, and revised your draft into a clear, succinct document, you will take pleasure in this experience. Your document will come back to you with remarks like "Excellent job, see my notes on page 14" or "Okay as written." In the worst possible case, where you miss the mark entirely, the only thing that happens is that your reviewers tell you about the problem and suggest changes.

The review cycle varies from organization to organization and also depends on the length of your document. It is not fair to ask reviewers to spend all their time on your work, so you must build into your schedule enough time for them to get to your document without being inconvenienced. One or two weeks works well in most cases; longer, and your document might get buried in someone's pile of "things to do"; shorter, and you are hurrying your reviewer. Generally an organization has a standard review sheet that you attach to the copies of your document. Each reviewer returns the sheet to you with his or her approval or changes noted. Keep a file of these for future reference (sometimes people forget that they approved documents).

When the review comments come back to you, it is your responsibility to evaluate and implement them as soon as possible. Sometimes reviewers who fancy themselves writing critics cover your document with red ink. If the stylistic or grammatical suggestions are worth heeding, do so. Usually, however, the only changes you are obligated to make are the substantive ones.

You may find that one reviewer's comment conflicts with a remark made by another. You must resolve such conflicts before you create the final document. In other cases, the reviewer might write "This is not true" or "Not so" without telling you what *is* true. It is up to you to chase down the right answer.

To implement changes in the most efficient manner possible, make a copy of your review draft and use it as a "change copy" onto which you transfer the remarks of all reviewers. Gathering all remarks onto a single document gives you a clear picture of what must be done and where reviewers concur or disagree. If necessary, color-code the remarks of each reviewer so that you can easily return to the source.

Evaluate the degree of change required. Is it a matter of fine-tuning, or is a major revision needed? If the latter is the case, briefly go through your situational analysis. Did you miss the mark at the outset? If so, make your corrections at the prewriting stage and carry them through all of the successive stages. Perhaps you misinterpreted some of your research. If this is the case, go back to the book or report involved and reread the pertinent passages. If you see things in a new light upon this second exposure to the materials, make your changes. However, if your original understanding still seems valid, set up a meeting with your reviewers and discuss their reasons for making the statements they made. Another possibility is that you overlooked some areas of discussion you hadn't realized were important. When this is the case, you may have to do more research before you can change your document.

You have to be the judge of how much change is required to get your document approved for production. Don't be afraid of change, however; you are well versed in the subject matter by now and should feel confident about your writing abilities. Step back into the technical writing process as far as necessary, and carry any required changes forward through the successive steps. If necessary, send your document out to a second technical review to ensure that the changes you have made are satisfactory. Then you can submit it for approval to the proper authority.

THE FINAL DRAFT

In business, research, and industry you submit the final draft of a document for approval both to your supervisors and to legal counsel. Most organizations have a standard sheet that is attached to the document for this purpose. If approved, the document is ready for duplication and distribution. Duplication may mean merely taking it to the copy center and having a certain number of copies made, or it may mean taking it to a printer for professional production. In either case, the final draft must adhere to all specifications and must be attractive, easy to read, and neatly bound, stapled, or clipped.

If you are working with a printer, it will probably be your responsibility to receive *galleys* (typeset proofs of the document), which must be carefully compared to the original document. You need to check for typographical errors, transposed characters, misspellings, and dropped lines or sections. After approving galleys, you might see *boards* of the document—a preliminary view of the layout with type stripped in. Finally, depending on the sophistication of the document, you might see *blues* (the printer's final proof before the document goes to press) and *color keys,* a multilayered film of colors as they will appear when printed in the document. The further along in the production process you are when you make changes to your document, the more expensive production will be, so take the initial proofreading seriously. It is best to have someone who has never read your document check it over to ensure that your familiarity with the work does not prevent you from picking up problems.

A WORD ABOUT WORD PROCESSING

In recent years, the word processor has quietly revolutionized the writing process. This specialized computing tool is designed to make the composition, revision, storage, and printing of documents simple and efficient. The word processor can be used at each phase of the technical writing process—from the situational analysis to the production of the final printed document. If you know how to type, you can use a word processor to improve your productivity and the quality of your documents.

Word processing capabilities are available as part of individual personal computers or terminals linked to a mainframe (i.e., large-scale) computer.

Each of these systems leads to advantages and disadvantages. Personal computers offer users a wide range of software applications; however, these individual units are isolated from each other and from corporate data bases. Terminals connected to a mainframe may not provide as many software applications; however, users can access data in corporate data bases, share information with other users, or transfer files to other users through their terminals.

A typical word processor includes a keyboard, a video display monitor, a central processing unit, software, a storage unit, and a printer. Let's look at how these components work together to make the writing process easy.

Just as if you were typing at a typewriter, you enter words into the word processing system by using the keyboard. Your words do not appear on a piece of paper, however; they appear on the video display monitor. A blinking light, called a *cursor,* indicates the next character space to be filled on the screen. Special keys on the keyboard control the movement of the cursor, allowing you to make editing changes with ease. Once the cursor is positioned where you want it to be, the software lets you perform such editing functions as inserting, deleting, or changing information; underlining, boldfacing, or capitalizing characters; repositioning information; merging information from a stored document into the one you are working on; sorting information on the screen; reformatting the document; and more. All this can be done automatically through a few simple commands. If you were using a typewriter, on the other hand, these changes would use up a great deal of time and effort.

The central processing unit translates each keystroke into digitized information that can be "understood" by the computer. When you press the RETURN key, your information is automatically saved in the temporary storage area of the central processing unit. Once you have finished your work, the software enables you to file your document in the main storage unit of the word processor or on a diskette. Then when you wish to work on the document again, you simply recall it into the central processing unit.

The software also lets you send your document to a printer, where it will be produced according to the format specifications you provide. For instance, margins can be justified right and left, and the spacing between lines can be adjusted. Tables can also be easily created with the help of most software.

Depending on the sophistication of the software, you can perform a range of additional automatic functions, such as making copies of your document, checking spelling, replacing a word or phrase with a substitute at each occurrence of the original, building an index, changing a typeface, and merging files. The software also lets you maintain your files more efficiently than you could with a standard filing cabinet and paper documents. You can rename documents and reorder them, receive a printed list of all the documents you have on file, keep track of the most recent date you worked on a document, and determine the number of pages in a document. Best of all, your documents won't get lost as easily, and their physical appearance won't deteriorate.

Once you master the operation of a word processor, its use will enable you to produce written work more quickly. But with increased speed come some potential pitfalls. For instance, when you move words, sentences, or paragraphs around on the screen, you may weaken your transitions between ideas. You may also delete information you did not intend to delete. To avoid these problems, always read the screen before you file the information on it. Additionally, always print a preliminary hard copy of your document and proofread it carefully before you produce the final version.

To avoid erasing newly revised material, file sections you have edited heavily before you move on to other sections. Use back-up files to avoid losing a document if the system "goes down" or a diskette is damaged.

Even though writing at a word processor is quick and easy, don't be lulled into carelessness. Follow the technical writing process faithfully so that you work from a plan. And beware of a tendency toward wordiness: revise your work carefully so that it is succinct and direct.

Above all, take the time to learn the particular system you are using. The user guide will provide you with tips that will make your work as simple and efficient as possible.

CHAPTER SUMMARY

In this final stage of the technical writing process, you draw up the rough draft on the basis of your outline and revise it numerous times before it is ready for review. Once your document is reviewed, you may have to go back several steps in the process to implement the required or suggested changes.

As a handy reference, the criteria you should apply during the revision process are summarized at the end of the chapter exercises. Use these criteria for the writing assignments you produce, which begin in Chapter 6. Soon these questions will come naturally to you, and you will learn to anticipate them as you develop your drafts, making revisions as you go.

At this point, the technical writing process may seem time-consuming and quite involved. However, if you follow the steps along the way carefully, you will soon be creating polished documents. The better accustomed to the process you become, the more confident you will be, and the less time you will require. At first the process may seem forced or awkward, but revising will become second nature, and you will be able to skip any steps that prove unnecessary in a specific situation.

EXERCISES

1. Write three sentences based on the information supplied. Each sentence should show a different sentence structure. For example:

Use the telephone. / Telephones save time. / Telephones save money.

SIMPLE:	Use the telephone.
COMPOUND:	Use the telephone and you will save time.
COMPLEX:	If you use the telephone, you will save time and money.

You may add subordinating or coordinating conjunctions or relative pronouns. You may focus on only a portion of the information if necessary.

a. Computers are tools. / Computer use is growing. / More nontechnical people are using computers.

b. Ivy is widely cultivated. / Ivy is in the ginseng family. / Ivy has evergreen leaves.

c. A rhombus is a parallelogram. / The parallelogram is equilateral. / The angles are usually oblique.

d. The confluence of the two streams is 3 miles south. / The streams form a tributary of the river. / The river feeds into the Mississippi.

e. An orthographic projection represents an object in several related views. / It is as though the views were all in the same plane.

2. The following paragraph contains many complex sentences. Rewrite the paragraph so a high school audience could understand it.

Responsibility for development of scientific-invention competence has been spontaneously assumed, in ever-increasing degree since 1961, by students in the professions—at first in architecture, engineering, and science, and now by students of all kinds. These students see that the world's prime, vital problem bears repeating a million times. It is: how to triple—swiftly, safely, and satisfyingly—the overall performance realizations, per pound, kilowatt, and man-hour, of the world's most comprehensive resources. To do so will render these resources—which at present design level can support only 44% of humanity—capable of supporting 100% of humanity's increasing population at higher standards of living than any human minority or individual has ever known or dreamed of. (Buckminster Fuller, "Report on the 'Geosocial Revolution' ")

3. Rewrite each of the following sentences to eliminate the problems cited. Underline the problem areas.

Unnecessary Repetition

a. Combine both of these features into one model.

b. Most of the test results had negative outcomes.

c. The Brooklyn Bridge required a strong foundation of heavy pilings to be driven deep into the ground to support it.

d. Certain crystals will exhibit the piezoelectric effect of generating electricity when they are subjected to pressure or of undergoing mechanical stress if these crystals are subjected to an electric field.

e. The sides of the canal were fortified with gabions to strengthen them.

Unnecessary Noun Phrases

a. The meal during the early part of the day is vital to our ability to concentrate and be productive.

b. Verifiability of information is desirable in a technical writing environment.

c. User dependence upon the computer is increasing drastically.

d. The performance of the vehicle was sufficiently impressive to sway the reporter from the magazine from his preconceived notion of the car's mediocrity.

e. A growing ambivalence toward technology and progress is apparent among people old enough to have witnessed dramatic changes in the past decade.

Unnecessary Prepositions

a. The reaction of the participants in the survey suggested that the latest economic policies of the White House are not of much help in the solution of the problem of unemployment.

b. Salt particles were suspended in a solution of water in a beaker from the laboratory to determine the amount of time it takes for them to dissolve.

c. Reggae is a form of music of Jamaican origin, influenced by calypso music and characterized by a strong rock and roll rhythm.

d. A secant is a straight line extending from the center of a circle through the end of an arc of its circumference to another straight line that is tangent to the radius at the other end of the arc.

e. Through a system of ducts and tubes in its body, the echinoderm (starfish) is able to filter food from sea water in order to eat.

Unnecessary Passive Voice

a. It was understood that no one was to be allowed near the controls that had been computerized on the pipeline.

b. The injection was administered by the medical assistant rather than by the doctor.

c. Passenger fares were reduced by the airline company.

d. The evidence was presented by the prosecuting attorney to the jury.

e. The conclusions were challenged by the research of Cannon and Paternik.

Unnecessary Relative Pronouns

a. Professor Tillman, who is a man who demands quality, insisted the experiment be conducted once more.

b. The shipment, which arrived yesterday, was not logged in until this morning.

c. The man in the corner, whose name is Constantin, is an exchange student who is studying nuclear physics in the special program that is sponsored by Cal Tech.

d. The emperor penguins, who get their name from the crown-shaped golden crest that is on their foreheads, put their young in a communal area, which is similar to a kindergarten.

e. Several penguin "matrons," who act like kindergarten teachers, keep the young entertained while the parent birds hunt for food, which is scarce in the waters of Antarctica.

Unnecessary Modifiers

a. The intensely cold weather caught the vast majority of people by surprise.

b. The garden snail moved at a particularly slow rate across the garden.

c. The Tower of Pisa, which had been continuing to slant at an almost imperceptible rate for centuries, suddenly increased its declining speed to as many as five inches in a single year.

d. The migration from one release of a computer operating system to another is relatively complicated if an organization has tailored the original software to its own very specific needs.

e. It is an especially interesting and challenging experience to participate in the development of this highly innovative machine.

Wordiness

a. Along the order of 15,000 people took part in the annual running of the Boston Marathon.

b. It was while he was on his way to his car that the lights streaked across the sky and attracted his attention.

c. Exactly what the strange markings around the Tycho Crater of the moon are no one can say.

d. What lights up the rest of the moon at this time is earthshine, or the very strong reflection of sunlight from the earth.

e. Beyond a shadow of a doubt, the H_2SO_4 will eat through the container.

Expletives

a. There are five species of plantago that are indigenous to the American Southwest.

b. It is likely that the next satellite launching will be in July.

c. There is one report that treats this topic.

d. At one time it was believed that massive doses of vitamin C helped a cold sufferer recuperate more quickly.

e. There is an animal in Australia called the echidna that produces a shelled egg, yet the animal is considered a mammal because it seems that its young hatch inside a pouch at an embryonic stage.

Jargon

This section should be conducted as an in-class discussion. Students in the fields represented should act as "translators."

a. *Agronomy:* Because the herbicide diuron moves acropetally through the apoplast via the transpiration stream, its phytotoxicity is directly affected by factors that affect the transpiration rate (such as increased temperatures or decreased moisture).

b. *Computer science:* The relational access interface of these user-friendly tools allows end users to access needed information in sequential or indexed files and display it in tabular arrays at their terminals.

c. *Botany:* The aerobic azotobacters are the most efficient in carbon utilization; they fix ten to twenty milligrams of nitrogen per gram of carbon compound.

d. *Electrical engineering:* A diode is an electron tube of the high-vacuum type with a cold anode and a heated cathode, used as a rectifier of alternating current, a demodulator, and so on.

e. *Nutrition science:* Digestion in humans involves the mechanical disruption and enzymatic breakdown of foodstuffs taken into the alimentary tract.

f. *Nautical science:* A yardarm is a tapered spar extending a square lateen or lug.

4. Provide a concrete statement to replace each of the following abstract sentences.

a. The fragile seedlings succumbed to the late frost.
b. The rats were more agitated than usual.
c. The children frolicked in the park.
d. The volcano was active.
e. The car had trouble driving up the mountain road.
f. The bridge was unstable in the high wind.
g. This computer program is easy to use.
h. Sweet-Aids come in a convenient dispenser.
i. The unusual sound was captured on tape.
j. The lawnmower is efficient.

5. The following is a list of emotionally charged words. In a technical writing environment such subjective words are usually inappropriate. They may cost a writer his or her credibility. Sort them into positive and negative categories. Then select for each an alternative word or phrase that is more objective. (Often a concrete description is the best solution.)

Example: The speaker was boring.
The speaker did not hold my interest.

frail	majestic	dumb
starvation	stubborn	hostile
confused	special	lucky
hero	illegitimate	pompous
morbid	outsider	exquisite
choleric	formidable	effortless
admirable	sacrosanct	pitiable
tedious	epitome	fruitless

EDITING CRITERIA

As you revise your rough drafts, apply the following revision criteria. (Additional lists of criteria appear at the end of Parts II and III.)

1. Is the thesis clear?
2. Do topic sentences relate to and support the thesis?
3. Do supporting sentences within paragraphs support the topic sentences?
4. Does each sentence use an appropriate structure, given the writing situation?
5. Does each paragraph deal with a single idea?
6. Is the discussion of each point complete?
7. Are transitions between sentences logical and clear?
8. Are transitions between paragraphs logical and clear?
9. Is the overall arrangement of information appropriate to the writing situation?
10. Is the arrangement of information within paragraphs logical and appropriate?
11. Does the writer meet the needs of his or her audience, as defined in the situational analysis?
12. Does the writer accomplish his or her purpose, as expressed in the situational analysis?
13. Is emphasis appropriate throughout?
14. Is the level of detail appropriate to the needs of the audience?
15. Is word choice appropriate throughout?
16. Is wordiness avoided whenever possible?
17. Is grammar correct?
18. Is spelling correct?
19. Is punctuation correct?
20. Have all format specifications been observed?

PART II

Techniques

C H A P T E R

6

Classification, Division, and Partition

CHAPTER PLAN

Technical subjects are often broad or complex, and it is up to the writer to analyze the topic and select expository techniques that explore in coherent, meaningful ways the relationships among ideas or parts. By grouping diverse elements in appropriate categories, you can examine the characteristics that are pertinent to individual aspects of the topic or to the whole.

This chapter presents an overview of the techniques you can use to organize your discussion of complex subjects, and it serves as a cornerstone for subsequent chapters on other techniques. To that end, this chapter will

- help you determine whether grouping is called for.
- present ways in which to create meaningful categories.
- examine methods of discussing elements within categories.

DETERMINING THE NEED FOR GROUPING

Three techniques for categorizing exist: classification, division, and partition. *Classification* is an inductive method of analysis that combines like elements into groups on the basis of characteristics that group members share. The result can be a clear picture of an overall, broad topic. *Division,* a deductive process, breaks down a broad topic into its subordinate elements, thereby enabling readers to understand the many aspects of a subject. *Partition* (either inductive or deductive) explores the specific attributes of individual parts of a whole in order to demonstrate the relationships among those parts.

We are all familiar with the use of such analytical techniques, although we may not have recognized them as such:

1. Classification helps us make sense of a telephone directory, wherein the names of residents (white pages) or products and services (yellow pages) are alphabetized. Think how difficult it would be to find the number of your favorite movie theater or a friend if this information were not organized. And think how confusing it would be if each city had its own unique method of organizing its phone book.
2. Car dealers use division when they organize their offerings according to model or price tag. Computer manufacturers sell their products as hardware, software, or networking technologies, rather than as a bunch of random parts. Restaurants employ division to present their appetizers, entrees, and desserts in meaningful arrays.
3. Partition is heavily used in descriptive documents such as manuals and product specifications to examine the inner workings of machine parts or designs. It is also used in organizational plans to identify functional units of the company and to demonstrate their relationships to the enterprise as a whole.

Your ultimate goal in using any of these three expository methods is to create manageable units of information, thereby improving the reader's comprehension and the accessibility of information.

In the creation of categories, a logical or inherent organizing *premise* serves as the ruling principle. Categories are clearly delineated because of shared characteristics and, in technical documents, are enhanced by the use of format conventions such as headings, indentation, underlining, roman numerals or decimal notation, page breaks, chapters, sections, volumes, or issues. The more appropriate this organizing premise is to your audience and purpose, the sounder the grouping system.

WHEN TO USE CLASSIFICATION TECHNIQUES

Classification is an expository method by which we begin at the specific level and work toward a general understanding (hence it is an inductive technique). It is especially useful to researchers and investigators who are working with fragments of information in an attempt to draw a complete picture. Consider the remarkable Swedish botanist Linnaeus (1707–1778), who personally analyzed the characteristics of thousands of plants, formed logical premises on the basis of which to categorize them, and initiated the plant classification system as we know it today. Although memorizing taxonomic names in a botany class may not seem particularly useful to the average student, Linnaeus's system enables botanists to identify plant varieties quickly and efficiently for medicinal, research, and industrial purposes.

A more immediate example is the student in a biology laboratory class who records the results of experiments. Initially there may be very little correspondence between one set of results and another, but as the experiment is repeated and tighter controls are imposed, patterns emerge that enable the student to group the results in categories. In both of these examples, patterns are inherent in the topic and it is up to the writer or researcher to discern them in order to develop categories and draw conclusions from them.

Hypothesis testing, a practice with which most scientists are familiar, is a subset of classification. A general rule (or hypothesis) already exists, and the writer or researcher explores the degree to which the individual case stands up to the general premise.

In this kind of inquiry, the writer usually reviews the characteristics of the general premise and its pertinent categories. She or he introduces the new item into the analysis and measures it against the established group by asking the following questions:

1. Does the item share *any* of the characteristics of existing group members? (If not, it does not belong in this classification system.)
2. If so, does it share *all* the characteristics of existing group members? (If so, the item is grouped in this category from that point on.)
3. If some characteristics are shared, does the new item have enough unique traits to warrant establishing a new group? Does the general premise itself require reconsideration on the basis of this new information?

For example, consider the recent discovery of a new chemical compound on earth. Scientists first had to determine whether this compound was indigenous to our planet. Through special chemical testing and precise dating techniques, they matched the new compound to other known compounds and determined that it was not native to earth. Rather, it was brought to our planet by a very large meteor during the Cretaceous Age (about 75,000,000 years ago). In this case, analysis of the new item proved that it was *not* a part of the existing classification system, nor was it necessary to devise a new system or alter the existing one to accommodate it.

However, the dating of the compound caused paleontologists (scientists specializing in past geological periods) to reexamine their theories. Because the huge meteor's impact coincided with the extinction of the dinosaurs, these experts tested the new information against the various (and sometimes conflicting) reasons thought to account for the disappearance of dinosaurs. Having applied this new information, many scientists now believe that the impact of the enormous meteor had several devastating effects on the animal and plant life of that epoch. Those who subscribe to this theory suspect that a shock wave of great magnitude occurred, which suspended a thick layer of dust in the earth's atmosphere. It also triggered tidal waves and earthquakes that immediately killed many animals. Plants, unable to carry on photosynthesis because of the dust screen, began to die off quickly. The animals depending on plant life became the next victims, until only the most adaptive organisms were able to survive.

WHEN TO USE DIVISION

Division is best used when you must simplify a complex topic by breaking it down into smaller, more manageable units of discussion. This process is deductive; you begin with the general topic and work toward specific elements comprising it. Unlike the preexisting, or inherent, relationships that the writer using classification must *discern,* the grouping mechanism used here is chosen by the writer on the basis of audience requirements and the document's purpose. For instance, if you were developing an article on microwave ovens for *Consumer Reports,* you might wish to break down your topic into a catalogue based on manufacturer, price level, options available, capacity, value, or efficiency rating. Each of these methods of dividing the topic is valid, but you would select only one as the essential premise of your categories. Because *Consumer Reports* prides itself on providing evaluations of efficiency, you would probably choose cost-effectiveness.

The following excerpt from a document on computer networking technology uses division to create understandable units of information.

PROCESSING RELATIONSHIPS IN A DISTRIBUTED SYSTEMS ENVIRONMENT (DSE)

DSE is designed to support three different types of processing relationships, or structures, between an organization's computer resources: horizontal, hierarchical, and hybrid. This flexibility enables the distributed system to parallel an organization's internal structure.

The horizontal structure is one of equal partnership between multiple host (mainframe) processors and is used to exchange data files between hosts and, in some cases, to exchange work for load-leveling purposes. In this structure, an organization may maintain two or more large data processing centers in different geographical locations. As information is gathered at individual sites, it can be transmitted periodically for consolidation (if desired). . . .

The use of a horizontal structure is illustrated by a typical service bureau. For example, two large computer centers may be interconnected between cities in a private communications network. If either computer becomes overloaded, its extra work can be moved to the other center. When the computers are compatible, many types of work can be processed interchangeably at either site. User files can be moved between computers if a user interaction or job must be moved. . . .

The hierarchical distributed structure is one in which a single host is interconnected with one or more satellite processors. The central processor is the focal point for centralized reporting while the remote processors support local work and even program development. The host and satellite processors can be the size and type appropriate to a specific organization's needs. . . .

An example of this structure is a public service system using three levels of intelligent components. Two of the levels may be large-scale computers such as the DPS 88 and the smaller DPS 8. One may be used as a host system and the other as a remote host processor in a regional computing center. . . .*

Another reason to use division techniques is to create a context in which to discuss a single facet of a broad topic. We saw this done in our ZSM-5 example (see the outline that starts on page 95). A graphical representation of the movement from general to specific would look like a series of concentric circles (see Figure 6-1).

In this case, the author assumed that the business reader would not appreciate or understand a discussion focusing just on ZSM-5. This audience required background information on catalysts in general—natural and synthetic. Second, the audience needed to realilze that synthetic catalysts have been used for a long time in industry for a variety of processes. Focusing further, the writer demonstrated the many ways in which synthetic catalysts are used to speed up petroleum processing. The most successful family

* *General Comprehensive Operating Supervisor & Summary Description* (Waltham, MA: Honeywell Information Systems, Inc., 1983).

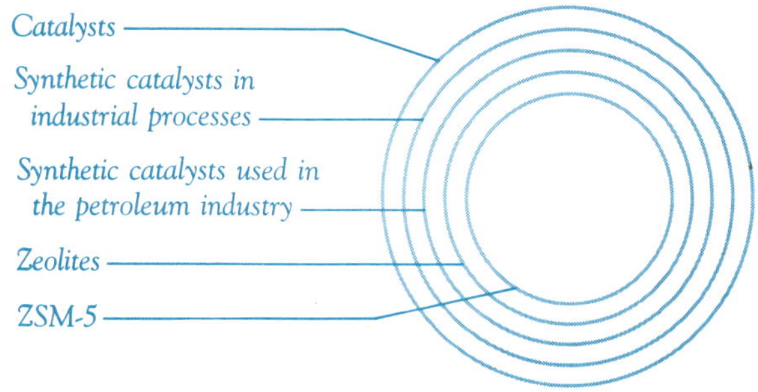

Figure 6-1 Division techniques move from the general to the specific.

of catalysts for this purpose is the zeolite group, of which ZSM-5 is the most resilient, flexible member.

WHEN TO USE PARTITION

Partition helps readers grasp the relationship of individual parts to a whole (inductive arrangement) or that of the whole to its parts (deductive order). Usually applied to a physical entity and its components, this technique makes both physical and functional connections clear. Here is an example of how this technique can be used to explain the inner workings of the eye.

AUDIENCE	Readers of a magazine such as *Today's Health*. They are health-conscious but do not have formal medical training.
ROLE	Eye surgeon.
PURPOSE	Describe the structure of the eye so that readers will appreciate its complexity. List several simple protective measures that readers can take in caring for their eyes.
FORMAT	Article in popular journal.

YOUR EYES ARE WORTH PROTECTING

Our eyesight is one of our most valuable possessions.[1] Without it, life cannot be experienced to the fullest. As prized as our eyes may be, few people know very much about these complex, delicate organs. By knowing more about the eye's structure, people may treat their eyes with greater respect.[2] As an eye surgeon,[3] I have seen too many tragic accidents that could have been prevented by just a little knowledge.

1. Introduction: subject is identified
2. Reason for discussion cited in thesis
3. Role

The eye is a hollow organ held within a bony socket of the skull. It is made up of four sections: the outer layer, the vision center, the inner chamber, and the optic nerve. 4

4. Outline statement: serves as organizing technique for entire document

The outer layer of the eye is a tough, thick layer of tissue called *sclera.* In addition to protecting the eye from injury, this layer exerts pressure upon the inner chamber to help the organ keep its shape. The sclera rests against a cushion of fatty tissue that lines the eye socket. It is also attached to a complex of muscles that allow the eye to move. Directly beneath the sclera is the *choroid,* a thin layer that houses a constantly moving stream of blood vessels that bring oxygen and other nutrients to the organ. 5

5. First category and points of discussion

The vision center is the most complicated section of the eye. It contains the *cornea,* a transparent bubble-like extension of the sclera that protects other parts of the vision center. The cornea acts as a window letting light from the outside pass into the eye. Directly beneath the cornea is a cushioning layer of liquid called the *aqueous humour.* Beneath this solution is the *iris* of the eye. 6

This portion of the eye is pigmented, giving the eye its color. More important than this aesthetic function is the role of the iris as a shutter-like device. The iris, attached to ciliary (hairlike) muscles, can adjust its size: when the iris is small, the hole in the center of the eye, called the *pupil,* is wide (dilated). When the iris expands, the pupil contracts. The iris acts as a light meter, opening and closing to adjust the amount of light that can pass through the lens.

The *lens* is a transparent, convex disc that focuses on the image from the external world. It passes the image from the pupil to the retina.

6. Second category and points of discussion.

The inner chamber is the largest section of the eye. It is filled with a fluid called the *vitreous humour.* This liquid exerts internal pressure against the outer layer of the eye. The amount of pressure is equal to that exerted upon the chamber by the sclera. Lining the inner chamber is a network of delicate nerve cells called the *retina.* These cells receive images that are let through the pupil and focused by the lens. The images projected onto the retina are inverted. It is this inverted image that is passed to the optic nerve. 7

7. Third category and points of discussion

The *optic nerve,* one of the most specialized nerves in our bodies, is located at the base of the eye, enclosed in a protective sheath. It rapidly transmits the inverted image received from the retina along to the sight center at the rear 8

8. Fourth category and points of discussion

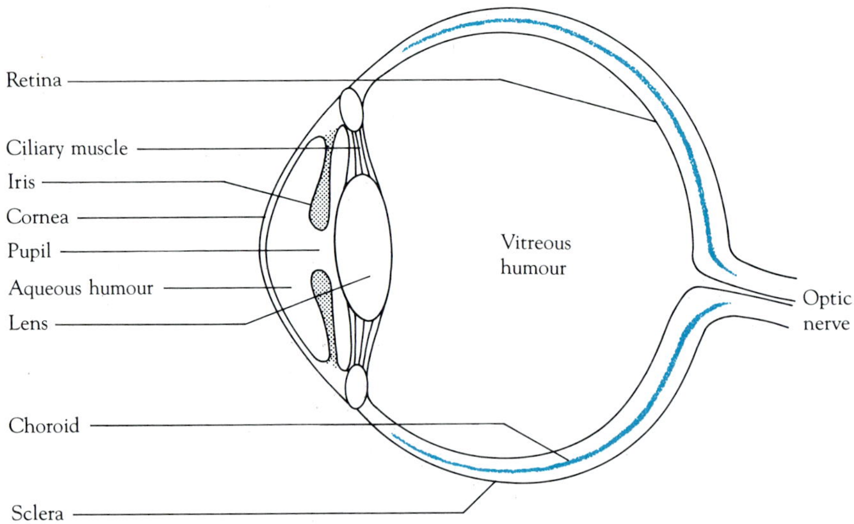

Figure 6-2 This diagram of the eye helps readers locate parts discussed in the partition document.

of the brain. The sight center then automatically adjusts the inverted image so that we can see the world right side up.

The accompanying diagram of the eye points out all these components.

9 The eye allows us to see by exercising all of its parts in a nonvarying sequence. With the aid of collaborative muscles and glands (e.g., eyelids and tear ducts), the process of sight is conducted smoothly. However, when any one of the eye parts is damaged or abused, the quality of sight can be threatened. All of the parts must work synchronously if sight is to be possible.

9. Reunification of "exploded" whole

10 Protecting the eye from damage or abuse is sensible and practical. Simple measures such as wearing a hat in the sun go a long way to enhancing the strength of our eyes. Protective lenses on the job help to prevent damage from flying debris or from sharp objects. Resting eyes after a long reading session refreshes the eyes as well as the reader. Any eye wash with clear, lukewarm water (no chemicals, please) relaxes the eye muscles as well as cleans out foreign particles that can irritate the sclera or cornea. Keeping eyeglasses and contact lenses clean cuts down on eyestrain and

10. Conclusion: supports thesis; catalogues preventive measures

the risk of infection. Providing ample light for close work can prevent eyestrain, too. Avoiding make-up minimizes the chance for irritation and disease. Eating foods rich in vitamin A (carrots, egg yolk, and butter) helps prevent night blindness.

Of all of these words of advice, none is more important than this: have your eyes checked by a reputable eye doctor or vision specialist at least once a year. If your eyes are red, itching, or tearing for no apparent reason, see a doctor immediately. The same is true for any abnormal conditions such as blurred or double vision, fading vision, eye aches, throbbing, or general discomfort.

NOTE: In the case of partition, another important dimension of grouping intervenes before the concluding section. During discussion the whole (eye) is divided into its component parts; that is, it is "exploded." Before leaving the categorization portion of the document, the writer ensures that the whole is reunited: "The eye allows us to see by exercising all of its parts in a nonvarying sequence." When the whole is left in its exploded state, the reader may have a fragmented understanding of the subject or may place undue emphasis on one of the parts.

CREATING CATEGORIES

Once you decide that you want to use grouping techniques in your discussion of a topic, you must choose the categories. In the case of classification, you will have to analyze the specific elements of information in order to discover shared characteristics (patterns). These patterns may suggest organizing options based on the requirements of your audience and your purpose in creating the document. In the case of division, you will probably have a choice among several appropriate organizing principles. Because it deals primarily with physical entities, you can use partition to examine the topic in terms of spatial or functional distribution of parts.

THE FOUNDING PREMISE

Categories are developed on the basis of a single criterion called the *founding premise,* which either collects like elements into groups or breaks down a larger item (topic) into subunits (categories). The following table lists some of the founding premises used thus far in this chapter.

Subject	**Founding Premise**
plant life	species
biology lab results	corroborative evidence
chemical compound	chemical make-up
Distributed Systems Environment	structure
ZSM-5	increasingly specific types of catalysts
eyes	components

The founding premise should organize the topic into categories pertinent to the audience. In structuring a discussion of atoms for a group of molecular engineers, you might select *behavior* as the criterion; an audience of school children might require a whole-to-parts discussion of each component; whereas students in a chemistry class would be most interested in groupings according to atomic weight.

Most topics can be grouped in several different ways, just as some numbers have many factors (24 can be divided evenly by 1, 2, 3, 4, 6, 8, 12, and 24). However, as in mathematics, only one factor—i.e., one founding premise—can be applied at a time.

Consider the broad topic *nests,* which can be organized into categories on the basis of such founding premises as *material used, animal creating the nest,* and *shape.* With such a topic, you must decide what criterion is most important to your readers. If they are to identify nests in the field, perhaps *shape* is best. If they need to learn that other animals besides birds build nests, *animal* might be most informative. If they are curious about the manner in which nests are constructed, *material* is the best organizing rule. Figure 6-3 represents the results of applying each of these different founding premises to the topic *nests.*

Unless it is obvious to your audience, you should state the founding premise clearly at the beginning of the document so that the relationships between categories and members of each set are apparent:

Organic compounds are classified on the basis of their composition.

Fatty acids can be grouped according to their degree of saturation.

Any bone in the human body may be classified according to its gross appearance: long, short, flat, sesamoid. . . .

The founding premise should be applied consistently throughout a technical document. In an article about computers wherein the founding premise is *manufacturer,* readers expect a discussion of International Business Machines, Honeywell Information Systems, Digital Electronics Corporation, and so on. It would be illogical and confusing to include the category *personal computer* because it is not arrived at by applying the premise *manufacturer.*

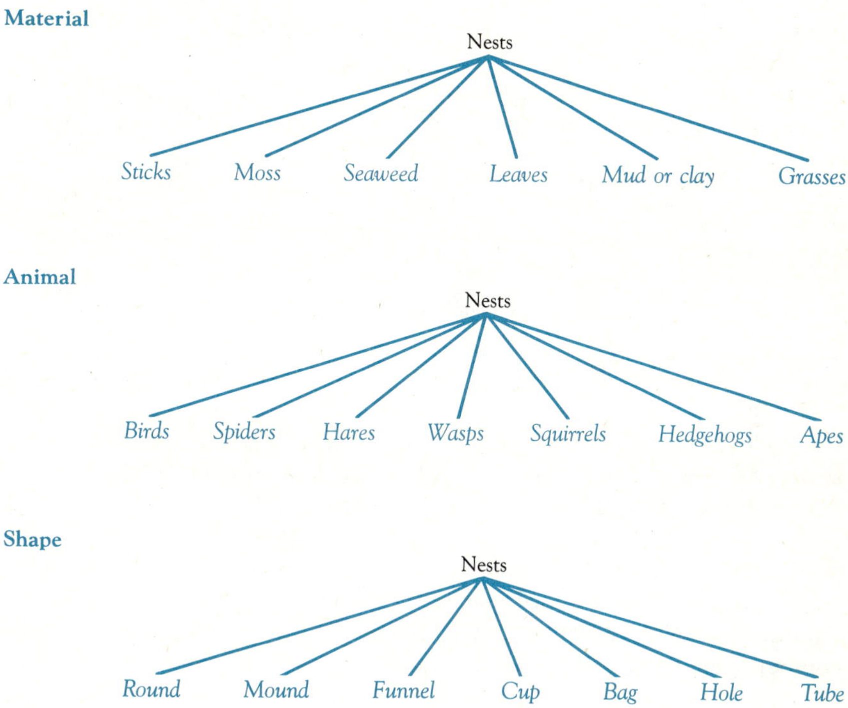

Figure 6-3 The founding premise determines the groups and the individual members of each group.

Although a broad topic can be focused through successive applications of founding premises, only one is applied at a time. And where the focus narrows, it is best to use a clear transitional phrase, sentence, or paragraph to define this next more specific level of discussion. In our nest example, it would be possible to survey the many animals building nests and then to focus further on birds by introducing a new founding premise, *type of bird* (see Figure 6-4).

At each stage of categorization, a sufficient number of groups should be represented (in the case of a highly focused document, that number may be only one). At the same time, a classification system should not include an overwhelming number of categories, unless the document is meant to be a comprehensive catalogue. Certainly hundreds of bird species exist, but the writer must decide which best represent various nesting habits or which are most important to readers. In Figure 6-4, then, it would be more accurate to describe the founding premise as *birds with representative nesting habits*.

Another important characteristic of the founding premise is its ability to yield categories that *do not overlap*. Though not all information falls neatly

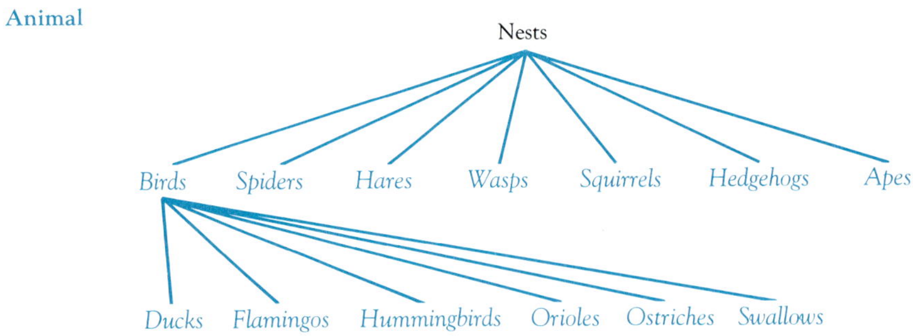

Figure 6-4 It is possible to focus by applying successive founding premises.

into categories, a truly logical founding premise should help you form relatively discrete units. When overlap is impossible to avoid, explain to your readers that the categories do share characteristics. For instance, in the nest example wherein the founding premise was *materials used,* some nests are created from a variety of materials (straw and leaves, moss and grass, sand and twigs). You may choose separate categories for each of these combinations, or you may find it easier to have a single unit called *combined materials.*

AN EXAMPLE

Now that we have seen the effects of a founding premise in a model, let us look at an extended example. The student paper we shall examine is similar to ones that will be assigned to you at the end of this chapter. The marginal comments pinpoint some elements that we have already discussed and some that we will explore soon. As you read, trace the writer's movement from general to more specific levels. Then answer the questions that follow the sample to determine whether you understand the application of these principles.

AUDIENCE Homeowners with little or no electrical engineering knowledge or experience. They know they should install some sort of fire detection equipment, but they don't know which kind to select.

ROLE Hardware salesperson.

PURPOSE To explain the differences between the two basic types of smoke detectors so that these homeowners can purchase the best type for their requirements.

FORMAT Flyer for point-of-sale display.

SAFE AND SATISFIED

Public service announcements, homeowners' insurance 1
plans, and pamphlets sent home with our children from school all remind us that a plan of action against a fire in our home is wise. The message is usually threefold:

- Plan escape routes.
- Have a fire extinguisher.
- Install a fire or smoke detector.

The escape routes must be based on the location and type 2
of fire as well as on the availability of exits. A fire extinguisher should be approved by the Federal Regulatory Commission Article 3444582, fully charged, and easily accessible. The smoke detector, a device that can save valuable time in reacting to a fire, is an important part of preparedness.[3]

There are basically two types of devices—one that 4
senses heat and another that reacts to smoke. The heat sensor takes longer to respond to a fire because the fire must be burning long enough to give off detectable heat in the area of the alarm. But many low-temperature, smoldering fires are just as dangerous. With a heat detector, these fires might go unnoticed until a great deal of damage is already done.

The smoke detector is far more reliable for home use. 5
Two basic types of smoke detectors are available. Each has its own method of operating that results in certain benefits and certain limitations. The selection you make depends on where the detector will be placed in the home.

The photo-cell smoke detector is composed of a small 6
dark chamber, a light source, and a photo-cell. A barrier separates the photo-cell from the light (usually a light-emitting diode). This barrier is designed to prevent light from activating the photo-cell. However, smoke diffuses light and allows it to pass around the barrier to trigger the alarm via the photo-cell.

This type of alarm reacts well to smoldering fires, yet it does not issue false alarms for cigarette or fireplace smoke. Smoke from a blazing fire, however, is smaller than the 1 to 10 micron range of this device (often as small as 0.1 microns) and will not activate the alarm.

1. Focus narrowed from general to specific
2. Focusing process taken to next more specific level
3. Thesis statement
4. Further focusing by eliminating category inappropriate to audience
5. Identification of discussion level
6. First category and points of discussion

These devices are priced from $15.00 to $25.00 and are available in most hardware stores. They can be installed easily on a ceiling or high up on a wall. They are most effective if placed in the main living area.

7 The second type of smoke detector ionizes air in a small test chamber by using a minute amount of Americum 241 (a mildly radioactive substance). If no smoke is present, the ionized air passes an electric current between two electrodes in the chamber. While the current is on, the alarm is deactivated.

7. Second category; parallel discussion of points

Any smoke entering the chamber will disrupt the current flow by resisting ionization. When the current drops, the alarm is automatically activated by a control circuit. This unit can detect very fine particles of smoke such as those produced by a blazing fire. However, this sensitivity can trigger alarms for cigarette, fireplace, and cooking smoke.

These devices cost between $20.00 and $40.00 and are also available in hardware stores or home improvement centers. They should be placed on a ceiling or high on a wall. They are best used in bedrooms or in hallways between bedrooms, where early warning is vital.

8 The best protection against fire would be a combination of these two types of detectors. False alarms can be minimized by placing the ionizing variety in rooms requiring greater sensitivity. Then, of course, test the alarms at least once a month (according to the manufacturer's instructions) to make sure they are in working order.

8. Conclusion: advice and action items

Consider the role the student has selected. Is it maintained consistently throughout the sample? If not, identify areas that need improvement. Is the topic focused at a level that is interesting and important to the designated audience? Is the founding premise clearly identified near the beginning of the document? Are discrete categories formed when the premise is applied? Have a sufficient number of categories been included to serve the stated purpose of the document?

DISCUSSING GROUP MEMBERS

Once you have chosen a founding premise appropriate to the writing situation, you can place each member element in its category. What you choose to say about the group members and the order in which you discuss them also depend on your analysis of the writing situation.

COMPARING MEMBERS OF GROUPS

Very often the points to be made in a document that requires grouping can best be made by comparing members of one group to members of other groups. You may be emphasizing how much more effective one product is than another, trying to determine the most practical theory, or recommending the most suitable plan of action. A controlled comparison of elements can help you demonstrate your conclusions to your audience and win their agreement.

To conduct such a point-by-point comparison, you should consider the same types of information within each category. For instance, in a discussion of fruit, you might compare the taste, texture, color, season, perishability, nutritional value, and cost for members of such categories as bananas, oranges, apples, pears, peaches, grapefruit, watermelon, and grapes. It would do readers very little good to find out how the texture, color, and perishability of bananas compare to the season, cost, and taste of grapes. (In this case it is possible to compare "apples and oranges," but it is not possible to compare color and cost.)

You should not only cover basically the same items of information but also follow the same order from category to category. This arrangement facilitates comparison. Only consideration of emphasis might cause you to change the order on occasion. With such a pattern, the nest example might cover the following points:

Nests

1. Cup-shaped
    1.1. Builders
    1.2. Construction
    1.3. Location
    1.4. Frequency of occurrence
    1.5. Building season
2. Tunnel
    2.1. Builders
    2.2. Construction
    2.3. Location
    2.4. Frequency of occurrence
    2.5. Building season

The points selected for discussion should be both interesting and important to your audience. A useful founding premise is vital, but unless you buttress it with a meaningful discussion of points within categories, your document will not be successful.

The author of the flyer on smoke detectors knew his audience required some discussion of context. He uses the first paragraph to move from the most general level of advice on fire safety to a more specific focus on detectors. The second paragraph rules out the heat sensor because it provides only limited protection. In the third paragraph we learn that two types of smoke detectors exist, each distinguished by its mode of operation. The discussion within categories covers only those items of most importance to homeowners: mode of operation, detection range, limitations, availability, and placement suggestions. The points are made in the same order for each category so that readers can compare characteristics easily.

ORDERING CATEGORIES AND INFORMATION WITHIN THEM

Creating proper categories and points of discussion takes a considerable amount of effort. Furthermore, for the document to be as effective as possible, these categories and their respective points should be presented in a meaningful order. Three arrangements discussed in Chapter 4 work especially well with grouping techniques: the graded, sequential, and spatial or functional arrangements.

Graded order allows you to rank information according to the needs of your audience and the purpose of your document. In a descending arrangement, the most important (or interesting or effective) category is placed first and is followed by categories of decreasing rank. The opposite arrangement is used when ascending categories are appropriate. To lend additional cohesiveness to your document, follow the same arrangement of points within categories as well.

The *sequential arrangement* of categories and points is appropriate when the information you want to convey follows a linear sequence of steps or reflects chronological development. Such is the case if you are writing a document on evolution, for instance. As we mentioned in Chapter 4, the arrangement can be either straight flow (from start to finish) or reverse flow (from finish to start). Which you choose depends on what you want to emphasize (first points generally receive more attention).

Spatial or functional arrangements are especially useful in a whole-to-parts discussion. The article about the eye uses this ordering strategy effectively, moving from the outer area inward.

Whichever arrangement you select, it is important to get off to a strong start that points your reader in the direction the discussion will take. You can do this by starting with an introductory overview statement that identifies the general subject and its categories. The listing of categories should reflect the order in which you intend to discuss them, thereby acting as an implied outline for the document. Look at our examples on the Distributed Systems Environment (DSE), the eye, and smoke detectors. You will find such "pointers" appearing within the first few paragraphs of each document. Similarly, each writer takes special pains to use clear transitional devices in grouping information. Readers remain aware of the progression from category to category and from parallel point to parallel point within groups.

The introductory portion of the document also contains the thesis that all this categorization is intended to support, and it motivates the audience to read the entire document. In the DSE example, that point is "This flexibility enables the distributed system to parallel an organization's internal structure." The point to be corroborated by categorization in the description of the eye is "By knowing more about the eye's structure, people may treat their eyes with greater respect." The smoke detector example explains that "The smoke detector, a device that can save valuable time in reacting to a fire, is an important part of preparedness." In each case the reader understands that the grouping—be it classification, division, or partition—exists to substantiate this thesis.

It follows, then, that the primary point should be reiterated *after* the discussion of categories. Those who wrote the description of the eye and the evaluation of the smoke detector do just that with wrap-up statements such as "Of all of these words of advice, none is more important than this . . ." and "The best protection against fire would be a combination of these two types of detectors." The thesis may be embellished at this point by calling for action or making recommendations.

CHAPTER SUMMARY

When the writing situation requires that you discuss a broad or complicated topic or one that requires a detailed close look, make good use of the grouping techniques recommended in this chapter. As an investigator compiling information, you will find that classification helps you make the patterns you have discerned in your data clear to your readers. When explaining a complex subject, you can simplify the exposition by dividing the topic into appropriate units of information. Partition, primarily a descriptive technique, will help you focus your discussion on specific parts within a larger context.

Categories are based on a founding premise that represents either patterns discernible in the data (in the case of classification) or appropriate shared characteristics (in the case of division and partition). Only one founding premise can be applied to the categorization process at a time, and it is always a good idea to let your readers know what principle of organization you are using. Categories and points within categories should be arranged according to the requirements of the writing situation, which generally calls for graded, sequential, or spatial order. Your discussions of points within categories should be consistent and parallel to facilitate comparison. You will undoubtedly find that most lengthy technical documents require some degree of classification. The longer the document, the more likely it is that you will want to use special headings, indentations, or sections to delineate categories.

EXERCISES

1. For each of the following topics, choose a founding premise that divides the subject into discrete categories. Then list as many of the categories as would be appropriate to a coherent, meaningful discussion. Identify the audience that would be most interested in such a perspective. What would the purpose of the discussion be?

insects	power resources	horses
cars	bridges	clothing
food	computers	students
circuits	furniture	fish
vitamins	armed forces	tools

2. Organize the elements in the following lists at least three times to form three separate classification systems. What is the founding premise for each system of categories? What audience would be interested in each? What would the purpose of each system be?

Cities	**Inventions**
Portland	telephone
New Haven	car
Nashua	camera
Montpelier	phonograph
Hartford	space shuttle
Bangor	television
Bennington	computer
Augusta	steam engine
Warwick	light bulb
Stamford	typewriter
Boston	airplane
Providence	telegraph

3. Return to the description of the eye (pages 134–137). Create categories based on function rather than location. Order the components within categories to reflect function.

WRITING ASSIGNMENTS

Select one of the following assignments. Preface the classification with a situational analysis specifying the audience, role, purpose, and format you have chosen.

1. Imagine that you are writing an article for a popular journal in your field. Select a topic that lends itself to grouping techniques, and present a classification of it that would be both interesting and helpful to the audience of such a journal. Select an appropriate founding premise for categories, pertinent points of discussion, and a complementary order of information. Limit: 500 words.
2. Imagine that you will be conducting a laboratory experiment or field study. Before you can start the experiment, you must write a document classifying all the materials, laborpower, and equipment needed for the project. Group these elements into meaningful categories; explain why they are needed, how they will be used, and so on.

NOTE: Apply the editing criteria at the end of Part II to your assignments before you submit them to your instructor.

CHAPTER 7

Definition

CHAPTER PLAN

Often the purpose of a technical document is to introduce new or unfamiliar ideas, theories, or technologies. Even more frequently, it is necessary to explain terms for other reasons. In a physics experiment, for instance, readers must understand the concept of *plasma* before they can conduct an experiment with electromagnetism. In a manual for inexperienced computer users, the term *command* must be explained before readers can implement any of the instructions provided. And so on.

One of the most important technical writing skills you can develop is the ability to *define* a new term or concept. To help you learn to write effective definitions, this chapter will

- examine the need for definition.
- assess various degrees of definition.
- survey the most effective techniques for explaining new information.

RECOGNIZING THE NEED FOR DEFINITIONS

Definitions of technical terms can be found all around us. The Sunday morning newspaper offers a discussion of organic mulches and the way they benefit various soil types. In a report to a budget committee, a project leader explains the properties of a new synthetic material so that board members will understand the project's importance and continue funding it. Introductory textbooks present terms basic to their respective disciplines (such as *Ohm's law* in electrical engineering and *technical writing* in the first chapter of this book). Joggers read about tendonitis in a medical guide for runners. Countless books explain phenomena, processes, and life forms.

As pervasive as definition may be, it should not be used without good cause. Rather, choose familiar terms and concepts as often as possible. Why use *pollex* when *thumb* will do? Is *monition* that much more impressive than *warning*? In fact, when writing is unnecessarily abstruse, readers can become confused.

Admittedly, variety in word choice can help maintain audience interest, but it is not worth sacrificing clarity or precision. Far too often, new vocabulary serves only to baffle rather than enlighten readers. If your reader is forced to concentrate on *how* you are saying things rather than on *what* you are saying, your message will not be as effective as it could be.

Of course, readers can benefit from learning new terms at times—especially in technical disciplines. In these situations, present both the word and a clear definition of it. One obvious reason for defining unfamiliar terms and concepts is that very few words look or sound like what they represent. Although in some oriental languages words do resemble their meaning (for example, the ideograms, or picturelike symbols, used for *horse, boat, man,* and *house* look like the thing they name), few languages exhibit this correspondence between symbol (word) and meaning. A classic joke makes this lack of correlation all too clear: Adam and Eve were instructed to name each animal in the Garden of Eden. Upon observing a large animal with a long, hair-covered neck, a broad back, and four sturdy legs, Adam said, "That is a *horse.*" When Eve asked him why he had chosen that particular name, he replied, "Because it looks like a horse."

Unlike Adam, your readers may not be able to determine what a word means because it looks like a *coulter* or a *diageotropism.**

Less obvious, but far more important conditions that call for defining terms may also exist:

* A *coulter* is a cutting tool that is attached to the beam of a plow and makes a vertical cut in the surface that allows the soil to be turned over effectively. *Diageotropism* is the tendency of growing branches or roots to extend their axes at right angles to the line of gravity.

1. The audience is unfamiliar with the term itself but may be well acquainted with the concept it represents. For instance, the word *acrophobia* (an intense fear of being in high places) may be new, but the reader can readily understand the fear it describes.

2. Two words are commonly mistaken for one another. A *bunion* is an inflamed swelling of the small sac on the first joint of the big toe. This term is used more widely than it should be—for instance, when the word *corn* (a local hardening or thickening of epidermis, as on a toe) should be used. Similarly, people unacquainted with engineering terms frequently confuse *stress* (a force exerted on a material) with *tension* (a condition existing within a material), because we use these two terms almost interchangeably in our daily lives.

3. A term has a unique application within a specific document. Such an explanation becomes a "working definition" and holds true only for the document involved. Defining such terms is similar to assigning values to symbols in an algebraic expression. An example is the word *barrio* used in a city planning document. Under normal circumstances, this anglicized word means simply "neighborhood." For the purposes of this planning document, however, the term is assigned the following working definition: "that economically deprived area bordered by Escondido Parkway and Chula Vista Road on the east and west, and by Central Avenue and Mission Street on the north and south." Each time the term is applied in the document, the notion of economic deprivation and the particular physical boundaries of the area are evoked.

4. A word or phrase is coined. As technology and research advance, new words are formed to describe new information. When a word is coined, it should be placed within quotation marks and explained. Until it becomes a generally accepted word, it should appear within quotation marks. Terms that have evolved recently but are now accepted as standard include *ergonomics, cybernetics, ethnobiology, downloading, end user,* and *user-friendly.*

5. A strong connotation associated with a word needs to be dispelled. Consider a word such as *discrimination,* which has become negatively associated with bigotry or unfair treatment. This word can also be used to mean the very positive quality of using good judgment. Certainly an audience must be told which meaning is intended in a document, if it is not immediately evident from the context.

6. A word's meaning varies in different contexts. Even though the word in question may not symbolize a complex concept, it is still best to let readers know which meaning is implied. Consider a simple word like *bit.* To a computer scientist it means binary digit; to a horseback rider it is a device placed in the animal's mouth; to an actor it is a very small part in a production; and to a mechanic it is an attachment for a drill. In many

technical writing situations, the context makes the meaning clear. However, in documents addressed to lay audiences, a brief, introductory explanation may be appropriate.

7. Both the term and the concept it represents are unfamiliar to the audience. In the other situations we have described, a brief explanation of the term may suffice. However, when *both* the term and the concept are unknown, readers deserve an extended discussion. Few readers unfamiliar with the concept would be satisfied with a one-sentence definition of *Boolean logic, laminitis, spectroscopic analysis, biomimetics, fermentation,* or *electromagnetism.*

DEVELOPING BRIEF DEFINITIONS

The writers of technical documents must evaluate the words and concepts they use to determine whether definition is warranted. In many cases, when the concept is relatively simple to grasp, a brief definition is adequate. Brevity is an advantage in these situations, because providing the necessary information does not interrupt the general flow of the discussion. However, when you decide to be brief, you must be sure that only minimal clarification is required; otherwise, brevity could be self-defeating.

This chapter has already given brief definitions of such terms as *ideogram, acrophobia, coulter, diageotropism, stress,* and *tension.* The sentence in which each term appears is not disrupted by the explanation. Skim the early part of this chapter to locate these terms. Note that clarification is provided as a service to readers, yet the point of the sentence depends on the audience's understanding the proper meaning. Defining is always better than risking audience confusion. In fact, you should always define a term if it may be misunderstood.

When it is necessary to include a brief definition in a technical document, you have several options, including the use of a word or phrase, sentence, or note.

Sometimes an additional word or phrase may be all that is required to clarify an unfamiliar term. The term *paleontologist,* used in the last chapter, is defined parenthetically by the phrase "historian of geology." In the following sentence, a *synonym* (a word that has the same meaning as another) is used: "The amputee has a large cicatrix, or scar, running down the side of his shoulder where his arm had once been." In this case, a phrase using *or* is employed to introduce *scar* and is set off from the rest of the sentence by commas rather than parentheses.

Not all single-word explanations are synonyms, however. The English language is particularly rich in words that have distinctive meanings. In fact, few words mean exactly the same thing as another. Sometimes, for instance, the relationship between the term and its one-word definition is that of specific case to general category.

The jassid (leafhopper) has caused significant crop damage this year.

The gum surrounding the lower right canine (tooth) is inflamed by calculus (deposits).

A *jassid* is only one type of leafhopper, a *canine* is a particular kind of tooth, and *calculus* is a type of deposit. By placing the unknown specific term within its general, more familiar category, the writer is able to introduce new information efficiently.

On other occasions, it is possible to clarify a new term's meaning by relating a general category to a familiar specific case.

A pompano (butterfish) population could be introduced into the reef system of the Marshall Islands to provide a larger food source.

A cicatrix (hilum) is left on the seed where the ovule was attached.

A *butterfish* is a type of pompano, whereas a *hilum* is a type of cicatrix, or scar. By mentioning a specific member of the general category, the writer helps readers understand the new term.

Whether you are using a word or a phrase to clarify a term, the punctuation you choose affects the amount of attention the explanation receives. Setting the definition off from the rest of the sentence by commas lets the sentence flow on with minimal interruption. Slightly more attention is paid to the definition when parentheses are used; a word or phrase within dashes enjoys still greater emphasis.

His acrophobia, or fear of heights, can be so intense that he cannot climb a staircase without experiencing anxiety.

His acrophobia (fear of heights) can be so intense that he cannot climb a staircase without experiencing anxiety.

His acrophobia—fear of heights—can be so intense that he cannot climb a staircase without experiencing anxiety.

If an audience is likely to know a term, use commas. If only half of your readers might know it, choose parentheses; and if most won't know it, dashes are appropriate.

For even greater emphasis, you can use a full sentence to supply a brief definition. The more complicated the sentence structure, the more attention the definition receives.

Acrophobia is a fear of heights. His condition can be so intense that he cannot climb a staircase without experiencing anxiety.

Acrophobia is such an intense fear of being in high places that it can make its victims feel helpless. In his case, he cannot even climb a staircase without experiencing anxiety.

A final possibility for the brief definition is a note or footnote. In the case of a note, text is interrupted, the word Note is used with a colon (:) to announce the explanation, and the definition is indented.

His acrophobia is so severe that he cannot even climb a staircase without experiencing anxiety.

NOTE: Acrophobia is an intense fear of being in high places.

He generally avoids activities that require him to be above ground level. His apartment and office are on the first floors of their respective buildings.

In the case of the footnote, the term has either an asterisk or a superscript (raised) number following the term. The explanation, like that of *coulter* and *diageotropism* earlier in this chapter, appears at the base of the page, while the text continues without interruption.

Through the use of a single word, phrase, sentence, or note, you can briefly explain an unfamiliar term to your audience when a definition is required. When several terms are introduced, vary the methods used according to the degree of emphasis required for each explanation. However, do not mix methods simply for the sake of variety. If a single method will work in all cases, use it consistently.

DEVELOPING DETAILED DEFINITIONS

In some technical documents, both a term and the concept it represents are unfamiliar to your audience and require an extended, or formal, explanation. Such a definition usually begins with a three-part introductory statement followed by a detailed explanation that may take several paragraphs—or even an entire document—to complete.

THE INTRODUCTORY STATEMENT

The first part of the statement used to introduce an extended definition identifies the term being defined. The second part places this specific term in its broader category (see the classification techniques discussed in Chapter 6). The final part distinguishes this term from other members of the broader category. The following chart demonstrates the use of this introductory statement for four terms that may be unfamiliar to you.

Part 1 (Term)	**Part 2 (Category)**	**Part 3 (Distinguishing traits)**
A windlass	is a hoisting device	that has two drums of different sizes on the same axis. A line wound around the larger and unwound on the smaller provides lifting power.
A piton	is a metal spike	[that is] driven into rock or ice as a support during mountain climbing. It often has an eye through which a rope may pass.
A geländesprung	is a jump in skiing	[that is] made from a low crouching position with the aid of both ski poles. It is usually made over an obstacle.
An atoll	is a coral island	that consists of a reef surrounding a lagoon.

To your readers, this three-part formula signals the starting point of an extended definition. It also acts as a disciplining device for you, the writer. It places the term in its proper context (category) and distinguishes it clearly from ideas, acts, or objects that might be considered similar. A well-devised introductory statement helps you elaborate on the points that are crucial to understanding the term.

The following sample definition shows how the introductory statement acts as a launching pad for the rest of the definition.

AUDIENCE New camp counselors in pre-summer orientation. They will be working in an environment where scabies is a possibility. Most are college students. Their camp's reputation could be damaged by an epidemic of the disease.

ROLE Camp director, just returned from health seminar.

PURPOSE To acquaint readers with the disease and its symptoms so that any cases can be detected and treated before widespread infection occurs.

FORMAT Page in a training notebook.

Scabies is a skin disease in which sufferers can develop severe itch, blisters, and skin scaling. This highly contagious disease is a special problem at summer camp because it is spread through direct skin contact, poor hygiene, and contact with contaminated bed linens, towels, or utensils. If an outbreak of scabies were to occur, it could seriously damage the reputation of this camp.

The term *scabies* comes from the name of the mite *Sarcopetes scabiei,* which causes the disease. The female mite lights upon human skin. When she is ready to bear eggs, she burrows a tunnel in the skin, deposits her eggs, and dies. When the eggs hatch, they continue the cycle of burrowing and procreating until the human host is infested.

The first symptom of scabies is a mild itch that worsens during the night. Mosquito bites can cause the same symptom, so campers who complain of itch should be checked carefully. If the disease has been contracted, you may be able to see faint lines of raised skin (the burrows). Send a camper with such lines to the clinic immediately.

If an isolated incident of the disease occurs, we will quarantine the camper for treatment. If several campers become infected, special bathing, disinfecting, and medicinal procedures should be followed by all campers. These include the use of strong soap and frequent baths, alcohol rubs, and the use of Eurax, a dermicidal lotion.

It is up to each barracks counselor to detect this disease and keep it from spreading. Helping campers maintain a reasonable level of hygiene every day will be an important preventative measure. Your complete cooperation is required in this matter.

You may have noticed in this example (and in the situational analysis preceding it) that the three-part formula is being applied to define *nouns.* It is always best to translate verbs or other parts of speech into their noun forms before creating the introductory statement, because most readers are accustomed to receiving explanations of nouns rather than of actions, adjectives, or adverbs. *To jump* can be expressed as *jumping, to dissect* as *dissection,* and so on.

Several other construction problems often plague introductory statements. Try to avoid each of the following pitfalls:

1. A category or context too broad to be truly informative. Always select a category that is as precise as possible yet not so limiting as to have no other members in it besides the term being defined. For instance,

A melanoma is a *usually malignant tumor* containing dark pigment.

is more informative than

A melanoma is a *growth* containing dark pigment.

2. Distinguishing characteristics that do not adequately separate the term being defined from other members of the category. For example,

Disaster recovery is a plan for dealing with a catastrophe (fire, flood, tornado, or sabotage) at a data processing center *aimed at restoring service as soon as possible at an alternative location.*

is more distinctive than

Disaster recovery is a plan for dealing with a catastrophe (fire, flood, tornado, or sabotage) at a data processing center *aimed at restoring service.*

In the second version, the recovery tactics being described could just as easily be system restart activities. To generate a set of distinguishing characteristics, first list all the traits of the term in question. From this list, choose those traits that are most important to your audience and purpose *and* that differentiate the term from other members of the category.

3. A definition that is not sufficiently concrete. Include characteristics that the reader can perceive via his or her senses (touch, smell, taste, sight, and hearing). Of course, this objective cannot always be achieved, but even relatively abstract notions can be translated into terms a reader can experience.

Reparation is the act or process of providing *money, goods,* or *services* so that *damages, injustices,* or *debts* are offset.

is preferable to

Reparation is the act of *making amends* or of *eliminating a debt.*

4. An introductory statement that uses a derivative of the term being defined. In deciding to offer an audience a detailed definition, you are deciding that the concept, not merely the term, requires explanation. Therefore, using a form of the unfamiliar word in its definition will confuse your readers. (Such "circular definitions" lead readers around and around and leave them pretty much where they started: with a term they don't understand.)

An optical scanner is a device that visually recognizes letters, digits, and other symbols legible to both a machine and a human.

is far more helpful than

An optical scanner is a device for reading documents that uses *optical character recognition*.

Each of us can recall the frustration of looking up a word in the dictionary only to find it explained by means of a term derived from the same root. Dictionaries do this because they don't have the space to define every form of a word at length; you can always track the full definition down quite easily. But don't expect the reader of a technical document to do that. Your definition is meant to be the only help the reader needs to understand the term.

5. The use of other, equally unfamiliar terms in the explanation. Readers who require an explanation of a term require that it be simple, direct, and clear.

A yard is a long, tapering pole made of wood or metal that attaches the upper edge of a four-sided or triangular sail to the mast of a sailing ship.

is far more helpful to a reader who does not understand nautical terminology than

A yard is a long, tapering *spar* attached at a right angle to a mast to hold and spread the *head* of a *square sail, lugsail,* or *lateen*.

EXPANDING THE INTRODUCTORY STATEMENT

After you have presented an introductory statement for your formal definition, it is necessary to explore the term or concept in greater detail. Numerous ways of expanding on the initial statement are possible, and each has its own special effect or purpose. These approaches include

authoritative source	purpose
etymology	negation
physical description	comparison
examples	causation
contexts	procedure or process

No single method is truly sufficient to expand a complex term or concept; rather, an appropriate combination of techniques is desirable. Understanding the needs and characteristics of your audience will help you decide what methods to use.

1. One way to expand a definition beyond the introductory statement is to quote from a reputable *source* such as a dictionary, law, or institute. Associa-

tion with an established source can help build credibility if the audience is skeptical or hostile. It tends to shift the focus from the writer to the authority; likewise, it helps eliminate any doubt about the term's "true" meaning.

The "institutionalized" definition should be used sparingly for two reasons. First, you don't want to pass the buck and depend on higher authorities to create credibility for you. That is your job as writer. Second, because the authoritative definition ought to be enclosed in quotation marks, properly cited, and indented, it is a major departure from the tone and general flow of the document.

The following definition of *toxic pollutants* is excerpted from the 1972 Federal Water Pollution Control Act.

"Toxic pollutants are defined as those which will on the basis of information available to the Administrator of the Act, cause death, disease, behavioral abnormalities, cancer, genetic mutations, physiological malfunctions, or physical deformities. . . . The term is further extended to include substances that cause toxic effects after concentration in a food chain or in combination with other substances."

2. Another possible expansion technique is using word history, or *etymology*, to clarify a concept. Word history provides interesting background information and is especially appealing to a well-educated audience. The following example illustrates this approach.

Kwashiorkor, a severe protein deficiency, derives its name from the Ga language used in Ghana, Africa. It literally means "first-second." The disease often shows up in a first child after the second child is born. The first child is weaned to a diet of starchy foods providing little or no protein so that the second child can be breastfed. Because of the poverty level of the parents, children are denied animal protein, which is too expensive. The starchy diet fails to supply an adequate source of protein to substitute for the protein of breast milk. An adequate protein intake during the early childhood years (1–4) is crucial for normal growth and development.

3. *Physical description* is another technique you can use to clarify a concept. This method helps you create a "word picture" for readers so they can envision the object, disease, creature, or mechanism being defined. Physical description is an effective means of making information tangible by concentrating on such features as dimensions, texture, color, sound, smell, responses, or symptoms. The following excerpt from a definition of *laminitis* is addressed to a group of future veterinarians.

Acute laminitis most commonly affects the front feet of a horse. Symptoms can include anxiety, trembling, increased respiration, and a general reluctance to move. Heat can be felt over the sole and wall of the hoof, and a uniform tenderness is present over the entire sole. In severe cases, the hoof wall may separate from the inner tissues. Death, although not common, may result from acute laminitis.

4. Perhaps the most effective means of making a concept or term tangible for your audience is the use of *examples*. This book relies heavily on examples to introduce new points. Seeing is not only believing; in a technical document is can also mean understanding. Of course, it is important to be sure the examples you use are typical and representative cases.

An atypical example can mislead an audience. In a definition of computer assisted instruction (CAI) for adults, it would be proper to discuss the deductive methods used in most courses as typical. It would be misleading to present the inductive approach to computer instruction practiced by Seymour Papert of M.I.T.'s Artificial Intelligence Institute, because, although it has been used extensively with children, it has been used very little with adults. It *would* be acceptable to use this example if the audience were told that the approach is new and innovative, and if this example were coupled with at least one other, more representative case.

Although not every definition requires an example, examples usually save time and make the meaning very clear. Consider the many ways in which examples have been used in this book and how helpful (we trust!) that approach has been.

On occasion, it may be best to avoid examples in order to leave the audience as open-minded as possible. For instance, in the mid-1970s the Supreme Court decided that individual communities should determine their own definitions of *obscenity*. In this landmark decision, Justice William J. Brennan, Jr., used almost every kind of definition tool imaginable—except examples. If Justice Brennan *had* used an example, he would have established a standard against which other situations could be tested. In effect, he would have been defeating the very purpose of the ruling by limiting the meaning of *obscenity*!

5. Giving the *context* of a definition means explaining the circumstances that surround the use of a term and affect the way it is understood. For instance, in a sociological investigation of 1987 attitudes toward conscription, a social scientist might need to define the phrase *draft opposition*. In recent history, this phrase was interpreted by different groups to mean different things. To many people of draft age, draft opposition meant a show of resistance. The phrase had relatively positive connotations for this group; resisters were men who had the courage of their convictions and were willing to suffer the penalty of imprisonment or voluntary exile in order to oppose the draft. To many World War II veterans (frequently the parents of the resisters), however, this phrase had a highly negative connotation. They viewed opposition to the draft as a form of disloyalty and

cowardice. To fully understand 1987 attitudes, the social scientist might need to trace these historical perspectives.

6. Yet another means of explaining a term is to present its *purpose.* A statement of purpose helps readers understand why something exists, how it is used, or what it is supposed to do. Statements of purpose are excellent tools for explaining terms that represent designs, devices, medicines, procedures, or structures.

Consider this example of a statement of purpose that follows the introductory statement:

LOGO is a programming language developed by Seymour Papert of M.I.T. for use in a computer-assisted learning environment. Designed chiefly for teaching young children the basic elements of computer programming, this simple and rather limited language is strongly oriented toward graphics and is of little value to general data processing professionals. It is highly interactive, inasmuch as it lets children learn quickly how to draw geometric patterns and pictures on the terminal or personal computer screen.

7. When one term might be confused in your readers' minds with another, *negation* is an effective definition technique. This technique pits the two terms against each other in order to eliminate the misconception. When this technique appears in your document, it usually calls for a variation on the standard introductory formula:

Part 1 (Term)	**Part 2 (Misconception)**	**Part 2A (Real category)**	**Part 3 (Distinguishing traits)**
Negligent homicide	is not murder;	rather, it is extreme carelessness or lack of consideration	that contributes directly to the death of an individual

The confusion between the term being defined (*negligent homicide*) and *murder* occurs because the reader is placing the term in an inappropriate category. *Negligent homicide* belongs in the category of *extreme carelessness* rather than in that of *willful act,* the category appropriate to *murder.* In either case, the distinguishing characteristics apply; that is, the act directly contributes to the death of an individual.

8. When confusion exists between two terms within the same category, another useful definition technique, *comparison,* should be used. This method compares and contrasts the distinguishing characteristics of the two

terms in question. The following definition of *diesel engine ignition* distinguishes it from the way in which a gasoline engine accomplishes this task.

> The major difference between diesel and gasoline engines is the method of ignition. A diesel has no ignition system as such; rather the air/fuel mixture is ignited by heat generated during the compression stroke. Heat can be thought of in terms of molecular motion—the hotter the substance is, the faster the molecules vibrate. When a gas is compressed, part of the energy of compression is transferred to the molecules as velocity increases. . . .*

Comparison is an especially useful technique for evaluation purposes (see also Chapter 6). By conducting a parallel examination of points, you can provide readers with data from which they can draw their own conclusions.

The *analogy* is a special kind of comparison. For our purposes, an analogy is a comparison drawn between two things, one unfamiliar (the term being defined) and one familiar (the subject of the comparison). Outside the confines of the analogy, these terms are not usually thought of as similar; however, there are sufficient likenesses to enable you to use the familiar term to clarify the unfamiliar idea. The following analogy is taken from a computer manual on a school scheduling system.

> The clustering process involves distributing the number of students in courses as evenly as possible, determining the demands placed upon rooms and faculty members, and analyzing the conflict/no-conflict relationships between courses. To help understand this grouping process, imagine the departmental mailboxes in the school office. A clerk has received a stack of notices for students from the mail room and must distribute them evenly to faculty members. When the system receives its course offerings, it is very much like the mail clerk with the stack of notices. The system takes each course (starting with those of least demand) and places it in the first available "mailbox" or time slot, just as the clerk places a notice in each mailbox. It places the next course offering in the next available time slot (mailbox) and so on, until each box has one course offering.**

9. Exploring the *causal relationships* between elements is another possible technique to use in extended definitions. Understanding the causes, readers can often understand the effect (the term being defined). Likewise, if the term being defined is the cause, a discussion of effects may prove useful. Here is an example of an effect explained by means of its causes:

> Structural dormancy is caused by physical characteristics of a seed. For instance, immaturity of the embryo can produce dormancy, for either the

*Paul Dempsey, *How to Repair Diesel Engines* (Blue Ridge Summit, PA: TAB Books, 1979).
** *SCRIBE Summary* (Waltham, MA: Honeywell Information Systems, Inc., 1981).

embryo is too small to germinate or its parts are not fully formed. Dormancy will also occur if the seed coat will not allow enough water to be taken into the seed, or if the seed coat resists seedling emergence. Seeds of this type are called "hard seeds." Exchange of oxygen and carbon dioxide is also necessary for germination. Seed coats that do not allow free exchange of these gases will cause dormancy.

Here is an example of a cause defined by means of its effects:

Inflation is an abnormal increase in available currency and credit beyond the proportion of available goods, resulting in sharp and continuing rises in price levels.

Note that these statements use words that indicate causal relationships, such as *resulting in, cause, produce,* and *allow.* Other transitional words that can be used are *because, by means of, despite,* and *occurring.* As we learned in Part I of this textbook, transitional words and phrases help keep readers on track. In causal explanations, such devices can be especially helpful.

10. A final definition technique is one that describes a *procedure* or *process.* (The difference between the two terms is explained at the beginning of Chapter 8.) In using this method, you trace a sequence of events that results in a determinable end. This mode of definition is closely related to cause-and-effect statements because it often encompasses a variety of causal relationships. In the description, however, emphasis falls on the entire process rather than on the result or on any of the individual steps. The following definition of *water cycle* illustrates this point.

The water cycle is a natural process that includes three phases: evaporation, condensation, and precipitation. The sun heats water to create water vapor; water vapor rises and forms clouds; water returns to earth in the form of rain.

To describe a process in order to explain a concept, first develop a list of all steps, stages, or points involved. Present the information in its natural or logical sequence. Use clear transitional devices to indicate which steps are the most important to the process (or which are the most complex, time-consuming, or difficult).

PUTTING IT ALL TOGETHER

Having surveyed the major techniques for defining terms, we need only consider the order in which information is presented. Just as in classification, the graded arrangement works very well when it is necessary to rank the elements of your definition according to the priorities your audience will assign to them. Sequential order is appropriate when you are using procedural or causal techniques within your definition. Spatial arrangement is useful when your definition consists primarily of physical or functional description.

Now let us examine a fully expanded definition that employs several techniques and presents information in a way that is meaningful to the audience. Earlier in this chapter we looked at a camp director's definition of *scabies* in order to see the introductory statement at work. A second sample definition of the same term follows; this one was written by someone working in the Department of Health. The audience, purpose, and format are different from those that shaped the first definition of *scabies*. Consider how these two versions differ in content, level of detail, emphasis, and language.

The annotations in the right column will help you identify the various definition techniques that we have discussed. Try covering the margin of the page the first time you read the same and identifying these techniques yourself.

AUDIENCE	Parents of elementary school children.
ROLE	Member of the Department of Health. Writer is trying to be informative and helpful.
PURPOSE	To acquaint readers with symptoms, causes, and treatment of scabies. These readers must determine whether their child has the disease.
FORMAT	A leaflet sent home with each child.

Recently, a student in your child's class developed scabies (also called "the itch"). This is a very contagious	1	1. Attention getter
skin disease. Scabies is caused by a mite (a tiny insect) that lives in human skin tissue. The presence of this mite can cause itching. If left untreated, scabies can cause blisters and infection. In extreme forms it can cause scaling of the skin and scarring.	2	2. Brief, formal introductory statement
The tiny mite likes to settle in skin folds such as the groin area, behind the ears, and between fingers and toes. It digs tunnels in which it lays its eggs. When the eggs hatch, the mites quickly reach maturity, burrow more tunnels and lay more eggs. The cycle continues unless all the mites are killed.	3	3. Procedure description
The first signs of scabies may be mild itching. The itching is stronger at night. If left untreated, scabies will advance to a second stage in which the mites' burrows appear	4	4. Physical description

as light, raised lines on the skin. The third stage is blisters at the edge of the tunnels. Serious infection can occur at this point.

Despite popular belief, scabies is not a problem of poor hygiene (although that fosters the mite population's growth). Rather, the disease is transmitted by direct skin contact with a person carrying the mites. For that reason, family members of a contaminated person often contract the disease as well. 5

5. Negation

To treat the disease, the first thing to do is visit your family physician. He or she will prescribe a lotion to help eliminate the mite. The contaminated person should be scrubbed with warm, soapy water and rinsed well to expose the mite burrows. After the person has dried the skin, the lotion should be applied. Clothing, towels, bed linens, and utensils should be kept separate and disinfected. Treatment takes from 10 days to 2 weeks. 6

6. Procedure description

If you think your child has contracted scabies, please follow our advice. Contact the school nurse's office at 555-9999 if you have questions. 7

7. Call to action (conclusion)

CHAPTER SUMMARY

As we have seen in this chapter, clarification and definition are often required in technical writing. However, it is up to you, the writer, to determine what degree of explanation is required and what techniques to use. Brief definitions can be presented as a single word, a phrase, a full sentence, or a note; formal, extended definitions can comprise an entire document. The formal definition usually begins with an introductory statement that identifies the term, places it in its proper category, and distinguishes it from other members of that category. At least ten techniques for defining are available to help clarify terms. Information should be ordered in such a way that it best meets the requirements of the overall writing situation.

EXERCISES

The following exercises reinforce many of the points presented in this chapter. The kind of practice each exercise affords will help foster skill in writing definitions.

1. The following definition of *scabies* is taken from a medical encyclopedia. The citation at the end of the example reveals that the authors are a doctor and a nurse, respectively. They bring years of expertise to this explanation. Determine what methods of definition are used. From the level of presentation, construct a list of characteristics that audience members are likely to exhibit.

Scabies (ska'bez) is a contagious skin disease caused by the itch mite, *Sarcopetes scabiei*. Scabies, sometimes called "the itch," is most likely to erupt in folds of skin, as in the groin, beneath the breasts, or between the toes or fingers.

The adult itch mite has a rounded body about one-fiftieth of an inch long. Scabies is caused by the female, which burrows beneath the skin and digs a short tunnel parallel to the surface, in which it lays its eggs. The eggs hatch in a few days, after which the baby mites find their way to the skin surface, where they live their brief lives until they too are ready to burrow and lay their eggs.

Symptoms

During the initial tunnel-digging and egg-laying the human host may be oblivious to what is happening. There is little itching. The very slight skin discoloration may be mistaken for any one of numerous other skin disorders.

In about a week, the itching becomes intense because of hypersensitivity to the mite. The itch is much worse at night. The tunnels in the skin can now be discerned as slightly elevated grayish white lines. The mite itself can often be seen—with the aid of a magnifying glass—as an infinitesimal white speck at the end of the tunnel. Blisters and pustules also may develop near the tunnel.

Transmission

Scabies is easily transmitted from person to person by direct skin contact or to a limited extent by contact with clothing of infected persons. Epidemics are fairly common in such places as camps, barracks, and institutions. It is unusual for one member of a family not to communicate it to others.

The period of communicability lasts until the itch mites and eggs are totally destroyed, a period of 1 to 2 weeks, depending on the effectiveness of the treatment used.

Treatment and Prevention

The usual therapy begins with a hot bath and thorough scrubbing to open and expose the burrows. This is followed by the application of some type of DDT preparation.

The patient's underwear and bedclothing must be changed and laundered daily until all the itch mite eggs are hatched out and the mites eliminated.

Good personal hygiene and wearing fresh and clean clothing are two most effective ways to prevent scabies.*

*Benjamin F. Miller, M.D., and Claire Brackman Keane, R.N., B.S. *Encyclopedia and Dictionary of Medicine and Nursing* (Philadelphia: W.B. Saunders, 1972), pp. 856–857.

Figure 7-1

2. Figure 7-1 presents a definition problem. What is the problem? Rewrite the cartoon to eliminate the difficulty.

3. Create three different informal definitions for each of the following terms: spelunking, crustacean, promontory, bus.

4. Look up each of the following words in the dictionary: *acme, compensation, cytolysin, group, spar*. Then create a chart like the one on page 155 and place each part of the dictionary definition into its appropriate column (*term, category, distinguishing traits*).

WRITING ASSIGNMENTS

Select one of the following assignments. Preface the definition with a situational analysis of audience, role, purpose, and format.

1. Choose a field-specific term with which you are very familiar. Assume the role of expert and explain this term to an intelligent adult who has no formal training in your field. Present this explanation in the form of a newspaper article worthy of publication in the local paper's science section. Select only the most appropriate methods of amplifying the term's meaning. Limit: 400–500 words.

2. Choose a field-specific term with which you are very familiar. Assume the role of project head and explain the term to the budget committee that supplies funding for the project. Select only the most appropriate methods of amplifying the term's meaning. Limit: 400–500 words.

NOTE: Apply the editing criteria at the end of Part II to your assignment before you submit it to your instructor.

CHAPTER 8

Describing Processes, Procedures, and Activities

CHAPTER PLAN

Scientific and technical environments frequently require documentation of such natural processes as circulation, crystallization, decomposition, digestion, meiosis, and photosynthesis. And just as important are clear descriptions of methods used in such engineering procedures as computer design and programming, bridge construction, effluent treatment, and pollution control. Likewise, technical fields depend on the accurate description of specific activities: operating heavy equipment, using a word processor, performing an experiment, and testing product quality, for instance.

In both your academic and professional careers, you will be required to describe processes, procedures, and activities. To equip you to meet these challenges, this chapter will

- help you determine the need for descriptions of processes, procedures, and activities.
- introduce the components of such descriptions.
- examine the writing techniques that are most effective for presenting each of these descriptions.

DEFINING PROCESS, PROCEDURE, AND ACTIVITY

This chapter concentrates on the several ways of describing actions and reactions in the scientific and technical environment. Depending on the subject matter and the purpose of the document, you may be called upon to describe a process, a procedure, or an activity.

A *process* is a natural series of actions and reactions that yields identifiable, verifiable results. All of us participate in a variety of processes every day. We digest our food, heal ourselves by generating new cells or producing antibodies, distribute resources through our bodies via respiration and circulation, and use our five senses. Natural processes go on all around us as well. Plants produce their own food via photosynthesis, yeast ferments, storms build and move, volcanoes erupt, fertilized eggs mature, and matter decomposes—the list seems endless.

In some cases, we can create or control the *conditions* that facilitate or impede a natural process, but rarely can we control the process itself. When we put a pot of water on a stove and heat it to boiling, we have created the right conditions, but we have not created the tendency of water to turn to steam when it is heated to 100°C. Similarly, we can chew our food 30 times before swallowing it so that enzymatic juices are secreted early in the digestive process. Or we can place sand bags on the banks of an overflowing river to slow the erosion process. However, in none of these cases are we creating or changing the tendencies inherent in the natural process.

Describing a process, then, is primarily a matter of observing and reporting. Our usual purpose in writing about a process is to *explain* it to an audience requiring that understanding. Descriptions of processes rely heavily on cause-and-effect analysis to explain their subject matter, and they can be found in a variety of technical documents. For instance, a report about the preparation of a site for a dam might have an introductory section explaining the forces of erosion. The textbook for a chemistry course might discuss molecular bonding and catalysis. An article in a scholarly journal might describe the ways in which rain becomes increasingly acidic as it absorbs sulphur particles from industrial pollution.

A *procedure* is a human-engineered set of actions and methods that leads to a predetermined, verifiable end. A procedure requires active human involvement both to create the necessary conditions and to carry out the action. We all perform basic procedures every day; they include laundering our clothes, driving a car, cooking dinner, and writing a document. In a technical environment, standardized procedures generally exist for such tasks as conducting experiments, forecasting conditions, and designing programs.

Some procedures can be performed handily without much forethought, whereas others require concentration and considerable expertise. Some consist of steps that can be conducted in a random sequence (though certain patterns might be more efficient than others); some depend on the precise execution of several steps in unvarying order.

Describing a procedure is primarily an evaluative job that relies heavily on classification and answers the basic question of "why" something is done. We group actions into logical or chronological categories in order to best examine the recommended ways and the alternative ways of reaching a goal. The rationale for doing things in a certain way (cost-effectiveness, efficiency, timeliness, or ease) is an important part of the document.

Generally, the purpose of the description of a procedure is *not* to have readers perform specific actions after reading the document. It is rather to provide an *overview* of activities so that readers can understand and appreciate the methods suggested. The most common document in which descriptions of procedures appear is the technical manual, wherein these sections provide important information for those who will ultimately be responsible for either performing the procedure or supervising those who carry it out.

The description of an *activity* is actually a subset of the description of a procedure. It discusses a specific task in terms of a well-defined set of steps. Rather than providing readers with an overall understanding of actions, the description of an activity is designed to enable readers to *replicate* the results put forth. The most common form such a description takes is the set of instructions. In this straightforward list, precise steps are presented in the imperative; the reasons behind each step get little or no emphasis. Most of the time, information is arranged in linear fashion with each discrete action following the basic pattern "Do this . . . to that . . . with this."

For each of these three types of descriptions, we will now examine the kind of analysis required, the necessary components, recommended ways in which to order the information you include, and a sample document or excerpt.

UNDERSTANDING THE CAUSAL RELATIONSHIPS

To develop an effective description of a process or a procedure, you must thoroughly understand the causal relationships inherent in the phenomenon or event so that you can best arrange them within the document and explain them to your audience. Almost all processes and procedures depend to some degree on cause-and-effect or action-and-reaction relationships; few can be classified as mere sequence. Even if one action does not directly cause another, it creates the necessary condition or prepares the way.

A *cause* is a person, thing, or action that initiates a specific result: overeating is often the cause of stomach upset. An *effect,* on the other hand, is produced by such an agent: stomach upset is often the effect of overeating. A *condition* is a situation or set of circumstances that allows a cause or effect to occur: a person is very hungry, many delicious foods are laid out, and the host insists on serving second portions. The condition is not the cause, however (it simply made overeating possible—and likely).

The *causal relationship* is the way in which causes and effects interact. In Chapter 4 we looked at several possible arrangements of information in various causal relationships:

single cause—single effect
single cause—multiple effects
multiple causes—single effect
causal chain ("chain reaction")

Each cause can also be classified according to the degree of impact it has on the overall effect. For instance, a *necessary cause* is a required agent—one that must be present if an effect is to occur. It is *necessary* for sperm to penetrate the ovum in order for fertilization to take place; this is true whether sexual intercourse or artificial insemination is used. However, a necessary cause is not always a *sufficient cause*—that is, an agent capable of initiating the expected result by itself, without aid from any other agents. Consider a homicide investigation in which the detective determines that the bullet wound to the brain was enough to kill the victim even though he was shot at least three other times.

A *contributory cause* is one of several causes that together produce an effect (for example, inflation results from a combination of too much currency, too little demand, and too much credit). No single contributory cause is able to produce the effect on its own. An *immediate cause* occupies the next-to-last position in a causal chain; it is the agent immediately preceding the final result. The *initiating cause* (sometimes called the ultimate cause) is the first cause in a chain of reactions. Finally, a *probable cause* is the agent that is likely to have been responsible for an effect, though it is impossible to be absolutely certain. Probable causes fall in the realm of hypothesis and theory. For example, paleontologists conjecture that a great meteor falling to earth 75,000,000 years ago caused an enormous dust cloud that blocked the sun and subsequently caused the death of less adaptive forms of plant and animal life.

Effects, too, can be grouped into categories in terms of their relationship to causes. The *direct effect* is one that results immediately from a sufficient cause. In the statement "The remark angered him," anger is the direct effect and the remark is the sufficient cause (the single-cause–single-effect relationship). A *secondary effect* is one produced tangentially or indirectly from a cause ("The chicken pox lowered his resistance to infection [a condition], and he subsequently developed pneumonia when he was exposed to bacteria"). A *triggering effect* is one effect in a causal chain that becomes the cause of yet another effect. For instance, in the chain "The storm caused the electricity to go out / the temperature in the refrigerator rose / the food began to spoil," the electricity going out is a triggering effect.

Categorize each of the causes and effects you discover when you examine your process or procedure and apply the proper label. If your audience is familiar with these labels, you can use them to make transitions from point to point. However, it is more likely that your readers will not know these labels. You can still provide guideposts, which place causes and effects in their appropriate categories, by using terms such as

first, second, third, finally
initially, then, last
minor, major
lasting, important, significant

Likewise, it is important to distinguish between conditions and causes. Ovulation, for instance, does not *cause* fertilization: it allows it to happen if other conditions are satisfied. It is equally important to distinguish between mere sequence and actual causation. For instance, in evaluating a landslide and a dam break in northern Italy, researchers learned that the landslide occurred first and then the dam broke. However, they found that both of these events were caused by an earth tremor; the landslide did not cause the dam to break.

Labeling causes and effects will also help you organize information in the order that best reflects the way the process itself unfolds. Your classification skills will prove useful as you group certain causes or effects into categories or phases of action. When grouping is necessary, use headings to make information more accessible and easier to follow.

DEVELOPING THE DESCRIPTION OF A PROCESS

The components of a process description can vary from document to document, but they generally include

an introductory statement
phases of action and reaction
summary
conclusion

The *introductory statement* follows a pattern quite similar to that used in definitions (see Chapter 7). Recalling that three-part formula, you will want to present the name of the process (*fermentation, peristalsis*), the process category (*biochemical, electromagnetic, physical, atomic, geological*), and an overview of the key action phases—in the order you intend to follow in your discussion. This statement not only previews the process for your readers but also acts as a disciplining device for you, the writer, because in it you commit yourself to follow the pattern you establish. In addition to this handy one- or two-sentence introduction, you should include a thesis statement or statement of purpose in the opening section of the document. The following sample, "How Food Is Digested," illustrates the kind of introduction that works well for the description of a process.

The bulk of the description covers the actual *phases of action:* the causes and their effects. These should appear in the natural sequence in which they occur. In the cases of simultaneous action (many causes and effects that "run parallel" or branch off rather than being linked in a chain of events), you may wish to use a visual aid to demonstrate the pattern. Clear transitions are also crucial here. In other cases, where order either is not apparent or is of little consequence, you may want to use a graded order to present the information in a manner most appropriate to your readers' interests or to the purpose of the document.

A lengthy process description (one that runs several pages) will benefit from your summarizing the phases or steps before you draw any conclusions. Rather than merely reiterating the information you have presented earlier in the document, however, bring new points to light or emphasize the importance of particular phases over others.

Like the documents relying on grouping and definition that we have examined in this textbook, the description of a process should have a concluding paragraph or two that reinforces the thesis, draws conclusions important to your readers, makes recommendations, or calls readers to action.

Let us consider how the following document illustrates the elements we should watch for in the description of a process.

AUDIENCE	Incoming students, who must understand the way in which food is digested before they can appreciate the kinds of foods that provide the best nutrition.
ROLE	Instructor in introductory nutrition class at a junior college.
PURPOSE	To provide an overview of the way in which food is digested. Students will be tested on the information they learn from the handout.
FORMAT	A handout for use in the introductory nutrition class.

HOW FOOD IS DIGESTED

Digestion is a biochemical process that breaks down food into simple nutrients usable by the host organism.[1] In the human body, digestion takes place in the digestive system, a collection of specialized organs and glands. Digestion can be 2
discussed in stages, each stage associated with a different portion of the digestive tract. The stages are chewing, swallowing, peristalsis, decomposition, absorbing, and excreting.

1. Introduction: three-part formula is distributed throughout the first paragraph (no purpose or thesis is provided)
2. Reveals the order that will be followed

The first stage occurs in the mouth, where food is chewed. The effect of chewing is the breaking apart of large pieces of food. Secretions from the salivary glands increase during chewing and are mixed with the food. The mixture is softer and smoother than the original. 3

3. Phase 1—chewing; cause and effect: chewing causes increased salivation and a change in food texture

The tongue pushes the food toward the back of the mouth, which induces a series of muscle contractions (swallowing).[5] As the food is swallowed, the soft palate closes off air from the nasal cavity above.[6] The larynx automatically rises, while the epiglottis lowers itself onto the trachea; these actions propel the food into the esophagus.[7] 4

4. Phase 2—swallowing; series of causes and effects:
5. Pushing causes contractions
6. Palate closes, restricting air from nasal cavity
7. Larynx raises and epiglottis lowers, propelling food

The esophagus, in turn, transports the food to the stomach by means of a series of contractions called *peristalsis*. When the food expands the walls of the tract, complex components in the autonomic nervous system begin to cooperate. Sense organs release a chemical called acetylcholine, which makes the muscles contract. In this way food is pushed into the stomach. 8

8. Phase 3—peristalsis; cause-and-effect chain: food enters/walls expand/acetylcholine is released/muscles contract/food is pushed into stomach

In the stomach, glands secrete acids and enzymes to break down the food into a fine mash-like substance.[10] It takes food an average of two and one-half hours of contact with these acids before it reaches this state. During this time, the enzyme pepsin is splitting protein into simple components.[11] 9

9. Phase 4—decomposition; simultaneous action:
10. Acids and enzymes break down food into fine mash
11. Pepsin splits protein into simple components

By means of stomach contractions, this fine mixture is passed into the small intestine. Only small quantities of food are moved at a time. In the duodenum (the first of three sections of the small intestine), the digestion process is almost com- 12

12. Phase 5—absorption; two stages; stage 1—activity of duodenum, simultaneous action:

pleted. Here bile breaks down the fat in the mixture into smaller bits so that enzymes can break it down further.[13] The enzymes of the duodenum convert the starch and carbohydrate into usable substances.[14] At this point food is at its most basic level of components. However, the nutrients have not been absorbed yet, or transported via the blood and lymph. Nor have wastes been removed from the body.

[15] The second part of the small intestine, the jejunum, is responsible for the absorption of these basic components. Its walls are lined with small projections called villi, which are made up of capillaries and lymph vessels. The lymph vessels remove the fatty substances from the nutritive substance and pass it to the main thoracic cavity for distribution throughout the body.[16] The capillaries absorb the nutrients and carry them indirectly to the right atrium of the heart, which sends them to all parts of the body.[17]

[18] At this point the digestion process is complete. Waste materials are sent to the large intestine for further processing and finally excretion.

[19] Without this natural process, our bodies could not make full use of the foods we eat. The nutrients absorbed into our blood are distributed to cells throughout our bodies. They replace energy lost through normal levels of activity. They help our bodies build new cells as well.

[20] When the digestive process goes awry, serious complications can result. Aside from obvious symptoms such as vomiting and constipation, we can experience more gradual and serious problems such as anemia, anorexia, and kwashiorkor, each of which, in its extreme form, can cause death.

13. Bile breaks down fat
14. Enzymes convert starch and carbohydrates
15. Stage 2—activity of jejunum, simultaneous action:
16. Lymph vessels remove fatty substances
17. Sequence: capillaries absorb nutrients; they then pass to right atrium, which distributes them
18. Phase 6—excretion
19. Summary point is made
20. Conclusion is expressed

Note that although our sample will probably accomplish its purpose (providing an overview of digestion to incoming nutrition majors), it would benefit from a direct statement of that objective in the introduction. The comments in the margins demonstrate the effective use of cause-and-effect analysis. Included are a single-cause–single-effect relationship, a series of causes and effects, a causal chain, simultaneous action, and sequence. In addition to these analytical techniques, the author brings division skills to bear as she breaks down the last phase (absorption) into two subcategories based on the area in which activity takes place. The summary toward the end of the document does not restate the overview sentence; rather, it points to the importance of the process and adds new dimension to the discussion. The conclusion helps the instructor prepare students for some of the problems that the course will address in the future.

DEVELOPING THE DESCRIPTION OF A PROCEDURE

Writing an effective description of a procedure requires you to be both familiar and comfortable with the procedure you are discussing. You must be acquainted not only with how to do something but also with why it is done a particular way. Your initial cause-and-effect analysis will serve you well here. In addition, you should include some contingency plans for your readers to use if they encounter problems. Assuming that you have this degree of knowledge and expertise, you are well prepared for the task of explaining the procedure to readers who are unfamiliar with it.

Before you actually begin writing, you should list the following elements:

purpose
phases of activity (including individual steps and the reasons for them)
materials, equipment, and/or expertise required
prerequisites
constraints
appropriate contingency plans

Remember that the primary purpose of the description of a procedure is to provide readers with an overview of the procedure so that they can understand *why* it is done the way you recommend as well as *how* it is done. Therefore it is important to develop a clear, concise statement of purpose for your introductory remarks. This statement can very closely follow that which you frame for your situational analysis before writing the document. The introduction should also include a brief three-part formula that identifies the topic, specifies the kind of procedure (hydraulic, manual, engineering, automated, computerized, or whatever), and provides an overview of the *phases* of activity to be discussed—in the order you will follow.

You can use your classification skills to form logical, functional, or sequential categories of activities to help explain the procedure. List all of the activities with which you are familiar (in much the same way as you brainstorm prior to developing an outline). Collect related steps into *phases* of action, and provide a descriptive label for each category. For instance, in a *Sports Illustrated* guide for runners, the author describes the racing start by organizing information into categories called "On your marks," "Set," and "Bang," the three stimuli runners receive just before they take off. In a landscaping manual, the author organizes information about transplanting trees that have been grown in cans by using the phases "Preparing the hole," "Cutting the can," "Loosening the roots," "Covering the roots," and finally "Supporting the tree."

The body of the description of a procedure should follow the order you previewed in the introductory statement. Within each activity phase you have formed, present the individual steps involved. Consider the sequence and provide appropriate transitions to keep readers on track. Must one step precede another? Are the actions simultaneous, sequential, or discrete? Cite reasons for performing each step if they are not readily apparent. Are there warnings, cautions, or notes that you should supply to ensure the safety and well-being of the person performing the procedure? Do you have any alternatives to suggest to the reader who runs into problems along the way?

List any required materials (soil, water, cardboard, tin cans), equipment (tractor, shovel, plow, Bunsen burner), or expertise (four years of field experience, a B.S.E.E.) within the appropriate phase. If any of these prerequisites applies to the overall procedure rather than to just a single phase of action, it should be discussed in the introduction.

Use the concluding portion of the document to summarize the phases of action (if the discussion is lengthy) and to evaluate the merits of the suggested methods by discussing such constraints as budget, schedule, personnel requirements, weather conditions, and resource availability. Make recommendations about implementation and suggest any changes you have in mind. And if a call for action from your reader is appropriate, insert it here.

The following excerpt from a manual on scuba diving discusses donning equipment underwater. It can rely heavily on classroom instruction to augment the information presented, because no one can get air for underwater diving without being certified and no one can be certified without passing a course. However, you will find that the document makes sense without extra help. Likewise, note how effectively graphic aids illustrate the information provided in the narrative.

AUDIENCE People about to start a scuba diving certification class. The book is required reading prior to the start of the course.

ROLE Scuba diving expert.

PURPOSE To familiarize students with the basic techniques for donning scuba equipment underwater.

FORMAT Textbook chapter.

Needs title

The following orientation, best practiced in a swimming pool, is offered as a suggested procedure. Instructors develop techniques and progressions that adapt best to their individual programs. This suggested procedure is, however, the product of many of these adaptations.

After completing predive checks and assembly of equipment, buddy teams should, in shallow water, each don all equipment with the assistance of the other. This initial donning should emphasize the BCD position, proper fastening of quick-release devices and the positioning of the weight belt so as to permit instant doffing if necessary.

Special care must be taken during the following exercises to prevent unvented ascent after breathing compressed air while submerged. Even in the shallow end of the pool there is potential hazard, and such an ascent could cause some form of lung injury.

Recheck the regulator function before submerging. With the mouthpiece in place, breathe normally. Lie on the bottom of the pool, face down. Alter the breathing volume and note the change in buoyancy which results. Roll to the side and to the face-up position. In using the single-hose regulator, free flow does not occur, regardless of position, so it is not a problem.

Moves to active to procedure

After completing the initial breathing exercise, the trainee should surface while breathing normally. Return to the bottom, remove the mouthpiece and surface while exhaling continuously.

Back to passive

Clearing the flooded mouthpiece and resuming breathing are the next phase. The flooded single-hose regulator may be cleared after inserting the mouthpiece by pushing the purge button or by puffing into the mouthpiece. Inhalation following either method should be of the cautious, trial type rather than a full breath.

Back to active

Clearing methods are simple and usually accomplished on the first attempt. Ridiculous as it may seem, however, the beginning diver must be impressed with the fact that inhalation cannot occur if the lungs are already full. Complaints of regulators not functioning, no air, or inability to breathe after mechanically purging the mouthpiece are, in many instances, traceable to this human failure. The trainee should practice the appropriate clearing procedures until complete ease is achieved with each.

After initial orientation, trainees will benefit from swimming several lengths of the pool while on the bottom, breathing normally. If they so desire and are comfortable enough, they should flood and clear the mouthpiece and mask several times during this exercise. Experimentation with and regulation of breathing volume during the exercise will demonstrate the desired buoyancy control.

Utilization of the foregoing skills (and their necessity) will be featured in the next exercises, involving donning and doffing scuba while underwater.

Figure 1 The five-step overhead donning method.

Donning

Before attempting don and doff procedures underwater, a full demonstration of both should be given on pool deck or shore. Again review the nature and function of all harness, because underwater don and doff practice will involve feeling, rather than seeing, much of the equipment. Initial performance of the exercises underwater will be easier if done in shallow water and while wearing the mask.

Overhead Method

After the scuba has been positioned on the bottom, with the weight belt placed over the tank near the base and the single-hose regulator attached, dive to the scuba [see Figure 1]. Grasp the valve, holding it close to the chest so as to maintain bottom position. Clear the regulator and breathe. While still face down, grasp the tank and move it, parallel to the bottom, to the area of the thighs. Roll to your back and sit up. If using fins, they may be donned before putting on the scuba harness to permit greater freedom of the body while leaning forward. Place the scuba on or between the thighs with the backpack on top. Transfer the weight belt to your lap. Make up the shoulder harness to fit the upper body. Insert the hands and forearms into the straps and grasp the air tank. Be sure that the right arm is *outside* the single hose before inserting the arm into the harness—otherwise the hose will be caught under the armpit when the tank is swung over the head. If hose is caught under harness or arm, this may be corrected by removing the mouthpiece and repositioning the hose. Next, insert and clear the mouthpiece and continue breathing.

The tank is now lifted through an arc into position on the back. Position the shoulder straps. Grasp and pull the waist strap into position and fasten it. If a crotch strap is part of the harness, it may be used to pull the bottom of the tank into position. The crotch strap may be held until needed by anchoring it beneath the buttocks or between the legs.

With harness in place, adjusted and fastened, the weight belt may now be donned. This can be done while still in the sitting position, but is better accomplished in a face-down "knee-chest" position which allows snug fastening without a lot of juggling.

The latter part of the overhead procedure may be practiced by standing erect in the shallow end of the swimming pool and swinging the scuba tank from pool edge over the head and into position on the back. This will permit easier adjustment of tank, hose and harness, and make these steps more natural when donning underwater.

Side Donning Method

The initial contact and clearing are the same as in the overhead method. While in prone position, draw knees forward to the area of the scuba valve, keeping the body inclined forward. Pull the tank and weight belt up between the thighs with weight belt in the right hand. Raise the right knee and swing it to the right. You are now kneeling on the left knee only. Place the weight belt across and high up on the right thigh for ballast. Mask and harness may now be donned. Next, insert the right arm into shoulder harness, making sure scuba hose is *above* the right arm. Using both hands, scuba tank is moved up and around the right shoulder to the back. Insert the left arm through harness and fasten shoulder harness. Kneeling forward, fasten weight belt around the waist. Don fins by raising each knee and bringing the foot forward.

Both the overhead and side donning methods work more easily when the diver is bare-skinned than when wearing a wet suit or other protective clothing. When such equipment is worn, shoulder straps and other harness may have to be maneuvered into position because they do not easily slip over rougher materials.

There are numerous variations of these two basic donning methods, including kneeling on both knees and lying on the back. All have merit and are dictated by physique, buoyancy, dress and equipment design. Individual preference and experience of the instructor will help determine which method best suits the individual diver.*

This author provides an extended overview of the scuba orientation and then applies it to donning equipment underwater. The description would have benefited from a listing of the five steps in the procedure; while reading, it is difficult to pick out the same five steps as the author illustrates in the graphic. Nonetheless, if they read this description several times, audience members should be able to perform the procedure under classroom conditions. The document does accomplish its purpose: *preparing* the reader to perform the procedure.

Note the use of warnings throughout the piece. The italicized warning affects the entire procedure and therefore appears in the introductory portion of the document. *Warnings* are messages about potential danger to individuals. *Cautions* (none are used here) indicate potential problems that could affect the performance of equipment. *Notes* are used to insert points of interest or background information but *not* to alert readers to potentially dangerous situations or actions.

The author employs the most common ordering technique for descriptions of procedures: a sequential arrangement of phases and actions within categories. After one entire method is presented, however, he presents a second possibility (with briefer coverage). When more than one method is described, this technique can be quite effective. Generally, the favored method is presented first, then the alternative method(s). Overall evaluation is usually postponed until the concluding remarks and is added only if the author wishes to help the reader decide which method to use.

DEVELOPING EFFECTIVE INSTRUCTIONS

Description of an activity generally takes the form of a list of instructions rather than a narrative. This precise medium is designed to enable readers to perform specific tasks in such a way as to achieve immediate and replicable results.

Ideally, technical instructions are self-sufficient. If they are well-developed, readers who follow them should have little difficulty implementing the actions described. However, if you have ever had to assemble a stereo, bicycle, or piece of furniture from the instructions that came with

*Council for National Cooperation in Aquatics, *The New Science of Skin and Scuba Diving,* 6th ed. (Piscataway, N.J.: New Century Publishers, 1985), pp. 101–105.

the product, you know that instructions often leave a great deal to be desired. The following model is also a lesson in that the subject you are being instructed in is writing instructions. Because you do not know very much about developing instructions at this point, your response will be the perfect test of whether instructions can be complete and self-sufficient!

HOW TO WRITE TECHNICAL INSTRUCTIONS

SUMMARY Writing technical instructions is a direct process of arranging information sequentially so that the audience can duplicate the described actions with replicable results.

MATERIALS Several pieces of notebook, typewriter, or printer paper.

EQUIPMENT One pencil or pen.
One elite or pica typewriter or one computer with word processing software.
One table or desk.
One comfortable, straight-backed chair.

WARNING *If you are using a computer, be sure to avoid eyestrain by taking a break after no longer than two hours of work in front of the terminal screen.*

CAUTION If you are using a computer, do not smoke, eat, or drink in the same room; air pollution can affect the performance of the system.

NOTE The writing of technical instructions has a history as long as that of written communication. Our civilization depends on written instructions as a method of preserving knowledge and avoiding duplication of effort.

Instructions

1. Adopt an omniscient writer's role.
2. Assume nothing about your audience except a moderate ability to read.
3. Perform steps 4 through 23 (draft preparation) on a piece of notebook paper with either pencil or pen, or on a word processor.
4. Devise a precise title that indicates the purpose of the instructions.
5. Write an initial confidence-building summary statement.
6. List all materials (expendable items) in precise kind and measure.
7. List all equipment or tools in precise kind and measure.
8. Write any overall warnings (to personnel) in italics, and use boldface if you are using a word processor with that option.

9. Write any overall cautions (about equipment), and use boldface if you are using a word processor with that option.
10. Write any overall notes (background, additional, or preparatory information).
11. Prepare a precise sequential arrangement of steps.
12. Number each step.
13. Limit each step to instruction in a single action.
14. Describe each action with a single, imperative verb.
15. Begin each step with an imperative verb.
16. Follow the basic sentence pattern "Do this to that with that."
    NOTE: Do not be concerned if one part of the pattern does not exist in any given step.
17. Keep sentences simple.
18. Write any warnings or cautions specific to a step in italics and boldface (if you have that word processing option) directly beneath the associated step.
19. Write any notes specific to a step directly beneath the associated step.
20. Design any figures appropriate to individual steps.
21. Refer to figures parenthetically within the associated step.
22. Cite points of repetition, when necessary, as a step.
23. Indicate the final step with a clear transitional device of completion.
24. Review your draft of the instructions for errors.
25. Test the instructions with persons representative of the intended audience.
26. Revise instructions as necessary on the basis of the results of the test.
27. Type instructions from pen or pencil draft, or run a final copy on the word processor.
28. Finally, repeat steps 1 through 27 for each set of instructions to be written.

Note how the emphasis within a set of instructions falls on the steps to be taken, not on the reasons for taking them. Apparently omniscient, the writer dictates what steps to take, even though several alternatives might be possible. The imperative mood of the verbs gives readers a set of commands rather than suggestions. Instructions of this nature appear in documents such as user guides and manuals and on the cartons of products requiring assembly.

The style seems clipped for several reasons. First, the sentence pattern rarely varies ("Do this to that with that"). Second, the sentences are short. Third, no transitions occur between steps beyond the sequential numbering; ideas seem to be fragmented. (This fragmentation is also a result of the lack of classification of associated steps.) With very complex activities, you may wish to group steps into phases of action and use clear, descriptive headings. Steps are numbered by the section in such cases (1-1, 1-2, and so on). Finally, only one action is described in each step.

Despite the seeming simplicity of a good set of instructions, writing them is an art. The ability to isolate steps so that the instructions are self-sufficient is a painstaking process in itself. Writers familiar with an activity may overlook actions (hence the testing mentioned in step 25 is essential to prevent any oversights). Selecting a precise verb that indicates *only a single action* is also quite demanding. Consider the following student example. The student assumes the role of a lab technician preparing incoming students to use microscopes. This set of instructions is intended for inclusion in a lab manual. Note the use of single verbs. In some cases, however, these verbs do not suggest single actions. Identify these complex verbs and divide them into the several actions they actually represent.

EXAMINING BODY CELLS WITH A COMPOUND MICROSCOPE

SUMMARY	Examining body cells with a compound microscope is a simple process of collecting the sample, preparing the slide, and focusing the slide under the microscope lens.
MATERIALS	Water
	Iodine
TOOLS	1 eyedropper
	1 wooden toothpick
	1 clean slide
	1 clean cover glass
EQUIPMENT	1 compound microscope
	1 table or large, flat surface
	1 chair

Instructions

1. Gently scrape the inside of your cheek with the blunt end of a wooden toothpick.
   CAUTION: Do not prick skin.
2. Spread collected material on a clean slide.
3. Using eyedropper, add several drops of water and a drop of iodine.
4. Place cover glass on slide over specimen.
5. Place microscope in good light near edge of table.
   CAUTION: Do not use direct rays of sun.

6. Sit close to table in a chair that brings your eye a little above the level of the ocular lens (eyepiece).

7. Adjust mirror to provide a clear, even light through the hole in the center of the stage.

8. Place prepared slide between the two clips on the stage to hold it in position.

9. Look into eyepiece.
CAUTION: Use either eye. Keep both eyes open.

10. Focus objective lens.

11. Finally, notice groups of five-sided cells, with the nucleus of each cell appearing as a small dot, the membrane as a thick dark line, and cytoplasm as interior, grainy material.

CHAPTER SUMMARY

Understanding processes, procedures, and activities is a key to understanding the nature of our world. Technical writing plays an important role in communicating knowledge about actions and causes to scientific, industrial, and lay audiences.

Descriptions of natural processes examine the cause-and-effect relationships between stages of action. In digestion, for instance, muscle contractions cause food to pass from the mouth to the stomach. Enzymes in the stomach contribute to the decomposition of food and the isolation of its basic nutritive components. However, chemicals secreted in the small intestine are necessary to make the breakdown complete. The purpose of such technical descriptions is to help readers understand the process or control the conditions surrounding it.

Descriptions of procedures concentrate on the stages of actions performed by humans. The actions may be described in great detail or in general terms, depending on the needs of the audience and the writer's purpose. Often the reasons underlying the methods suggested are emphasized, rather than the actions themselves. However, in technical instructions (the most common form of a description of an activity), the emphasis falls solely on the precise steps to be implemented.

Descriptions of processes, procedures, and activities appear in a variety of technical documents, including manuals, encyclopedias, textbooks, reports, proposals, guides, and articles. Often these descriptions serve as the foundation for recommended methods of control.

EXERCISES

Perform several of the following exercises in order to practice the skills required in writing descriptions of processes, procedures, and activities.

1. For each of the following processes, procedures, and activities, develop a three-part introductory statement. Incorporate the stages of action as the distinguishing traits. Here is an example: "Removing a defective transistor from a printed circuit board is a simple manual procedure of fluxing and positioning desoldering braid, applying heat, absorbing solder, and lifting the transistor in question away from the board." Before you begin, classify each item as a process, a procedure, or an activity.

changing a light bulb	hydrolysis
cleaning a handgun	programming
drafting	sailing
driving a car	sleeping
fermentation	swallowing
gestation	typing

2. Examine a textbook in your major field. List a representative sample of the processes and procedures described. Bring the list to class to compare with the lists that other students made up.

3. Select a description of a procedure given in a textbook you are using in one of the technical courses in your major. This description is probably written for someone who has had some experience in your field. Suggest how it should be changed if it were being presented to an intelligent adult who had had no formal training in your field. How would it change if it were being presented to your instructor in that course? How would it change if it were being presented to an eight-year-old child?

4. The following set of *instructions* is excellent for helping someone perform each of the steps involved. However, it is of little use to a gardener who wishes to understand the reasons for the actions indicated. From these instructions, develop a *description of a procedure* that classifies steps into related actions, raises the level of detail to that of an overview, and supplies the reasons why the actions are included.

HOW TO REPOT A PLANT THAT HAS OUTGROWN ITS ORIGINAL CLAY POT

SUMMARY Repotting a plant is a simple manual procedure of transferring the plant that has outgrown its original pot into a larger pot in which it will have room to grow.

MATERIALS 2 quarts of water in a pitcher
3–4 sheets of newspaper
1 old coffee can (or container of similar size)
¼ cup bleach
10 rocks (½″ to 2″ in diameter)
1 pound potting soil

TOOLS	1 clay pot (2″ larger than diameter of original clay pot)
	1 old, dull knife
	1 2-cup mixing container
	1 old toothbrush
EQUIPMENT	1 table or counter top
	1 sink
WARNING	*Soil has a tendency to spill on clothing. (An apron or old clothing might be best to wear.)*
CAUTION	Plant should be handled gently to avoid bruising leaves, roots, and stems.

Instructions

1. Clean table or counter-top work space.

2. Spread sheets of newspaper over this surface.

3. Place original pot, new pot, bag of potting soil, rocks, coffee can, pitcher of water, old toothbrush, and old knife on newspaper.

4. Water plant in old pot until soil is soaked.

5. Let original plant drain in sink.

6. Clean new clay pot with a mixture of ¼ cup bleach and 1 cup water.
   NOTE: The easiest way is to mix the bleach and water in a container and use the old toothbrush to scrub the pot with the mixture. Do this over the sink.

7. Rinse new pot thoroughly with tap water.

8. Place new pot on table.

9. Set rocks in bottom of new pot until the depth of rocks is about 1–½ inches, allowing for some air space and drainage.

10. Open bag of potting soil.

11. Place 4 handfuls of potting soil in the coffee can.

12. Add small amount (¼ cup) of water from the pitcher to soil in coffee can.

13. Mix soil and water with hands.

14. Repeat steps 12 and 13 until the mixture sticks together like mud.

15. Place some of this mixture into new pot until rocks are covered by an inch of soil.

16. Take old knife in hand.

17. Place knife between soil and original pot.

18. Simultaneously, loosen soil from the edge of pot with knife and rotate pot with the other hand.

19. Simultaneously, place hand flat over soil of original plant with stalk of plant between fingers and gently turn pot upside down.

20. Simultaneously, loosen and lift pot with hand that is not holding onto the plant, while the other hand catches the plant and ball as they come out of the pot.
    NOTE: If plant does not come out easily, repeat steps 16 through 20.
    CAUTION: Do not pull plant out of pot.

21. Turn plant and root ball right-side-up over the new pot.

22. Place plant and ball of soil in new pot.

23. Simultaneously, hold plant at center of pot and fill in space between plant and edges of pot with mixture from the coffee can.

    NOTE: If soil mixture runs out, repeat steps 11 through 14

24. Simultaneously, hold plant in center of pot with one hand, while patting down the soil with the other hand.

    NOTE: Soil level should be ½ to 1 inch from the top of the pot. If more soil is needed, repeat step 23. If the level is too high, remove some soil until it is correct.

25. Finally, water plant in new pot with water from pitcher until water thoroughly filters through the soil and comes out the drainage hole.

WRITING ASSIGNMENTS

Select one of the following assignments.

1. Select one of the following activities and generate a set of instructions in the format presented in this chapter.

a. How to change a flat tire on your own model car (specify the model).
b. How to change a cloth diaper on an infant.
c. How to make a twin-sized bed so it will pass military inspection.
d. How to finish a raw pine table.
e. How to wash and wax a car.
f. How to set a table for a full-course, formal meal for four people.
g. How to polish leather shoes to a high gloss.
h. How to build a campfire in the wilderness.

NOTE: If you are not familiar with any of these activities, select one of like simplicity that you prefer from your major course of study.

2. Select a natural process (with five or six phases of action) that is related to your major field of study. You are to write a description explaining this process to a colleague so that he or she can control the conditions under which it takes place. Consult a scholarly journal in your field (such as IEEE *Spectrum*) for a format model. Submit sample pages of this document with your written assignment so that your instructor can assess the degree to which your description conforms to the format requirements of the journal. Assume your reader has as much field-specific knowledge as you do. Limit: 800–1000 words.

3. Select a procedure (with five or six phases of action) that is related to your major field of study. This procedure should be one with which you are quite familiar at

your current level of expertise. Assume the role of supervisor (newly promoted) who is training your own replacement. Explain the procedure to this person. Assume *only the amount of knowledge you had about your field when you first learned the procedure*. Present this description in the format of one of your field handbooks (one that has narrative descriptions rather than instructions only). Submit sample pages with your assignment so that your instructor can assess the degree to which your description conforms to the format requirements of the field handbook. Limit: 800–1000 words.

NOTE: Apply the editing criteria at the end of Part II to your assignment before you submit it to your instructor.

C H A P T E R

9

Describing Mechanisms

CHAPTER PLAN

Technical writing, as a means by which people communicate their experience and expertise, helps prevent subsequent generations from having to reinvent the wheel. Nowhere is this more true than in the description of mechanisms. We have detailed descriptions of medical apparatus and weaponry dating from the Roman Empire and ancient Greece. And since the Industrial Revolution, the documentation of designs, inventions, patents, and prototypes has been an increasingly important part of the technical writing discipline.

This chapter examines the special skills required to describe mechanisms and their use. It will

- help you determine when descriptions of mechanisms are needed.
- enable you to distinguish between the physical characteristics and the function of a mechanism.
- examine the various components of effective description.

DETERMINING THE NEED FOR DESCRIPTIONS OF MECHANISMS

A *mechanism* is any device used to accomplish work. Mechanisms can range from a simple sewing needle to a sewing machine, from a plowshare to a tractor, from a piston to an entire automobile. It is also possible to approach less "mechanical" structures in the same light, as in anatomy (see the description of the eye in Chapter 6) or the study of archeological artifacts. Each structure, be it mechanical or otherwise, performs a specific kind of job. The description of a mechanism is as concerned with the *ways* in which that job is accomplished (functional analysis), as it is with the job itself and the physical characteristics of the device.

The degree to which a device is described depends on the audience and the writer's purpose. A history student needs a far less detailed description of the "cotton gin" than Eli Whitney may have supplied to the U.S. Patent Office.

At least six reasons for describing mechanisms exist. You will write such a description when you are introducing a new invention or design, explaining a mechanism's function, describing the way in which the tool is used, promoting a mechanism, analyzing a mechanism's construction, or describing the way in which the tool is maintained.

Often the description of a mechanism accomplishes several or all of these purposes, but we will discuss each one separately.

INTRODUCING A NEW INVENTION OR DESIGN

Inventors and product engineers often have cause to describe their creations. They usually do so in formal ways via proposals, patent descriptions, or reports. The proposal may be an appeal to a manufacturer to accept a new tool for production; the emphasis in such a case may be on profitability, innovation, and competitiveness. A patent application is made in a standardized format that requires precise drawings, a description of the mechanism's function, and extensive discussion of its distinguishing characteristics. (Skill in definition will serve you well here.) A report, on the other hand, usually informs the writer's supervisor about ongoing activities on a project and cites any conclusions that can be drawn from the work.

EXPLAINING A MECHANISM'S FUNCTION

In order to explain a mechanism's function, you must distinguish between its physical characteristics (form) and its functional capabilities. *Physical characteristics* include size, weight, density, material, sound, odor, color, and the like. Also included are the number and dimensions of parts and subparts. Here is an example that catalogs a variety of physical characteristics:

> The folding pocket can opener is made of 1/32-inch thick, rigid stainless steel. It is 2 inches long, 5/8 inches wide, and 1/8 inch thick when folded. This compact opener weighs less than 1 ounce. Its blade is shaped like the dorsal fin of a shark. . . .

Functional capabilities, on the other hand, are those things the mechanism can accomplish—the work it does and the *ways* in which it performs these functions. This category includes the overall purpose of a mechanism.

> A lawnmower cuts grass and other light vegetation uniformly to a predetermined height along a plane surface by means of horizontally rotating blades mounted on an engine-driven shaft.

Discussion of function also focuses on the work accomplished by individual components of a mechanism and the relationships between working parts.

> [Separatory Funnel] . . . The stopcock is a hard plastic that completely fills the "bubble," allowing no liquid to pass through when the stopcock is closed. It has a slit in the center that can be lined up with the funnel to allow liquid to flow out the bottom of the cylinder. The part of the stopcock that extends to the outside of the funnel has a knob. This knob, when turned, allows or stops the flow of liquid by positioning the slit with or against the liquid in the funnel.

At this point a word of caution about the names of mechanisms is appropriate. Many mechanisms have labels that reflect the tool's function. Consider such tools as the *wastepaper basket, coat hanger,* and *eggbeater.* The inventors and designers of such devices had very definite functions in mind when they christened their creations. However, such names *could* limit a user's perception of the device. For instance, it is quite possible to discard things other than wastepaper in a basket with that name. (In a recent classroom discussion, students decided that the mechanism called a wastepaper basket could be used as a bucket to bail water from a flooded room, as a percussion instrument in a band, as a footstool, and even as headgear to protect its wearer in the event of an earthquake.)

Similarly, it is possible to hang shirts on a coat hanger, dry delicate clothes on it, or use it to enhance television reception. Cooks frequently use eggbeaters to mix the ingredients for cake batter (sometimes no egg is involved at all). Do-it-yourselfers use eggbeaters to stir paints and chemicals to the proper consistencies. The point is that a good functional analysis is not restricted to the meaning conveyed by a device's label. It should *explore* as many uses for a device as are important to the audience. Remaining flexible and perceptive is the key to successful description. Such qualities are particularly important to the inventor or designer, and they should be reflected in technical writing that describes mechanisms.

Explore all the possible uses for a device. Inventors, especially those seeking patents, must have as flexible a definition of their creations as possible. If the inventor of a toaster had defined the device as merely a machine to brown sliced bread, the first person who decided that the device (with only a minor modification in design) could be used to warm frozen French toast or pop tarts might have been able to patent the revised mechanism.

DESCRIBING A MECHANISM'S USE

As we learned in our discussion of how to write instructions (see Chapter 8), activities can be documented in a list of simple sentences that conform to the pattern "Do this to that with that." To provide readers with a basic understanding of the tool's use, instruction manuals often include a description of the mechanism to be used during an activity or procedure.

For instance, the following excerpt from the introduction to a user's guide discusses the basic parts of a computer terminal (keyboard and display screen) and the terminal's relationship to the main system. Because the audience consists of novice computer users, the description of the mechanism is not very detailed.

A computer is a tool. As such, it performs only as it is instructed. When it is instructed correctly, it performs with predictable results. When misinstructed, it will not perform and, in many cases, will reply with an error message explaining why the instructions cannot be carried out.

Instructions are given to the XYZ system by pressing keys on the typewriterlike keyboard of the computer terminal. These terminals present "screens" to their users that either request or supply information. The terminal is connected by cable or telephone line to a central computer residing in the data processing department of your company. The terminal transmits your input (data you supply) to the main computer and then returns the output (the computer's results) to you. When you sit at a terminal, it may seem as though you have the computer all to yourself; however, many users are probably accessing the computer at the same time as you. The ability of a computer to handle instructions simultaneously from many users at various locations is an important factor in accounting for the success and popularity of data processing in recent years. . . .

The display screen looks very much like a small television set. It is also known as a CRT (cathode ray tube) because of the electrical components enabling it to display information. Depending on the type of computer procedure being conducted, the information will appear line by line as it is typed at the keyboard (normal mode), or it will appear as a single preestablished form that fills the screen (forms mode). . . .

PROMOTING A MECHANISM

A surprising number of descriptions of mechanisms are in the form of promotional writing. Advertisements for products too numerous to mention abound in newspapers, magazines, and brochures, on billboards, and on television. Although we might not consider the discussion of these products particularly technical, very often technical personnel have had to analyze the mechanism in detail in order to describe it at a basic level. Understanding the features, advantages, and drawbacks of a product can help you emphasize its strong points and make the product attractive to the public.

A promotional description of a mechanism uses a combination of elements to attract readers and sustain their interest. For instance, headlines and graphics provide visual stimuli. Drawings, photographs, models, diagrams, tables, and other visual aids are also important to promotional descriptions of mechanisms. They save the writer copy space, enable the reader to see the product, and create a stronger impression than words alone can. The level of detail in promotional descriptions is general. However, important points must be discussed or the reader will not pay attention.

DESCRIBING A MECHANISM'S CONSTRUCTION

Product designers may need to write about the construction of a device in order to train assembly workers, to propose a building technique, or to justify a design change. The emphasis in these cases falls on the physical characteristics of the mechanism. Simple dimensions, though important, must be augmented by discussions of such characteristics as strength, reliability, flexibility, modularity, and efficiency. The degree to which these aspects are discussed depends on the needs of the audience.

The following description is taken from a specifications sheet for a mainframe computer. It is little more than a list citing the computer's features and its physical, electrical, and environmental characteristics.

Description

Large-scale, multidimensional information system that acts as a host computer in a distributed processing environment. It supports 32M bytes of memory and is designed for use with the XYZ operating system.

Features

- Simultaneous and asynchronous execution of components
- Byte-oriented ASCII
- Virtual memory addressing
- Cache memory for access to most recent instructions and data
- Concurrent operation of many peripherals of differing speeds
- Automatic Error Detection and Correction (EDAC)
- Hardware-enforced security and integrity protection, multistate ring mechanisms, and access control list

Physical Characteristics

CPU: Height: 73.0 in. (185.4 cm)
Width: 41.5 in. (105.4 cm)
Depth: 32.0 in. (81.3 cm)
Weight: 1100 lb. (499 kg)
Clearances: 36.0 in. (91.4 cm) front, 30.0 in. (76.2 cm) rear

IOM: Height: 73.0 in. (185.4 cm)
Width: 41.5 in. (105.4 cm)
Depth: 32.0 in. (81.3 cm)
Weight: 1100 lb. (499 kg)
Clearances: 36.0 in. (91.4 cm) front, 30.0 in. (76.2 cm) rear

Electrical Characteristics

Voltage: 120/208 ± 10% Vac, 3-phase
Frequency: 60 ± 1 Hz (50 ± 1 Hz available)
Power consumption (at 208 Vac, 60 Hz):
CPU — 3.66 kVA, 3.48 kW
IOM — 3.2 kVA, 2.88 kW

DESCRIBING THE MAINTENANCE OF A MECHANISM

Once a mechanism is built and used, it may need routine maintenance or repair. If you are in the service side of a technical field, you may be called on to describe a mechanism so that your readers can keep it in working order or fix it if it malfunctions. This kind of description incorporates both instruction writing and the description of a mechanism. The following excerpt explains how a motorcycle owner can check the condition of the cycle's front fork and repair it if necessary.

The Honkawzuki 950 has a front fork assembly . . . with the following components:

1. Caps (2)
2. Air valve
3. O-rings (2)
4. Spacer
5. Spring guide
6. Springs (2)
7. Inner tube
8. Slide metal
9. Pistol ring
10. Cylinder
11. Oil lock piece
12. Dust seal
13. Oil seal stopper ring
14. Oil seal
15. Outer tube

The tightening torque should be as follows:

A	1.5–2.5 kg/m	11.0–18.0 lb/ft
B	2.0–2.6 kg/m	14.5–19.0 lb/ft

In order to check the status of each of the parts, remove the front wheel (see procedure X.17), disconnect the speedometer cable guide, and remove the fender. Loosen the upper and lower front fork clamp bolts, so that the front fork can be pulled free from the body of the cycle. . . .

Inspection

The service limit of the fork spring is 516 mm (20.3 in.). If the spring measures longer than this amount, it must be replaced. Inspect the inner tube sliding surface for any scuffing and check the bent. Inspect the outer tube sliding surface for any scuffing as well. The damper rod ring must also be checked for wear and damage. . . .

COMPONENTS OF EFFECTIVE DESCRIPTIONS

As the examples in this chapter illustrate, the description of a mechanism can take many forms. However, certain components should be included in most descriptions; this section discusses them. On the basis of the situational analysis you prepare, you will have to decide what points to emphasize, how to order the information you present (see Chapter 4), and how much detail to offer. All of these decisions depend on the purpose of the description and the needs of the audience.

INTRODUCTION OF THE DESCRIPTION

The description of a mechanism rarely exists by itself. It is usually incorporated into a larger document such as a manual, report, plan, proposal, or advertisement. The length of the description may vary from only a paragraph to dozens of pages. The introduction of a description varies as well. It may set the scene so that a reader can understand the "state of the art" in a certain kind of technology. It may present problems that the mechanism has been designed to eliminate. Again, these factors depend on your purpose and your audience.

No matter what the length of or amount of detail in a description, it should include an introductory statement based on the three-part formula we discussed in Chapter 7. The specific term in this case is the mechanism itself. The category is the overall class of mechanisms into which the device falls, and the distinguishing traits are the primary components and functional characteristics of the device. Begin with as precise a name for the mechanism as possible. Use *Smith Corona portable electric typewriter* rather than *typewriter.* Use *folding pocket can opener* rather than *can opener.* If you are concentrating on a very specific device rather than the generic form, a name as detailed as the following is appropriate: The *500-ml. common standard tapor ground glass separatory funnel.*

Then identify the larger category into which the mechanism falls (labels such as *tool, device, product,* and *mechanism* are too broad to be informative). Some of the many possible categories are

apparatus	engine	switch
appliance	fixture	system
component	instrument	valve
container	machine	vehicle

It may be necessary to further qualify the category by means of a modifier or subcategory, such as *chemical, electrical, electronic, hydraulic, manual, mechanical,* or *thermal.*

Select an accurate, active verb to describe the function (distinguishing traits) of the mechanism. Consider such possibilities as

agitates	injects	renders
bakes	jams	stabilizes
controls	loosens	turns
decreases	manipulates	undulates
eliminates	narrows	vaporizes
fuels	opens	warms
grinds	pushes	wipes
hardens	queries	yields

In many cases, the active verb may not be enough to communicate the function of the mechanism; it may be necessary to add an object to the verb. A lawnmower doesn't simply *cut;* it *cuts grass.* And you may wish to further modify this action with additional information: the lawnmower *cuts grass to a uniform length.*

The final component of the introductory statement is an analysis of the way in which the mechanism accomplishes its function. This portion of the statement classifies the parts of the device into functional categories. These are then arranged according to their functional relationship in one of two possible ways, which are based on their spatial orientation. The first arrangement is called "straight flow." It presents the parts from the first working part (or general group of parts), to the next working part, through to the last working part. The following discussion of the centerfire metallic cartridge illustrates this overall arrangement.

The centerfire metallic cartridge is a metal container that houses all required firing components in a metal cylinder. The complete cartridge is approximately 1½ inches long and can be used in rifles and handguns. *The four main components of this device are the primer, metal case, powder, and bullet.*

The second arrangement, called "reverse flow," works backwards from the last part to the first. The following statement about a lawnmower illustrates the reverse flow of information.

A Davison 1401 gasoline-powered lawnmower is a self-propelled garden machine that cuts grass to a uniform length *by means of horizontally rotating blades mounted on a drive shaft directly connected to a 1-cylinder engine.*

The order of presentation in this part of the definition should reflect the order in which the components are discussed in the body of the description.

NOTE: It is not necessary to include all the components of a definition in a single sentence. Don't let the definition become unwieldy; use two or more sentences if necessary.

The introductory section of the description may include a discussion of the physical characteristics of the mechanism as well. It may illustrate the components by means of an exploded diagram. In particularly complex mechanisms, it may even introduce a comparison to help familiarize the reader with the nature of the device ("a computer terminal's display screen is like a television screen," "the can opener's blade is like the dorsal fin of a shark"). We will see this tactic used in the sample description of a microscope presented later in this chapter.

BODY OF THE DESCRIPTION

The body of the description expands on the overview statement of the introduction. Like a document that uses grouping techniques (specifically partition; see Chapter 6), it discusses the parts of the mechanism in relation to the whole device. The order of that discussion coincides with the functional flow presented in the introductory statement. The amount of discussion devoted to any one category depends on the needs of the audience. It is *not* necessary to give equal treatment to each category; however, the parallel discussion of points facilitates understanding.

The basic points of discussion within each category might be

physical characteristics of each part
functional characteristics of each part
relationship of the category (or part) to other parts
importance of the part to the overall function of the mechanism
special features
strengths and weaknesses

In examining the physical characteristics of the part, you should use visual aids. And, by incorporating measurements into the figure, you can save time and space. Discussion can then focus on only those features and dimensions crucial to the topic you are writing about, such as tensile strength or conductivity.

The functional characteristics can be presented by means of a brief description of the procedure. Consider these processes:

> Control rods are inserted through spaces in the reactor core to control the rate at which heat is produced. These rods are neutron absorbers. When the rods are completely inserted, more neutrons are absorbed; the chain reaction is reduced, and the reactor is out of service. As the control rods are withdrawn from the core, they absorb fewer neutrons, allowing a controlled chain reaction to resume and to produce more heat.

Functional analysis requires you to examine relationships between parts. Ask yourself: "How does the part (or group of parts) cooperate with others?" "How does it contribute to the overall function of the mechanism?" Note how clear transitional words and phrases are used throughout the description of the microscope (presented later in the chapter) to help make these connections between parts clear to readers.

Of course, most mechanisms require all of their working parts in order to accomplish their function. However, some parts may be more important than others. For instance, depending on the purpose of the document and the needs of the audience, an engine may be far more important to a description of a car than a windshield, trunk, roof, bumper, or door. In an example as obvious as this, an audience need not be told outright that the engine is more important, yet discussion of it should be far more detailed than discussion of the body parts.

In descriptions of unfamiliar mechanisms, the importance of a part may not be so obvious. In these cases, tell the audience explicitly how significant the part is.

The input/output multiplexer acts as a traffic controller for the entire computer system. It channels all input to the central processor from peripheral devices such as mass storage and terminals. It then sends the output to the proper location (to the user's terminal, to a printer, or back to storage). Without an efficient input/output multiplexer, a bottleneck can be created that slows down the response time for the system. If this component is poorly designed, data streams and instructions can be interrupted (thereby jeopardizing the integrity of the information), jobs can be lost, programs can be damaged, and resources can be underutilized.

The special features of a mechanism may make it particularly innovative, competitive, convenient, or durable. Therefore a discussion of special features is often warranted. Topics to be considered might include:

availability	luxury
convenience	mobility
cost	national reputation
durability	origin
ease of use	performance
efficiency	potential
flexibility	reliability
guarantee	service
horsepower	test results
integrity	value

Such analysis can often be incorporated into an overall evaluation of the product. The more promotional the document, the more extensive the analysis of its features ought to be. The specifications sheet excerpted earlier in this chapter (page 198) *lists* special features; the following brief sample accomplishes this same objective in a narrative format.

> Particular attention was paid to the ST-14's interior. Extended leg room and supple bucket seats allow you to sit comfortably for hours. The dashboard is a logical arrangement of push-button controls—all within easy reach. An adjustable steering wheel, with three thick spokes for a secure grip, helps you maneuver even the tightest corners. With the rear seats folded down, the ST-14 provides nearly 20% more storage space than most other cars in its class. A power-operated sunroof and an AM/FM radio and cassette player are all standard equipment. . . .

When it is appropriate to evaluate a mechanism, carefully analyze any strengths or weaknesses of a particular part or group of parts. If the handle of a golf club requires the user to hold it awkwardly, the golfer will not play comfortably or well. If the brake drums of a car are particularly well constructed, their strength should be noted. Likewise, if any special handling is required or any warnings necessary, be certain to inform your readers.

CONCLUSION OF THE DESCRIPTION

The conclusion of the description of a mechanism can be short and to the point. A summary of its function, main parts, and primary features is usually appropriate. An overall assessment of the tool's value or performance is also important. If operating instructions include warnings or cautions, the writer does well to repeat them, in the final remarks, for emphasis. The following conclusion illustrates these elements.

> The rotary beater, which is made up of a cog wheel assembly, beaters, and a handle, is a very efficient means of mixing and whipping liquids and semi-liquids. One revolution of the handle is converted to four revolutions of each beater, thereby saving the user considerable time and energy. Its simple construction makes it cost-effective and easy to clean. The rotary beater is generally very easy to operate; however, the user should avoid placing fingers or utensils in the path of the beaters.

The following student sample is based on one of the writing assignments at the end of this chapter. It effectively uses all of the techniques for describing mechanisms that we have discussed. The notes in the margin identify these components in action.

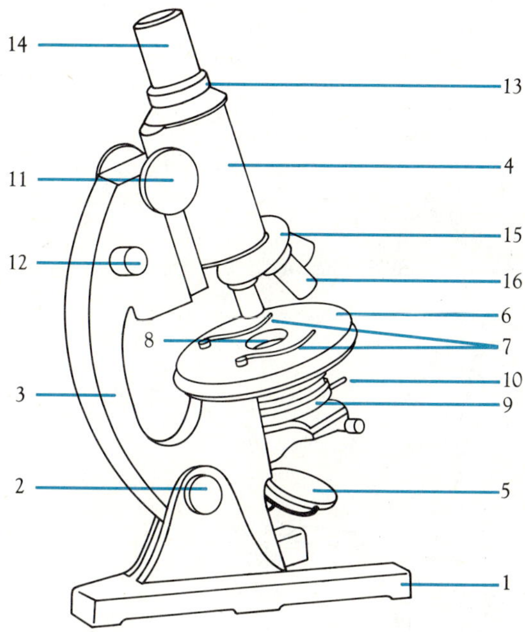

1. Base
2. Movable joint
3. Handle
4. Body tube
5. Mirror
6. Stage
7. Spring clips (2)
8. Hole
9. Diaphragm
10. Lever
11. Adjustment knob (coarse)
12. Adjustment knob (fine)
13. Extension cylinder
14. Ocular lens
15. Nosepiece
16. Objective lens

Figure 1 The 16 main components of a compound microscope.

AUDIENCE	My supervisor, a research biologist.
ROLE	Inventor of the compound microscope.
PURPOSE	To introduce my new invention and convince my supervisor that it would be worthwhile and profitable to market.
FORMAT	Informal report.

COMPOUND MICROSCOPE

[1] Research biologists currently study small plant and animal life through the use of the single-lens magnifying glass. Visibility is affected directly by the steadiness of the researcher's hand, the lighting conditions, and the power of the lens. Variance of any of these factors often results in considerable distortion and inconsistency of data. This severe limitation is now overcome by a new kind of magnifying device called a compound microscope. The compound microscope is an adjustable scientific instrument that enables a researcher to secure material to be studied and make it visible to the human eye.[2] The five primary components of this device are the frame, stage, mirror, lenses, and adjustment knobs.[3] (See Figure 1.)

1. Introduction: problem with current technology
2. Definition overview
3. List of parts

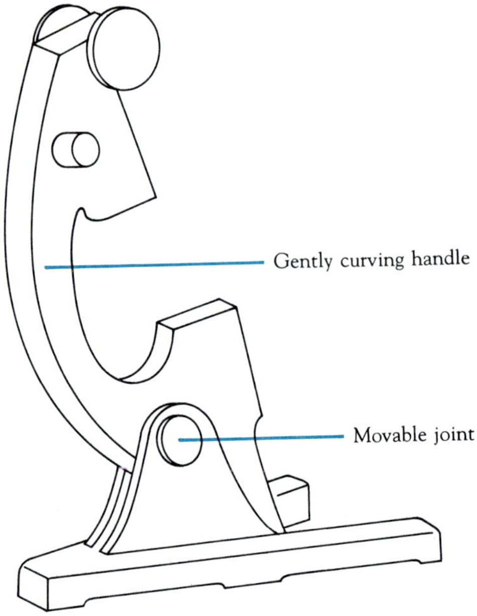

Figure 2 The microscope handle is built into the frame.

The Frame[4]

4. First part

The frame of the compound microscope is made of hard metal and is painted black so as to absorb light. It supports the moving parts of the instrument and provides a stable
research environment. The handle of the microscope is 5 5. Subpart 1
built into the frame. (See Figure 2.) It curves gently to fit the user's hand and allow for a firm grip. A movable joint connects the handle to the base and allows the user to adjust the handle to almost any appropriate angle.

The base of the frame is horseshoe-shaped. It acts as a 6 6. Subpart 2
three-point plane that firmly plants the microscope on the researcher's laboratory table. Small rubber cups are secured underneath each of the three contact points for even greater stability and adherence. (See Figure 3.)

The Stage[7]

7. Second part

The stage of the compound microscope is located in the middle of the microscope and serves as the foundation for the specimens to be observed. It is a round platform about

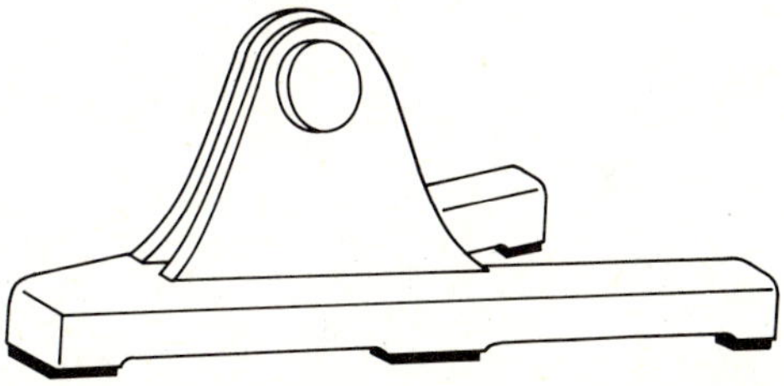

Figure 3 Rubber stops provide additional traction.

2 inches in diameter and is made up of several subparts, including the plane, spring clips, a hole, a diaphragm, and a lever.

The plane holds the object to be examined. The object is placed on a slide (a small piece of glass) and centered on the platform. Two spring clips on the stage hold the specimen slide in position. A hole in the center of the plane allows light to reflect from the mirror (see Figure 4) through the hole and then through the specimen on the slide. The diaphragm controls the amount of light that can pass through the opening. The lever, in turn, controls the opening and closing of the diaphragm.[8]

8. Subparts 1–5

The Mirror[9]

9. Third part

The illuminating mirror is a two-sided component of the compound microscope. It focuses light from a light source

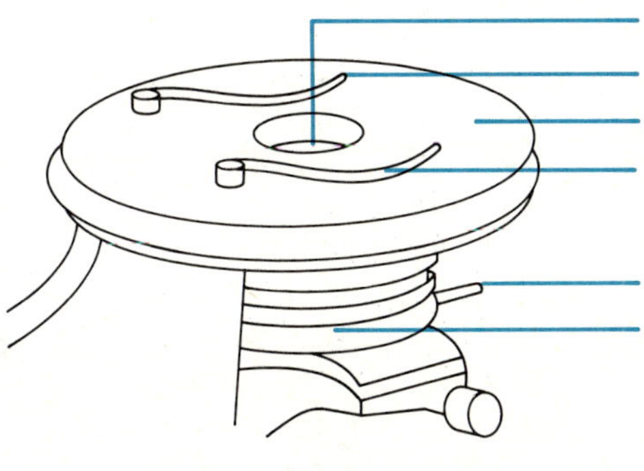

Figure 4 The five parts of the stage.

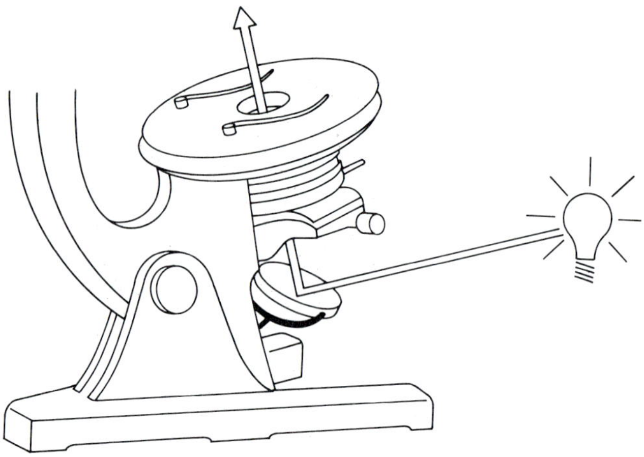

Figure 5 Light is focused from a source onto the specimen slide.

through the diaphragm and hole and onto the specimen slide. (See Figure 5.) Depending on the light source, it may be best to use the concave side of the mirror rather than the plain surface. The mirror is a very important part of the microscope. Unless it is adjusted properly, you will not see the specimen clearly. To adjust it, look through the eyepiece and turn the mirror until you have a clear, even light.

Lenses[10]

10. Fourth part

The lens mechanism of the compound microscope is quite complex, consisting of an eyepiece, smaller lenses, a body tube, and a movable nosepiece. The purpose of this mechanism is to magnify the specimen secured on the stage. The eyepiece, or ocular lens, is the place through which the researcher looks. (See Figure 6.) At the end of the body tube are situated a movable nosepiece and several smaller, or objective, lenses. The magnification power of the ocular lens is multiplied by that of the objective lens in use for a composite magnification power. The researcher can choose the appropriate power for each specimen by swiveling the nosepiece to position the appropriate objective lens. It is essential to keep the lens mechanism clean, or the specimen will appear distorted.

[11] 11. Subparts 1 and 2

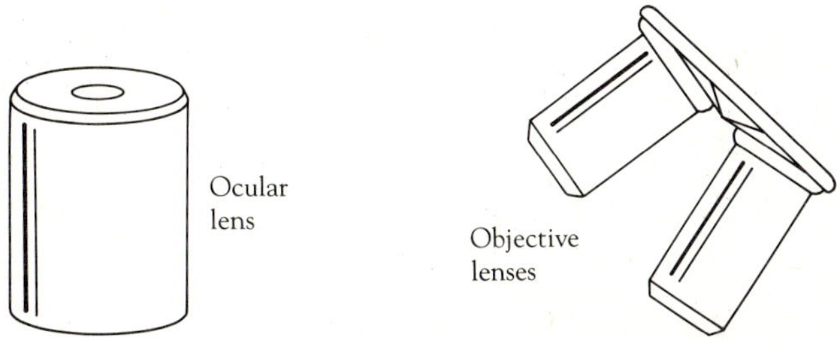

Figure 6 Two types of lenses.

Adjustment Knobs[12]

12. Fifth part

Two types of adjustment knobs facilitate clear focus and accurate research. The coarse-adjustment knob permits quick focusing on the object. The fine-adjustment knob permits more delicate focusing on the slide. The direction in which the coarse- and fine-adjustment knobs are moved determines the movement (up or down) of the body tube.[13]

13. Subparts 1 and 2

Conclusion[14]

14. Benefits-oriented conclusion

Exploring by microscope rather than via a magnifying glass should enable modern researchers to make great strides in the isolation and identification of new forms of plant and animal life and their behaviors. The instrument is highly marketable in today's economy, where research grants and laboratories abound.

Although there is some degree of artificiality in this sample (the student has supposedly *invented* the compound microscope), it demonstrates some key aspects of a description designed to introduce a new invention. The first paragraph identifies the problem that scientists have with existing technology (the single-lens magnifying glass) and then presents the new device as a solution to that problem. Within this framework, the inventor adapts the three-part introductory definition statement to name the product, place it in its proper context, and delineate five key components. Parts are mentioned in the same order in which they are described in the body of the document.

The order is based on a functional rather than a spatial orientation: the frame supports the working parts, the stage is the focal point, the mirror projects light onto the stage, the lenses enable users to see what is lit up on the stage, and the adjustment knobs allow the user to focus the lenses. Information covered within each category follows a parallel arrangement, which facilitates comparison of form and function from component to component. The writer mentions the purpose of each part and its benefits, as well as describing its physical characteristics.

The use of Figure 1 early in the document enables readers to understand the configuration of the mechanism even before extended discussion begins. Subsequent figures highlight the individual components being discussed. The use of figures saves both the reader and the writer time and enhances the clarity of the description.

CHAPTER SUMMARY

Descriptions of mechanisms appear in a variety of documents ranging from advertisements to manuals and from news articles to patent applications. Their form depends heavily on the needs of the audience. A detailed description is required for people who are going to construct or maintain the tool. Very precise description and graphics are required for patents. Only general discussion is necessary for procedures manuals.

The introduction of a description defines the mechanism by placing it in an appropriate category, identifying its function, and citing its main components. The body examines each of the components as they are related to one another and to the overall function of the mechanism. The conclusion evaluates the product in light of the audience's needs.

EXERCISES

In order to practice the skills discussed in this chapter, perform a selection of the following exercises.

1. Examine a variety of literature in your major course of study. Locate at least five descriptions of mechanisms. Then answer the following questions.

- What is the purpose of each description?
- How is each description incorporated into the overall document?
- How does the level of detail vary with the purpose, the nature of the document, and the audience?

2. Create a situational analysis for any three of the following mechanisms. Write a three-part introductory statement for each device you have selected. Include as many components of functional analysis as are appropriate. Expand to more than one sentence if necessary, but do not create unwieldy, single-sentence definitions.

.22-caliber handgun	shopping cart
iron	toaster

manual can opener	stapler
incandescent light bulb	briefcase
filing cabinet	stereo system
eyeglasses	screwdriver
coffee pot	razor
35-mm camera	fabric loom

3. Create an extensive parts list for your own automobile. Then organize these parts in logical categories by function.

4. Examine three brands of the same device (such as toothbrushes, hose nozzles, or pepper mills) and evaluate each variety according to those of the following characteristics that apply.

availability	ease of use
balance	flexibility
cost	handling
quality of construction	packaging
design	reliability
durability	safety

5. Using the editing criteria given at the end of Part II, edit the description of the compound microscope that appears on pages 205–209.

WRITING ASSIGNMENTS

Select one of the following assignments. Preface it with a situational analysis of audience, role, purpose, and format.

1. Imagine that you work in a manufacturing firm and are the designer of a new device (real or fictitious). Address your immediate supervisor in a brief report that introduces the product, describes its function, and assesses its marketability. Incorporate as many components of the description of a mechanism as are necessary to *convince* this person that your design is efficient and significant. Limit: 600 words.

2. Imagine that you are writing a procedures manual in your field. As introductory information, describe a field-specific mechanism that readers of the manual will be using. Emphasize preparedness to use the mechanism rather than technical detail. Limit: 600 words.

3. Imagine you are the quality assurance officer in a large corporation. In a brief report, evaluate a mechanism to determine its efficiency, durability, cost-effectiveness, and reliability. Provide an overall description of your findings to the product review committee so that it can investigate the source of any flaws you may have discovered. Limit: 600 words.

NOTE: Apply the editing criteria at the end of Part II to your assignment before you submit it to your instructor.

CHAPTER 10

Argumentation

CHAPTER PLAN

Although we generally think of scientific and engineering disciplines as totally objective, professionals in these fields often perform relatively subjective work. Such tasks include interpreting the results of experiments and drawing conclusions, testing theories, inferring causes of phenomena, proposing plans of action, recommending policy, and deciding among alternatives. In all of these cases, they deal not with truth or absolutes, but with *probabilities*.

In your own careers, you will undoubtedly be required to argue a cause—that is, to convince others that yours is the best answer to a given question. Your success will depend on the quality of your conclusions, the merits of your evidence, and the degree to which you anticipate and refute opposing points of view. To help you meet the many challenges of argumentation, this chapter will

- discuss the circumstances that call for argumentation.
- identify the components of an argument.
- examine the logical requirements of convincing arguments.
- introduce effective structuring techniques.

WHEN IS ARGUMENTATION APPROPRIATE?

In the technical writing environment, argumentation is an appeal to your audience's understanding and reason that secures agreement on how facts should be interpreted or what actions should be taken. Therefore, the purpose of an argument is to *convince* your readers that your perspective or the actions you suggest are the best. Unfortunately, argumentation has come to be associated with manipulation and appeals to emotion. But although these persuasive tactics may be legitimate in some forms of argumentation, only appeals to reason are acceptable in the scientific and engineering arena. In this chapter, you will learn how to develop effective evidence to help you win the causes you must argue.

Before we can appreciate the circumstances in which argumentation is appropriate, we need to rule out two possibilities. First, argumentation is not appropriate in cases where the issue at hand is *entirely verifiable.* When one engineer maintains that a structural beam is strong enough to withstand the stress of 400 pounds per square inch, and another engineer insists that the beam cannot handle such stress, it is quite possible to resolve the issue by using laboratory experiments or an accepted formula to determine what degree of stress the beam actually can withstand. That is, the matter can be settled solely by objective means. If the beam withstands the test, the first engineer is right; if the beam begins to buckle under the stress applied, the second engineer carries the day.

Second, argumentation is equally inappropriate for dealing with purely *subjective matters,* such as issues of individual taste. When one architect believes a new building is a striking addition to the city's skyline, whereas another believes it is an eyesore, no amount of argument will sway either architect from his or her position. No measurement can be applied and no evidence presented that will make a significant difference in the thinking of either person.

Suppose, however, we change the scenario involving the engineers and the structural beam to one in which the first engineer believes that beams produced by Acme Iron Works should be used exclusively in a structure, whereas the second believes that beams produced by Forthright Iron Works should be used instead. Now we have the basis for argument. We are faced with a series of *issues* to be resolved: "Which manufacturer produces better beams?" "Why would one manufacturer's beams be better for this structure than the other's?" "What are the structural requirements, and does one manufacturer meet them more closely than the other?" It is up to each opinion holder to demonstrate—that is, prove—the reasoning behind his or her conclusion. And although each might marshal verifiable facts and expert testimony to support her or his point, the conclusion remains a matter of interpretation rather than absolutes.

Argumentation, then, is appropriate only outside the totally verifiable and the utterly subjective realms, in contexts wherein conflict or potential conflict exists. *Conflict* under these circumstances means simply that more than one interpretation or option is possible. You will need to use argumentation to write effective recommendation reports, grant and bid proposals, feasibility reports, articles, editorials, essays, papers, cases, and promotional materials. And those of you who enter research-oriented fields such as biology, chemistry, geology, and physics will rely heavily on these techniques to present your future discoveries and theories.

The degree of advocacy you bring to an argument can vary from the mere explanation and evaluation of several options to unswerving, hard-hitting support of a single alternative. In all cases, however, you must maintain your sense of fairness, objectivity, and responsibility. Points must be "won" on the basis of substantive evidence rather than emotional appeals. You can gain ground only by proving that your conclusions and recommendations are the most defensible. Flamboyance and emotional appeals are absolutely unacceptable; in fact, they make readers suspicious of your arguments. The more reasonable and exacting your presentation, the more likely it will be accepted. That is not to say that persuasion (appeal to emotion) should be eliminated entirely; rather, it is important that your tone reflect an accurate understanding of your audience's needs, attitudes, and temperament.

Some words of caution to help ward off potential disappointments:

> Two important principles in the tactics and strategies of science are as follows: first, a useful concept may be a barrier to the acceptance of a better one if long-intrenched in the minds of scientists. Second, experimental discoveries must fit the time; facts may be at hand for years without their significance being realized; the total scientific situation must be favorable for the acceptance of new views.*

In science, a conceptual scheme is never discarded merely because of a few stubborn facts with which it cannot be reconciled. A concept is either modified or replaced by a better concept—never abandoned with nothing left to take its place.

Later in this chapter, to demonstrate the ways in which new ideas are argued, we will look at a theory currently posited by some eminent astronomers and paleontologists. The degree of acceptance that this theory has enjoyed is tempered by the constraints mentioned in the foregoing

*James B. Conant, "The Overthrow of the Phlogiston Theory," *Science: Method and Meaning*, ed. Samuel Rapport and Helen Wright (New York: New York University Press, 1963), pp. 88–107.

quotation. Someday you too may be faced with a discovery of great magnitude—a theory you just *know* is right—yet your colleagues may be reluctant to accept it. Perhaps, armed with the skills and techniques explained in this chapter, you will be better equipped than most to communicate your position.

COMPONENTS OF AN ARGUMENT

In order to develop an effective technical argument, you must be familiar with three key components: propositions, issues, and evidence.

PROPOSITIONS

An argument is about a *proposition*—a statement, conclusion, recommendation, theorem, or solution to a problem that is to be believed, doubted, or denied. Someone (usually the writer of the document) affirms it; another (usually the audience) denies it, doubts it, or is simply undecided. The two basic kinds of propositions are propositions of fact and propositions of policy. A *proposition of fact* maintains that something is true or highly probable. It frequently takes the form of a question containing such verbs as *is, was, does,* or *did:* "Did comets kill the dinosaurs?" "Is oxygen absorbed by a metal during combustion to form a calx?" A *proposition of policy* maintains that a particular line of action should be followed. It employs the auxiliary verb *should:* "Should Metropolitan Gas and Electric establish a computer center to serve the needs of end users?" "Should research into toxic weapons be curtailed?" In both kinds of propositions, the question posed is answered "yes" by those supporting the position and "no" by those opposing it.

We argue a proposition of fact primarily by deductively testing the statement against all available evidence, both pro and con. On the other hand, we usually begin discussing a proposition of policy by inductively examining and evaluating the possible alternatives. Both techniques rely on the grouping skills you acquired in Chapter 6, and both will be illustrated in our examples for this chapter.

Propositions are not always presented explicitly in a document; sometimes it is more appropriate to guide your readers step by step until they draw your conclusion for themselves. Nonetheless, it is essential that you be able to state your proposition in a *single, clear, unprejudiced* statement. In preparing to write a technical argument, allow your thesis statement to play this role. As you frame the proposition, identify any key words requiring definition. Be certain to provide that definition early in the document. Then stick to it as the argument progresses.

Often it is necessary to investigate the history of the question as part of the argument to be sure that both you and your audience are familiar with past interpretations of the issues involved and the circumstances that brought the argument into being. For instance, if you are arguing in favor of the mass production of electric cars, you should trace the history of the fossil fuel shortage as well as the history of such vehicles. Lavoisier (1743–1794) used such a tactic in his argument about the nature of "calces" (now known as *oxides*), when he detailed the then-popular "metalizing" principle of phlogiston held by most of his colleagues. (See the Conant article cited earlier.)

In addition to reviewing the history of your proposition, it may be important to explain why the question is particularly pertinent to current circumstances. In the case of electric cars, you might cite increasing levels of cancer-causing agents accumulating in the atmosphere and attributable to fuel-driven vehicles. Or, at the time of your writing, a new oil embargo might be causing economic disaster in the United States. In the case of Lavoisier, he was faced with mounting evidence that the weight gained by a metal as it formed a "calx" during combustion was equal to the weight of oxygen lost from the air in the experiment chamber.

ISSUES

Although there is just one proposition for a given document, the arguments supporting and those opposing it may be many and varied. The questions raised by a proposition—that is, the points of controversy—are called *issues*. These must be identified clearly, and they are usually presented in descending order of their importance to the proposition. Certain issues are considered crucial: their defeat means the defeat of the entire proposition. In the case of the dinosaur extinction theory expounded in the following extract, it is crucial for readers to accept that multiple mass extinctions of species other than the dinosaurs have occurred since the beginnings of life on earth. Similarly, it is vital for readers to appreciate that such extinctions exhibit a cyclic pattern. If either of these issues remains contested after argumentation, the proposition itself will remain contested.

Other issues are not so important; if they are not won, the proposition may still seem probable despite these few "stubborn facts." For instance, the theory regarding the extinction of dinosaurs posits that a 26-million-year cycle of mass eliminations is demonstrable. The opposition questions the accuracy of the figure 26 million, pointing to evidence for 28- and 33-million-year periods as well. The proponents do not deny this evidence; rather, they dismiss it as irrelevant, explaining that geologic time involves such unimaginably long periods that a discrepancy of several million years is inconsequential.

In order to discuss issues in greater detail, let us examine our excerpt from the article about dinosaur extinction. It is taken from the cover story of the issue of *Time* dated May 6, 1985. The marginal notes point out only those elements of argumentation that we have covered thus far.

DID COMETS KILL THE DINOSAURS?*

[1] . . . Every 26 million years or so, the theory holds, a rain of comets that lasts hundreds of centuries bombards the earth. The impact of the larger comets spews enough debris into the atmosphere to block the sun for months. As the skies darken, temperatures on the ground plummet and the majority of existing plant and animal species perish. . . . Researchers are hunting for an agent that could explain the apparent clockwork regularity of the celestial barrages.[2] Some suggest that a companion star to the sun periodically comes close enough to nudge comets gravitationally out of their natural habitat—a cloud of comets that circles the sun far beyond Pluto—sending them hurtling toward earth.[3] Others assign that role to Planet X,[4] while some insist that the slow, bobbing ride of the sun and its planets around the Milky Way galaxy is responsible.[5] Whatever the details, declares paleontologist J. John Sepkoski Jr. of the University of Chicago, the evidence for periodic mass extinctions "very strongly implicates an extraterrestrial mechanism."

1. Proposition: cyclic mass extinctions caused by extraterrestrial agent
2. Issue: what is agent?
3. Companion star?
4. Planet X?
5. Bobbing effect?

[6] . . . If the cyclic theory is true, the biologists argue, many assumptions about the course of evolution on earth—and even the likelihood of finding complex life forms on other planets—will be overturned. Says Daniel Whitmire (an astrophysicist with University of Southwestern Louisiana): "Just the possibility that life here has been controlled by an astronomical event is very far reaching."

6. Possible implication: assumptions about evolution in jeopardy

[Since the early nineteenth century,] a growing corps of paleontologists and geologists had determined that the world is not the static Eden-like meadow of legend.[7] At least intermittently, they concluded, it is an unstable, dangerous place, where vast numbers of species, like the

7. History of question: dates to early nineteenth century

*Natalie Angier with Robert Buderi and Christine Gorman, "Did Comets Kill the Dinosaurs?" *Time,* 125 (May 6, 1985), 72–74+.

giant mastodons, mysteriously disappear. . . . Some Victorian scientists viewed the discoveries as evidence of a primordial flood, possibly Noah's. . . . In the end, how- 8
ever, most of the scholars responded to the lure of Darwin, insisting that creatures die out because they are no longer fit to survive and must give way to the supremacy of the new. That argument seemed to apply particularly well to dinosaurs, which were denigrated as being too big, too slow, too pea-brained, and too cold-blooded for their own good.

As geology became more precise, scientists determined from the fossil record that at least five great dyings—and numerous smaller events—have occurred in the past 600 million years. Among the more significant: the Cambrian disasters some 500 million years ago, when many species of segmented creatures called tribolites disappeared from the seas they once dominated; the biggest extinctions, the Permian cataclysm of 248 million years ago, when up to 90% of all marine species died; and the late-Cretaceous event 65 million years ago, which saw the destruction of dinosaurs and many other groups of species. . . .

Yet for all the evidence of destruction, scientists could
not [determine the cause.[9] They] looked to the earth for 10
explanations, citing changes in climate or sea level. . . . When sea levels change, they reasoned, animal habitats in low-lying areas may be destroyed, or the climate farther inland may grow more extreme. . . . Those species that cannot adapt to the earth's erratic behavior simply succumb. To many paleontologists, as well as geologists, it seemed to make sense.

Serious questions remained however. Why, for example, were extinctions so selective, devastating some species while leaving others virtually unscathed. . . ?[11] The revolution[ary answer] began with an unassuming element known as *iridium,* a rare and hard silvery-white metal related to platinum and gold.[12] In the spring of 1977, Geologist Walter Alvarez of the University of California, Berkeley, was carefully chiseling through the rocks outside Gubio, Italy. . . . Alvarez was struck by a configuration of sediment layers, which resembled a sandwich in stone. [The clay was analyzed by Luis Alvarez, a Nobel-prize-winning physicist also at the University of California.] To everyone's surprise, it turned out to be 30 times as rich in iridium as normal rocks. The Berkeley team knew of only

8. Definition of Darwinian evolution; applied to extinction of dinosaurs
9. Repetition of issue: what is the cause?
10. Possible answer: earth's erratic behavior
11. Issue: why are extinctions selective?
12. Instance that makes revisiting the question timely: discovery of concentrated iridium

a few places where such high concentrations of this rare element might occur: in the earth's core, perhaps 2,000 miles below ground; in extraterrestrial objects like asteroids (or their fragments, meteors) and comets; or in the cosmic dust drifting to earth from a nearby supernova (exploding star).

The core material seemed too deep to come to the surface, and further analysis ruled out a supernova as the source, so the Alvarez team concluded that the iridium had been left by a giant asteroid hitting the earth at the end of the Cretaceous period.

Basing their calculations on the atmospheric consequences of the explosion of Krakatoa, they roughly estimated how much dust the impact of the asteroid would have thrown into the atmosphere, how long sunlight would have been blocked from the earth's surface, and what kinds of life would have been the most gravely affected by the climatic changes. They decided that plants, totally dependent on the sun for photosynthesis, and a variety of marine organisms would have died first, followed by the land animals highest on the food chain, the dinosaurs. . . .

Journeying to Denmark, . . . Walter Alvarez gathered corroborating samples of iridium and received still more . . . in New Zealand. . . . In 1980 the Alvarez team published its results in the journal *Science*. . . .

. . . Several researchers rushed forth to deny any extraterrestrial origins for iridium, attributing it first to a gradual process of sedimentation that concentrated the metal.[13] [Others declared that] volcanic eruptions might have forced the iridium from the mantle to the surface.

13. Opposition: iridium does not have extraterrestrial origins

. . . William Clemens, a paleontologist at the University of California, after systematically sampling parts of the eastern Montana area, declared that the layer of iridium and the bones of the last surviving dinosaur were too far apart to share any meaningful connection. Besides, he asked, why should the mammals have survived any Cretaceous catastrophe? Says he: "If you're going to have a nuclear winter killing off the dinosaurs, why didn't it kill off everything else?"[14]

14. Reiteration of selectiveness question

Nonetheless, the evidence in favor of an impact was rapidly accumulating. Other geologists uncovered similar iridium deposits just above the Cretaceous rock beneath the floor of the Atlantic Ocean and under the Raton basin in northeastern New Mexico. Additional analysis showed that these samples contained ratios of gold and platinum nearly identical to those found in meteorites. . . . Sediment layers containing abnormally high amounts of iridium [were found dating] more than 30 million years ago, at the end of the Eocene epoch. . . . The sediments harbored microscopic balls of glass called *microtektites,* which form in rock when something hits it with great force—further evidence of a major collision.

. . . In 1983 . . . Sepkoski and a colleague, paleontologist David Raup of the University of Chicago, presented a paper . . . at a symposium on mass extinctions. The pair had examined data on marine organisms that had become extinct over the past 250 million years, a list of about 3,500 families. . . . These two scientists found that most of the dyings fell into as many as a dozen events of varying severity, some of which had previously gone unrecognized as being mass extinctions at all. . . . The catastrophes were separated by periods of 26 million years. . . . [To explain the regularity] Raup and Sepkoski declared in their paper, "*We favor extraterrestrial causes.*"[15]

15. Reiteration of proposition

. . . Among the first cyclic mechanisms considered was one already familiar to astronomers: the sun's undulating route around the crowded Milky Way galaxy. . . . As it moves [the sun] bobs up and down through the central plane of the galaxy, where most of the stars and dust clouds are concentrated. . . .[16] Significantly, the sun and its accompanying planets hit the dustiest sections once every 33 million years or so, a figure that does not quite jibe with the apparent extinction cycle of 26 million years but is at least within the margins of astronomical and geological error.

16. Alternative 1: bobbing effect of sun and planets disrupts comets held in Oort cloud, thereby causing showers of comets to fall toward earth

. . . [Michael] Rampino [a geologist] and astrophysicist Richard Strothers [both of the Goddard Institute for Space Studies in New York City] suggest that the swirls of dust the sun and its planets encounter would gravitationally disrupt the Oort cloud, a vast bubble of comets . . . [thereby] sending showers of comets falling toward earth. . . . [The opposition to this theory states] that the sun is close to the middle of the galactic plane right now, and yet there has been no major extinction occurrence recently (the last one was 11 million years ago, wiping out some marine protozoans and mollusks). . . . Patrick Thaddeus, also of Goddard, pointed out that dust clouds are so widely dis-[17]

17. Opposition: if the first alternative were true, earth would be showered with comets now, because we are at about the

tributed that the Oort cloud should be encountering them practically all the time, not just once every 33 million years. . . .

middle of the galactic plane

The article goes on to consider other theories attributing extinctions to extraterrestrial causes. There is the theory of Nemesis, a supposed sister star to our sun that somehow disrupts the solar system's asteroid belt. And there is the Planet X theory, which posits that Planet X (as yet undiscovered) follows a sharply inclined, elliptical orbit that continuously shifts because of the gravitational tug of the other planets. Every 28 million years the shifting orbit intersects a disk of comets that lies beyond the orbit of Neptune. As Planet X passes through this disk, it dislodges comets, sending them toward earth.

Interestingly enough, however, the authors really take no stand in this argument; rather they report the issues and the evidence on either side of the proposition, leaving the readers to draw their own conclusions. And who are those readers? The average reader of *Time* is a college-educated, business-oriented person with a casual interest in science and technology. In this case, the authors realize it is not necessary to provide technical details. It is more important to summarize the two sides of the proposition and evaluate the merits of each stand.

Note that this article spends considerable time discussing the history of the question; it traces possible explanations dating as far back as the early nineteenth century. It also supplies the specific, compelling reason for revisiting the question of mass extinctions—that is, the discovery of high concentrations of iridium. You might also have noticed that the authors feel it is appropriate to offer a brief explanation (definition) of the Darwinian theory of natural selection, because certain implications of the proposition bring the validity of this controversial theory into question.

Before developing the article, the authors doubtless created a list of the relevant issues or controversies suggested by the proposition. As part of your preparation for writing an argumentative document, you should also create a comprehensive set of issues. If we were privy to the list of supporting and opposing views of the issues for the article on extinction of the dinosaurs, it might look like this:

PROPOSITION: DID COMETS KILL THE DINOSAURS?

Affirmative	Negative
1. Multiple mass extinctions have occurred in the history of the earth.	1. This issue is admitted.
2. Those extinctions follow a regular cycle of 26 million years.	2. Extinctions are random.

Affirmative	**Negative**
3. The cause is extraterrestrial.	3. The cause is terrestrial.
4. A possible cause is a shower of meteors or comets.	4. Other possible causes are earthquakes, volcanoes, or floods.
5. A possible cause for comet showers is the earth's bobbing route through the Milky Way.	5. If that were the case, showers would be almost constant.
6. Another possible cause for comet showers is a binary star to our sun called Nemesis.	6. If that were the case, the binary star would be acting differently from any other known binary star.
7. Another possible cause for the comet showers is Planet X.	7. Planet X has not been observed.

Although this list is not comprehensive, it reflects many of the key questions that must be resolved in order to convince readers that the proposition is likely to be true. In developing your list it is important to create as nearly complete a set of issues as possible. The list need not be systematic at first. But in revising, you should place together those points that are related to one another. In reviewing your list of issues, ask the following questions:

Do all issues bear on the proposition? (If not, eliminate the tangential issues.)

Does the list of issues fully cover the subject? (If not, add the missing points.)

Do issues overlap? (If so, create clearer divisions or plan to point out this overlap in your discussion of the topic, so that readers will not be confused.)

Do any of the issues really contain more than one idea? (If so, isolate the discrete points and create a new issue in your list.)

Which issues lend the most critical support to the proposition? (Arrange them in descending order of importance.)

Note in the foregoing list that the opposing viewpoint admits or concedes only the first point (that multiple mass extinctions have occurred in the history of the earth). Most of the other points are challenged. The structure of the article expertly plays one idea against the other, and the writer is careful to let the reader evaluate the merits of the evidence supplied.

This same list-building strategy works well for matters of policy (the extinction argument is one of fact). In this case, however, you must raise several different questions as you review the issues generated:

Is there a need for some change? (If so, why?)

Will the policy (plan of action) suggested by the proposition be effective?

Are the possible benefits of this policy greater than any new disadvantage it may create?

Is the proposed policy better than any alternative policy?

EVIDENCE

Evidence is whatever can be offered to support or confirm a proposition. Evidence can take a variety of forms, including facts, analogies, opinions, testimony, and reports.

A *fact* is something that can be verified or can be attested to by an unimpeachable source. Certain facts, such as the rising and setting of the sun, can be established through primary observation by the individual presenting the proposition. Facts can be measured, as can the stress capacities of a steel beam and the loss of oxygen from the air during combustion. Some of the facts supporting the proposition that comets wiped out the dinosaurs are as follows:

Fact	**Issue**
1. Fossil records indicate at least five great dyings.	1. Multiple mass extinctions have occurred during the history of life on earth.
2. During the past 250 million years, over 3500 marine families were eliminated in cycles of approximately 26 million years.	2. Extinctions follow a regular cycle of approximately 26 million years.
3. High concentrations of iridium were found in Italy, Denmark, and New Zealand.	3. The cause is extraterrestrial.

An *analogy* is based on the idea that, because two instances are alike in a number of important ways, they are alike on the point in question. A common example of this tactic is the use of animals in laboratory experiments. The results are often extrapolated to indicate similar effects in human beings. In order for an analogy to have merit, the two instances must be compared in important respects. The more similarities, the stronger the argument tends to be. Likewise, the differences between the two cases must be relatively unimportant. We can use analogy only to support a proposition, not to confirm it. The only supporting analogies in our excerpt from the article on dinosaur extinction are the following:

Analogy	**Issue**
1. Concentrations of gold and platinum were discovered near the iridium sites in proportions analogous to those found in comets and meteors.	1. The extraterrestrial cause is comets.
2. Sediment layers harbored microscopic balls of glass called *microtektites,* which form in rock when something hits it with great force.	2. The extraterrestrial cause is comets.

Opinions are the comments or remarks of others besides the writer of the argument. They should be used as evidence in a technical writing environment only when the authority of the witness is established clearly. The experience of an expert carries far more weight than that of the mere user, owner, or chance observer, provided that he or she is an authority on the subject at hand. (Hence Benjamin Spock, a renowned pediatrician, is not necessarily a good source of expert opinion on military and political philosophy.) Authority can also derive from the relationship of the source to the audience. The boss, company president, or team leader may have absolute authority for the immediate audience of his or her employees but may have none at all for other readers. If you use an authority the audience knows and respects in developing your arguments, you will gain an initial advantage over your opposition.

Several *opinions* are offered in the article on extinction of the dinosaurs. In each case the authority of the speaker is cited:

Opinion	**Issue**
1. J. John Sepkoski, a paleontologist from the University of Chicago, says the evidence for periodic mass extinctions "very strongly implicates an extraterrestrial mechanism."	1. The agent of disturbance is extraterrestrial and the extinctions are periodic.
2. Daniel Whitmire, an astrophysicist with the University of Southwestern Louisiana, says, "Just the possibility that life here has been controlled by an astronomical event is very far reaching."	2. The possibility of extraterrestrial agent is very significant.

Testimony is a documented comment based on immediacy to the issue. In the case of technical writing, it should also be based on the expertise or authority of the witness. Expert testimony is used successfully several times in the article on dinosaur extinction; however, the most compelling instance is that of Professors Sepkoski and Raup in their formal paper delivered in Flagstaff, Arizona, on mass extinctions. Here, after studying the last 250 million years of the history of life and observing the mass extinctions of 3500 families of marine life, they suggest that the cause of these 26-million year cyclic catastrophes is extraterrestrial.

This particular example of testimony results in conjecture. However, it is possible to establish facts via testimony in the form of a *report* (observation of an event) if the following conditions are met:

The fact is readily attested to by someone with authority.

That person had the opportunity to observe the fact directly.

That person was physically and intellectually able to observe and understand the event and report it accurately.

That person is unbiased (does not have a vested interest in the issue at hand).

Every time you offer evidence, you should tell your readers how the data were acquired, in order to enhance your credibility. The authors of the article on dinosaur extinction go to great lengths to indicate the way certain information was compiled. For instance, we learn that Dr. Alvarez found iridium concentrations in Italy and Denmark and then received confirming evidence from colleagues in New Zealand. The article spells out the means by which the comet theory was developed in the following passage:

> Basing their calculations on the atmospheric consequences of the explosion of Krakatoa, they roughly estimated how much dust the impact of the asteroid would have thrown into the atmosphere, how long sunlight would have been blocked from the earth's surface, and what kinds of life would have been most gravely affected by the climatic changes.

EFFECTIVE REASONING: INDUCTION AND DEDUCTION

Because argumentation presupposes conflict, your readers will naturally challenge your ideas and scrutinize your evidence. The way in which you attempt to prove your points is the way in which you reason—and no argument can be convincing unless the reasoning behind it is both accurate and effective.

Reasoning is the process by which the mind moves from certain data (evidence) to a conclusion that was not known prior to examination of the data. In our everyday thinking, we continually draw conclusions from cause-and-effect relationships. (The car will not start; perhaps the battery is dead. If you go swimming immediately after eating a heavy meal, blood will be drawn away from your stomach [where it is needed for digestion] and toward the arm and leg muscles, potentially resulting in stomach cramps.) Likewise, we use reasoning about subjects and their attributes to support our conclusions. (Cognac has a high concentration of alcohol, so it is best to drink it only when you will not be driving. Electric burners on stoves remain hot for several minutes after they have been turned off; therefore, do not touch one until it has had a chance to cool.)

Behind this ordinary sort of reasoning lies a formal structure—logic—that has been carefully documented since the days of Aristotle. Bear in mind that this structure has merely been *recorded* by logicians; they did not *create* it. It reflects the ways in which human beings develop and test their conclusions. The structure of logic has two basic forms: induction and deduction.

In Chapter 4, when we discussed planning and arranging steps within the overall technical writing process, we touched on inductive and deductive ordering techniques. *Induction*, you will recall, works from specific cases to a general statement, whereas *deduction* works from a general statement and corroborates it with specific instances. The same holds true when it comes to reasoning (rather than merely ordering) by these principles of logic.

Induction moves from a number of particular instances to a generalization (a probability of sorts). To be considered valid, induction must be based on the observation of a fair number of instances, and the instances must be typical—that is, truly representative of the whole. Finally, negative instances must be explained as atypical or somehow inconsequential (as in the case of 28-million-year rather than 26-million-year cycles).

The generalization is formed by means of a cognitive phenomenon known as the *inductive leap*. This synthesizing moment, which is characterized by the *recognition of patterns or trends*, occurs at different points, depending on the writing situation. One audience may require many more instances than another before it is willing to take the leap; one topic may be considerably more complex than another and require frequent demonstration of a pattern in the data.

Inductive reasoning is especially useful in arguments of policy wherein several alternatives are examined before a decision or recommendation is made. By "piling up" evidence, the writer hopes to make the reader recognize the same patterns or conceive of the same generalization that the document posits in its proposition. However, because it is impossible to observe all cases, induction can lead only to a statement of probabilities.

Deduction is better able to provide readers with certitude, because it begins with an axiom: an unquestionably verified statement of observed fact. You are probably most familiar with axioms in the context of mathematics (especially geometry), wherein statements such as "Things equal to the same thing are equal to each other" are never doubted. There is no attempt to prove such points. They are accepted as given—as "axiomatic." There *are* times when such basic assumptions are challenged (consider the cases of Copernicus or Darwin), but for the most part we regard these statements as *truth*.

Propositions of fact rely more heavily on deductive reasoning than on induction, simply because the writer must provide certitude, rather than probability, in order to be convincing. Deduction itself depends not on a compelling list of instances, but rather on a three-part formula for presenting a proof. This formula is called a *syllogism,* and the most famous example is

MAJOR PREMISE	All human beings are mortal.
MINOR PREMISE	Socrates is a human being.
CONCLUSION	Socrates is mortal.

We draw the conclusion by combining the subject of the minor premise (the specific case) and the predicate of the major premise (the axiom). This combination works well, provided the parts of the syllogism are all true; that is, all human beings really are mortal and Socrates really is a human being. It would be possible to engage in valid reasoning by means of a syllogism and still be wrong if the information within the structure were untrue. Hence it would be possible to deduce as follows:

MAJOR PREMISE	All astronauts are men.
MINOR PREMISE	Sally Ride is an astronaut.
CONCLUSION	Sally Ride is a man.

Needless to say, it is not true that all astronauts are men.

Four basic kinds of major premises exist:

1. All X is Y. *Example:* All human beings are mortal.
2. All X is excluded from Y (or no X is in Y). *Example:* No frogs are mammals.
3. Some X is in Y. *Example:* Some horses are thoroughbreds.
4. Some X is excluded from Y. *Example:* Some arguments are not won.

The article on extinction of the dinosaurs made use of the syllogism in several instances, one of which follows.

MAJOR PREMISE	All heavy concentrations of iridium in sediment are caused by extraterrestrial formations, such as comets and meteors.
MINOR PREMISE	The sediment of Gubio, Italy, exhibits heavy concentrations of iridium
CONCLUSION	The sediment of Gubio, Italy, is caused by extraterrestrial formations such as comets and meteors.

This exact phrasing did not appear, of course. Rather, the passage read

> To everyone's surprise, it [the sediment] turned out to be 30 times as rich in iridium as normal rocks. The Berkeley team knew of only a few places where such high concentrations of this rare element might occur: in the earth's core, perhaps 2,000 miles below ground; in extraterrestrial objects like asteroids (or their fragments, meteors) and comets; or in the cosmic dust drifting to earth. . . .

The syllogism underlies this passage. But what about the two other possibilities—the earth's core and cosmic dust? What we face here is known as a *disjunctive syllogism* (also called an *either/or syllogism*)—one in which alternatives exist but are ruled out. If the authors did not provide us with the reasons for ruling out these alternatives, we might consider the major premise untrue. Here is how these writers handled the disjunction:

> The core material seemed too deep to come to the surface, and further analysis ruled out a supernova as the source, so the Alvarez team concluded that the iridium had been left by a giant asteroid hitting the earth at the end of the Cretaceous period.

In an argument addressing readers more technologically oriented than those targeted by *Time,* the authors would have been required to provide the reasons why the core material "seemed too deep to come to the surface" and why the supernova was ruled out as a source.

An argument that does not follow from the evidence is called a *fallacy*. Induction is fallacious when too few cases are offered as the basis of a generalization. In an analogy (a kind of induction), a fallacy can occur when the two cases differ in important respects. In the case of deduction, fallacies occur when the major and minor premises are not related properly to one another. Recognizing fallacies will help you both in designing your own arguments and in countering those of the opposition.

An excellent discussion of logic as it is related to written argumentation is found in the section entitled "Argument" in Cleanth Brooks and Robert Penn Warren's composition textbook *Fundamentals of Good Writing: A Hand-*

book of Modern Rhetoric (New York: Harcourt, 1950), pp. 125–194. It covers the finer points of syllogistic reasoning and logical fallacies.

STRUCTURING THE ARGUMENTATION DOCUMENT

Although no prescribed format for an argument exists, we will suggest a structure to use in technical writing. The level of detail, order, and emphasis depend on the following considerations:

- the audience's attitude toward the subject.
- the writer's attitude toward the subject.
- the kind of treatment the subject permits (such as article, essay, report, or proposal).
- the audience's attitude toward the writer (if pertinent).
- the writer's attitude toward the audience.

For instance, in a recommendation report about developing an end-user computer facility, the audience of high-level executives has considerable interest in the issues. Some time should be spent on the history of such facilities so that audience members have something to compare with what is being proposed. However, very little time has to be spent on the relevant point that brings the question to a head; all readers are familiar with the problem of inaccessible data. First and foremost, readers want to know what is involved in establishing such a facility. Then they want to know what the experience of other companies has been and, finally, what other alternatives exist.

In most technical writing situations, the writer should maintain an objective, fair attitude toward the proposition. The proposition itself should be the result of careful, logical, unbiased investigation of the issues and evidence.

The following outline is used in a recommendation report, but it could also be used in a proposal or a report on an experiment. The same basic need to discuss these elements would exist in essays, articles, and promotional materials, but the arrangement of elements might differ.

Through careful consideration of the audience's attitude toward the writer and the writer's attitude toward the audience, it is important to find common ground and to overcome prejudices, personal hostilities, and (perhaps) habits of thinking. The writer of the recommendation report, for instance, must spend some time talking about ease of use, because the readers are somewhat cyberphobic (they have a fear of machines).

Here is the outline for the recommendation report that appears later in the chapter:

Proposition: Should Metropolitan Gas and Electric Establish an End-User Computer Facility?

I. Introduction
  A. Statement of the problem
    1. Relevant instance
    2. Proposition
  B. Overview of issues
    1. Purpose of facility
    2. Facility requirements
    3. Anticipated costs and paybacks
    4. Implementation strategies
    5. Experience of other companies
    6. Alternatives to establishing such a facility
  C. Method of acquiring data
    1. Secondary materials
    2. Interviews
    3. Site visits

II. Discussion
  A. Purpose
    1. Supply end users with data processing resources
    2. Help minimize applications development backlog
    3. Increase overall productivity of work force
    4. Optimize use of existing computer resources
    5. Help contain proliferating information management costs
  B. Facility requirements
    1. Space
    2. Staffing
    3. Product offerings
    4. Location
  C. Costs and paybacks
    1. Initial outlay
    2. Anticipated growth expenditures
    3. Maintenance-related expenses
    4. Quantifiable increases in productivity
    5. Payback period

- D. Implementation strategy
    1. Pilot program
    2. Full program
- E. Experience of other companies
    1. Carmel Candy Company
    2. South Dakota Utilities
    3. Prism Engineering and Construction
    4. Seaforth Telephone and Communications
- F. Alternatives
    1. Microcomputers
    2. Time-sharing services
    3. Status quo

III. Conclusions

- A. That Metropolitan Gas and Electric would benefit from such a facility
    1. Increased access to information
    2. Consistency of information
    3. Timeliness
    4. Decreased data processing backlog
- B. That such a facility is a better alternative than the status quo, time-sharing services, or the proliferation of microcomputers

IV. Recommendations

- A. That such a facility be established
- B. That it be situated in the west wing of the Phillips building
- C. That a pilot project be used initially

Here are some salient excerpts from this recommendation report:

AUDIENCE Top managers of the Metropolitan Gas and Electric. Such a facility would constitute a major financial investment, so they must review any recommendations made before they act.

ROLE Director of data processing department, who has been asked to review the issues related to establishing an end-user facility at Metropolitan Gas and Electric. Remains an objective investigator who approaches the problem with no preconceived opinion or bias.

PURPOSE To determine whether such a facility would be useful to the company.

FORMAT Recommendation report.

SHOULD METROPOLITAN GAS AND ELECTRIC ESTABLISH AN END-USER COMPUTER FACILITY?

I. Introduction

Statement of the Problem

At Metropolitan Gas and Electric potentially 6000 employees wish to use computerization to help them do their jobs better. Right now the data processing department serves 24 different departments within the company, including accounting, personnel, billing, benefits, marketing, competitive analysis, and financial planning. If the data processing center were required to support the needs of individuals within each of these many departments, our workload would be so great that very little could get done at all. Something must be done both to reduce the DP workload and to better satisfy end users.[1]

1. Relevant instance that makes question important

The purpose of this report is to investigate some of the alternatives—specifically, the establishment of an end-user computer facility.[2] The question to be answered is: Should Metropolitan Gas and Electric establish an end-user computer facility?[3]

2. Purpose of report

3. Proposition

Overview of Issues

The issues of greatest concern to both data processing professionals and the executive management to which this report is addressed are

1. What is the purpose of such a facility?
2. What are its requirements?
3. What are the anticipated costs and potential paybacks?
4. How can such a facility best be implemented (if it is decided that one is appropriate to the company)?
5. What are the experiences of other companies that have implemented such a facility?
6. What alternatives to an end-user computer facility exist?[4]

4. Overview of six issues to be covered in report (the order established here is the order the discussion follows)

Method of Acquiring Data[5]

5. Effort to build credibility to help make evidence convincing

Information presented in this report has been compiled by several means, including:

1. Review of existing literature carried in reputable trade journals such as *Computerworld, Datamation,* and *Computer Decisions* on the topic of end-user computer facilities.[6]

2. Review of reports by noted authorities in data processing such as James Martin, J. F. Rockhart, M. Treacy, S. Friedman, and P. B. Bittner.

3. Interviews with current users of microcomputers within our company and with data processing managers of such facilities.

4. Site visits to Carmel Candy Company, South Dakota Utilities, Prism Engineering and Construction, and Seaforth Telephone and Communications—all companies in which such facilities have been resident for at least two years.

6. Important for reader to believe that inductive process is based on a sufficient number of instances

II. Discussion

Purpose of an End-User Computer Facility[7]

7. First issue: definition (using purpose technique) of key term—vital to understanding of the proposition

The intention of an end-user computer facility (sometimes called an *Information Center* or *Solution Center*) is to provide non-DP business professionals with easy-to-use tools for accessing vital information, yet simultaneously provide a form that helps maintain data consistency and control. Such a facility ideally can enable personnel throughout an organization to do their own information processing quickly and easily. It enables the DP department to give them direct *but controlled* access to on-line information. Generally, such a facility features a variety of software and hardware ranging from personal productivity and decision-support tools to business-specific solutions. End users, with little or no data processing experience, can build their own spreadsheets, forecast trends with color graphics, browse through on-line files, and create or revise reports.

Such a center is meant to improve data management by providing consistent information to end users and by eliminating unnecessary application development. It can help introduce personal computers in a controlled way. And all indications are that such a facility can be tailored to meet the information processing demands of individual companies.

Facility Requirements[8]

8. Second issue: detailed examination of physical requirements

The size and location of the end-user facility should be determined by the number and distribution of the end users to be accommodated. The best strategy, it would seem, is to start small and build incrementally. Long-range planning is required to help ensure that the center can serve both current and future demands. The facility can range from something as simple as a remote or local telephone service, through walk-up windows, to a full-fledged consultation and training suite with terminals, personal computers, printers, and plotters available for use. It is important that surroundings be "friendly" and that the center be accessible to as many users as possible.

Brief history demonstrates (as do interviews) that such centers work best when they are physically located outside the actual DP department (although responsibility for the facility may remain with DP jurisdiction, if desired).[9]

9. Evidence

One version of an end-user computer facility is depicted below. Here, separate sections are designated for staff offices, classrooms, user cubicles, a printing center, and a reference library. The center is stocked with an ample supply of terminals, graphics devices, printers, and personal computers.

. . .[10]

10. [Report goes on to examine points listed in outline.]

Implementation Strategy[11]

11. Fourth issue: implementation

Based on interviews with managers of such facilities and on trade press coverage of the subject, it seems that the soundest strategy for implementing an end-user computer facility is "Start small."[12] Careful planning, testing, and evaluating can pay off in the long run. Initial focus should be on a single department with limited, well-defined needs. Once this department integrates its new access and computing power into daily procedures, the focus of the center can broaden to multiple departments or groups.

12. Evidence

. . .[13]

13. [See outline.]

Alternatives[14]

14. Sixth issue: alternatives

Some companies have tried to improve the situation by purchasing personal computers for end users. The languages are easy to learn, and the users work independently of the data processing department. In fact, personal computing has become very popular. Although these small computers let end users fend for themselves, it now seems that the proliferation of microcomputers may have serious repercussions. For instance, a recent *Business Week* article[15] points out that the spread of these units can hamper effective information management. It cautions that micros can "backfire." The article goes on to say, "If you are allowing people to manipulate data from their own data base, you don't have a microcomputer problem, you have a management problem." Data are inconsistent, results are suspect, and integration is just about impossible.

15. Evidence

Other companies have subscribed to time-sharing services to meet the end-user demand. But a *Computer Decisions* article[16] points out drawbacks here as well. Such services are usually quite expensive—$80,000 per year on the average. Their management is often outside the jurisdiction of the data processing department. This arrangement can make information management difficult; data are not integrated with the main system; new applications are hard to implement. Redundancy of effort can also result.

16. Evidence

. . .[17]

17. [See outline.]

III. Conclusions[18]

18. Logically based conclusions drawn from sufficient inductive discussion

After careful consideration of the issues involved in establishing an end-user computer facility, the data processing department draws the following conclusions:

1. That Metropolitan Gas and Electric would benefit from such a facility in terms of increased access to information for end users, increased data consistency, improved timeliness of information, and decreased backlog in data processing program development.
2. That such a facility is a better alternative than the status quo, subscription to a time-sharing service, or the influx of additional personal computers.

IV. Recommendations[19]

19. Recommendations based on agreement of audience

If the executive committee shares the conclusions we have found, we propose the following implementation strategy:

1. The facility be located in the west wing of the Phillips building since this position would be the most convenient to the largest number of potential users. The amount of space allocated should be at least 24′ by 80′ to accommodate the equipment and carrels we will be recommending.
2. That a pilot project involving the accounting department be selected initially to launch the program. This department's needs are immediate and well-defined. Additionally, it has some familiarity with data processing technology and would require less training than other departments. Such a selection would allow the data processing department to "learn the ropes" of operating such a facility with minimal risk of alienating an end-user department.

. . .[20]

20. [See outline.]

Although this example is only an excerpt from the lengthy report, it accurately reflects the unimpassioned sort of technical argumentation you should try to achieve. As the comments in the margin indicate, the writer is well aware of the audience and pinpoints their concerns. He takes pains to create credibility for his findings, provides background information, and compares alternatives with the recommendation he will ultimately make.

The difference between this proposition of policy and a proposition of fact is that here the writer can only recommend; there is no way to prove his points beyond all doubt. He must convince his audience that he *probably* is right. The more evidence he can cite from other companies and trade press coverage, the stronger the case he can build. At some point, his readers must say, "If these patterns are apparent in other companies, they are likely to hold true for ours as well."

CHAPTER SUMMARY

The occasion that calls for argumentation in a technical writing environment is one in which conflict is involved. However, the situation cannot be resolved by investigation alone—that is, by measurement. Nor can the situation be merely a matter of individual taste. Rather, argumentation involves situations in which the writer demonstrates the *probability* of his or her findings. The aim of such a document is to convince readers that the proposition put forth is correct.

The documents for which technical argumentation is appropriate include reports, proposals, articles, essays, and promotional materials. Two basic kinds of propositions exist. Propositions of fact maintain that a certain interpretation of data is true, whereas propositions of policy recommend that specific actions be taken to solve a problem. In neither situation is the proposition verifiable. It is arguable only. To support a proposition of fact, we rely primarily on deductive reasoning, wherein evidence corroborates or refutes the proposition. To support a proposition of policy, we depend on inductive reasoning techniques, wherein a sufficient number of convincing instances are presented and evaluated.

EXERCISES

Perform each of the following exercises to help you polish your argumentation skills.

1. Locate a scholarly journal in your field in which a researcher presents findings derived from an experiment. Are argumentation techniques used? If so, locate the proposition, the key issues, and the evidence on both the supporting side and the opposing side.
2. Review the article on dinosaur extinction. For each bit of evidence, develop a complete syllogism (like the example appearing in text) that demonstrates either the logic or the fallacy of the reasoning involved.
3. Consult the letters to the editor of a popular journal in your field (such as *Today's Health, Scientific American,* or *Popular Electronics*) to determine the argumentation techniques used. Because letters are usually so brief, the opposing arguments are often omitted and only the writer's point of view remains. Play "devil's advocate" and supply the opposing arguments. How do the statements hold up under this kind of scrutiny?

WRITING ASSIGNMENTS

1. Select an unresolved topic of interest to you in your field of study, and develop an argument supporting your views on it. Be sure to use acceptable logic, credible evidence (some research may be required), and compelling proof. The argument

may be based on either a proposition of fact or a proposition of policy. Create a complete situational analysis that justifies the format you choose, the level of detail, and the examples. Include a single, clear, unprejudiced statement of the proposition as a final part of the situational analysis. Limit: 1500 words.

2. Select an unresolved topic of interest to you in your field of study, and develop an argument supporting the view you *oppose*. All other constraints remain the same as in Assignment 1.

NOTE: Apply the editing criteria at the end of Part II to your assignment before you submit it to your instructor.

CHAPTER

11

Graphic Aids

CHAPTER PLAN

Graphic aids reinforce the written word and are often essential to technical documents. They help emphasize key points, clarify complex concepts or data, and summarize details.

Familiarity with a variety of graphic aids can help you develop more effective technical documents. This chapter will

- explain how to determine the need for graphic aids.
- examine the purpose and benefits of various types of graphic aids.
- introduce basic guidelines for using graphic aids.
- discuss computer graphics.

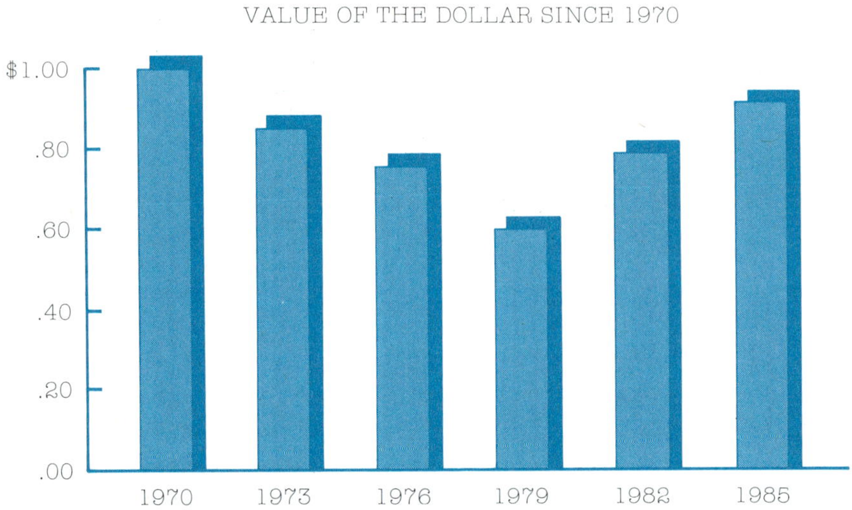

Figure 11-1 Graphs enable readers to see trends at a glance.

DETERMINING THE NEED FOR GRAPHIC AIDS

Technical information is often difficult to understand. To make that information more readily comprehensible, we include definitions, classification systems, causal analyses, procedural descriptions, summaries, and examples. Graphic aids can augment these written elements. They offer an added dimension through which we can convey complex information, make a point, or motivate readers to action.

Graphic aids can be used to demonstrate trends, establish relationships, compare or contrast elements, illustrate a point, present a specific or representative case, chart the flow of activity, or list relevant factors.

Though not all information requires graphic representation, including them improves most documents. As we will see, certain kinds of graphic aids lend themselves better to the representation of quantitative data, and others to qualitative and conceptual information.

THE BENEFITS OF GRAPHIC AIDS

The benefits of graphic aids can be fourfold: they can enhance clarity, emphasis, credibility, and interest. Just like providing concrete examples, *showing* a reader what you are talking about enhances understanding. From a mere glance at the bar graph in Figure 11-1, readers can determine the downward trend from 1970 through 1979; the upward swing since that time is likewise apparent. Although it is accompanied in its original document by a discussion that explains the various factors influencing the value of the dollar during these years, this graph simplifies the information and summarizes it visually.

Graphic aids can place greater emphasis on key points. In Figure 11-1, for instance, the writer reinforces the notion that a tremendous downswing in the value of the dollar occurred in 1970 and the years following because of the end of U.S. involvement in Viet Nam and the beginning of oil shortages. The graph also underscores the writer's observation that the dollar has been increasing in strength at a steady rate since the early years of the Reagan administration (1982).

Graphic aids constructed from research materials have the added benefit of "borrowing" the authoritativeness of the source and hence lending credibility to the points you are making. Certainly, you should never manipulate or create data in order to develop an impressive graph or table. However, if information lends itself to visual representation, including it may make your document carry more weight in the reader's estimation.

Reader interest can also be established and maintained through the use of graphic aids. Even a device as simple as an informal table makes reading easier by breaking up a page of text. Whenever you insert graphic aids, you give readers a chance to pause and take stock of the information.

CONSIDERATIONS IN SELECTING GRAPHIC AIDS

There are several factors to consider as you select an appropriate graphic aid to represent information.

1. *The needs of your audience.* Throughout this textbook, you have considered the needs of your audience in each writing assignment. The situational analysis you have learned to develop in preparation for writing projects will serve you here as well. In addition to enhancing comprehension, graphic aids enable you to stress key points, add flavor to the document, or summarize extensive data.
2. *The purpose of the document.* It is important to consider the purpose of the document. Will a graphic aid help you instruct, motivate, or persuade your audience? Will graphic aids help you explain, compare, or report information?
3. *The constraints of the writing situation.* It is important to take into account such constraints as time, budget, skill levels, and availability of resources. Consider the production of a repair manual, for instance. Because users of the manual should have as accurate an understanding of the equipment as possible, a clear photograph might be required. As the writer of a manual,

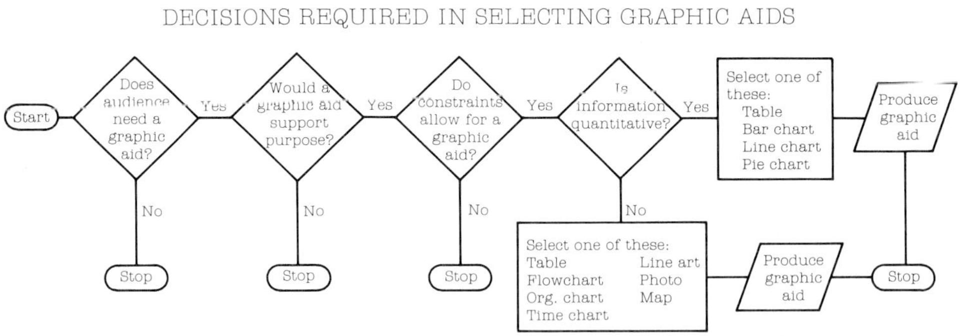

Figure 11-2 Qualitative information is represented symbolically.

you will have to find out whether such a photo already exists. If so, you must arrange to secure it for use in the manual. If the photo does not exist, you may have to take the picture or hire a professional to do so. In these cases, allow sufficient time, make sure the equipment to be photographed is available for the photo session, and be certain your budget can accommodate such an expenditure.

4. *The kind of information.* The final step before selecting a graphic aid is deciding whether the information is quantitative (measurable) or qualitative (conceptual). *Quantitative information* is numerical and is expressed as units of measure, such as dollars, weights, volumes, degrees of temperature, or velocities. The different vehicles for presenting this kind of information include bar graph, line graphs, tables, and pie charts.

Qualitative information, or conceptual information, can be represented in a variety of ways, including flowcharts, line art, organization charts, time charts, photographs, diagrams, and maps. Representing qualitative information can be somewhat more complicated than representing quantitative data, because it requires you to translate concepts into meaningful visual symbols. Figure 11-2 is an example of this translation process. It uses a flowchart to depict the four decision-making steps involved in selecting graphic aids. Diamonds represent decision points, rectangles symbolize action, and parallelograms stand for output (in this case, the actual graphic).

The action rectangles in Figure 11-2 include lists of graphic aids. As we proceed in this chapter, we will determine the appropriateness of each of these media to specific writing situations. Most important to remember, at this point, is that a graphic aid must indeed be an *aid;* it should help readers think about the information it offers. If it does not help clarify or emphasize the information being discussed, it may not be appropriate.

ARRESTS AND CONVICTIONS IN SALEM

Year	Total arrests	Total convictions on original charge	Total plea to lesser charge	Total dropped charges
1980	3500	2000	1000	500
1981	3300	1800	1200	300
1982	3600	2800	500	300
1983	3200	2500	450	250
1984	4200	3000	900	300
1985	3200	2000	600	600
1986	3800	2800	200	800

Table 11-1 Tables can summarize a great many data.

CONVEYING QUANTITATIVE INFORMATION

Numerically based information, or quantitative data, can be represented best in one of four ways: tables, bar graphs, line graphs, or pie charts. Each of these modes of display has unique characteristics that can help you reinforce different kinds of messages. By examining the purpose and effect of each type of graphic aid, its components, its varieties or subcategories, and the methods of developing it, you will become able to select the graphic aid best suited to whatever writing situation you encounter.

TABLES

Tables provide writers with one of the most important forms of "shorthand" possible. Through the use of columns and rows, tables can accommodate vast stores of quantitative data. Glancing from top to bottom and from left to right, readers can grasp relationships, perceive totals, and analyze trends. The summarizing power of a table is demonstrated in Table 11-1. The only real shortcoming of this kind of graphic aid is its inability to provide emphasis.

The components of the table include the rows and columns themselves, column headings, the stub heading (the leftmost column heading, which defines the line headings), and the line headings. All headings should be succinct descriptions of the data they represent. Subheadings (not illustrated here) represent subcategories within major headings. Subheadings may be necessary in a complex table, but they should be used with discretion for the sake of clarity.

In all cases, the columns and rows should be arranged in a way that seems logical for a particular table—for example, chronologically, in terms of descending quantities, in ascending order of importance, alphabetically, or by geographic location. Numerical data within columns should be aligned at the decimal point (if one appears) or at the units level.

The table title should be in a larger typeface than that used in the table itself. Formal tables are numbered consecutively throughout a document or within chapters or sections of longer documents. Tables are listed by number and title in the document's prefacing materials. Note that tables are numbered separately from figures.

BAR GRAPHS

A *bar graph,* such as the one shown in Figure 11-1, offers an excellent way to summarize numerical data for audiences who are unfamiliar with statistics or who do not need to know *exact* numerical values. Its primary purpose is to indicate patterns and trends or to compare elements. Bar graphs can also be used effectively to clarify, simplify, or reinforce key points.

A glance at a bar graph drives home the significance of the numerical information represented. Readers can tell whether increases, decreases, erratic development, or flat periods have occurred.

A drawback of this valuable tool is that the bar graph is relatively imprecise. You may have noted in examining Figure 11-1 that it is difficult to assess the exact value of the 1973 dollar ($.83) or the 1985 dollar ($.92). Rather, emphasis falls on the downward movement of the dollar's value from 1970 to 1979. This imprecision occurs because the structure of the bar graph is based on regular, or periodic, scales of measurement instead of absolute numbers. The basic elements of the bar graph are illustrated in Figure 11-3.

The essential components of a bar graph are the vertical axis, which contains the values of the unit of measurement; the horizontal axis, which contains appropriate variables; and the bars themselves. The base value is usually zero, as it is in Figure 11-3. Gradients can be clarified by using hash marks on the vertical axis. Bar widths should be constant, as should the spaces between bars. With complex bar graphs, you may wish to use shading to distinguish elements so that comparison is made easy. In Figure 11-3, the darker bars represent 1983 and the lighter bars represent 1984.

As in the case of all graphic aids we will examine, a succinct, concrete description of the information should appear as the title. Perhaps "Year-by-year Comparison of Rainfall in Albuquerque during 1983 and 1984" might

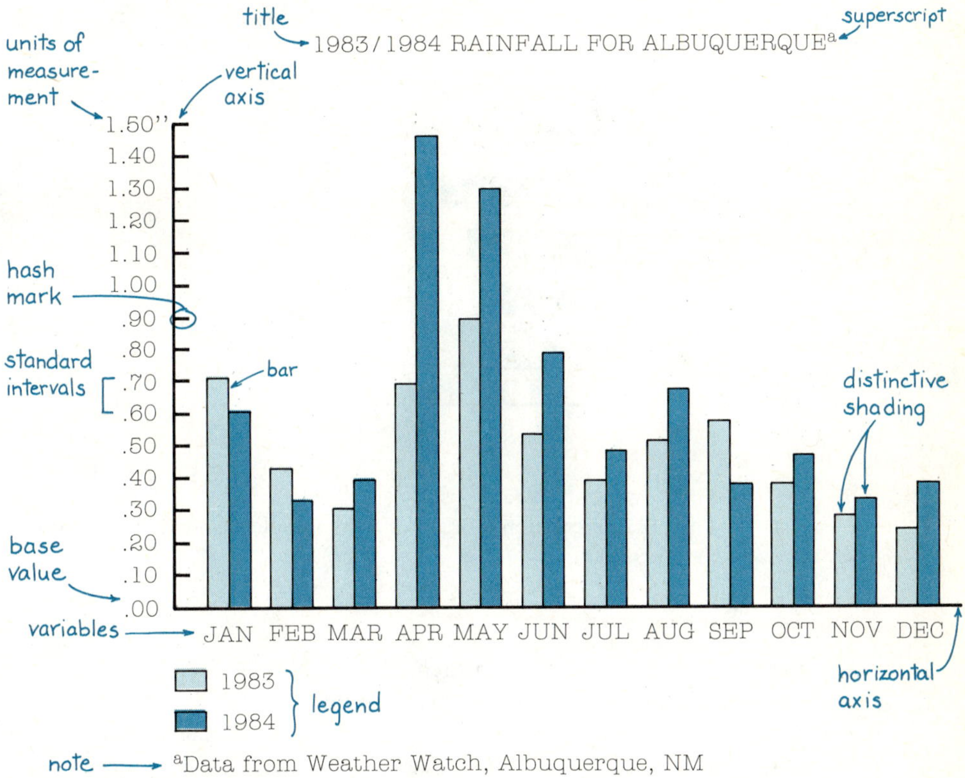

Figure 11-3 Bar graphs emphasize trends.

be a correct title for Figure 11-3, but "1983/1984 Rainfall for Albuquerque" is descriptive and succinct.

Note that in Figure 11-3 a superscript *a* appears after the title and a corresponding footnote citing the source of the chart appears at the bottom of the graph. It is common practice to use letters rather than numbers to designate notes and footnotes in charts, in order to avoid confusion with numerical scales or endpoints. In all graphic aids, notes belong at the bottom.

Most bar graphs use a vertical orientation such as that demonstrated in Figure 11-3. However, a horizontal orientation is useful when you are dealing with multiple factors reflected for a *single period.* As shown in Figure 11-4, this orientation requires that you place the variables on the vertical axis and the values of the unit of measurement on the horizontal axis.

The variations on bar graphs are many, but each choice should depend on the requirements of the audience, the clarity with which data can be con-

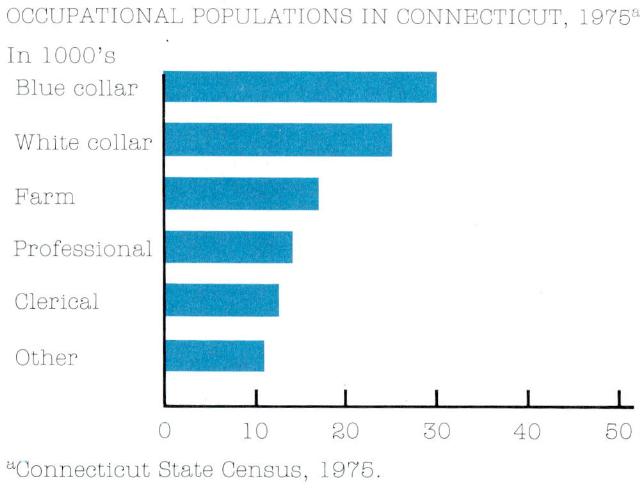

Figure 11-4 Horizontal orientation works best for multiple elements compared in terms of a single criterion.

veyed, and the purpose of the document. You probably are familiar with a quite popular variation of the bar graph, which is called a *pictogram* or *ideograph* because it is used frequently in newspapers and magazines when a technical subject is presented to a nontechnical audience. As Figure 11-5 shows, this type of chart uses pictorial symbols to convey quantitative values. A pictogram can add interest to a document, but it is far less precise than the traditional bar graph.

Use pictograms with younger audiences, with lay audiences, and when *only* gross patterns are important. Choose a symbol that is relevant to the data (barrels nicely portray oil consumption, and cows can represent livestock).

LINE GRAPHS

Line graphs, like bar graphs, are used to demonstrate trends and patterns. However, line graphs are the best choice when numerical precision is required. Line graphs are also the best choice when you are measuring several different factors simultaneously against the same vertical and horizontal axes.

To demonstrate the usefulness of this type of graphic aid, *glance* at the following series of numbers.

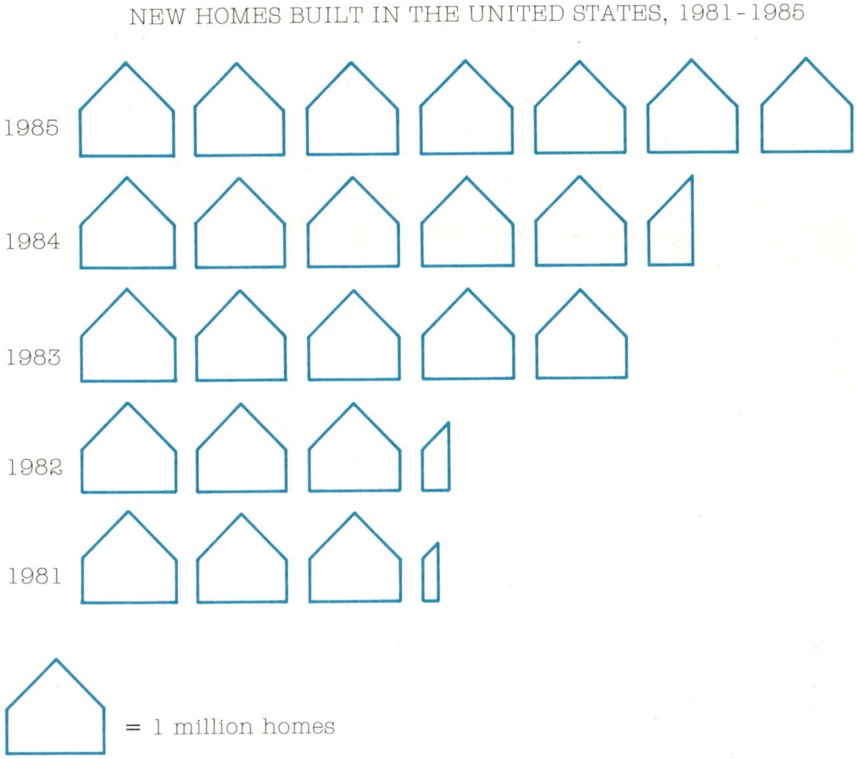

Figure 11-5 A pictogram depicts gross patterns.

Monday	**Tuesday**	**Wednesday**	**Thursday**	**Friday**	**Saturday**	**Sunday**
113,580	225,808	190,180	299,530	262,009	435,070	368,009

Are you able to discern a pattern easily? Probably not. Now look at Figure 11-6, which plots these values on a grid.

As you can see for yourself, the visual representation of these data speeds communication and improves understanding. It is now impossible *not* to recognize a pattern of upward development, interrupted by dips that do not halt the general upward trend.

This type of graph generally has a vertical orientation: the scale of measurement runs up the left side, and the variables run along the horizontal axis. Each plotted line represents the value of a factor that is being measured. Unlike bar charts, these graphics can have intersecting lines (see Figure 11-7), because multiple elements can be charted within the same

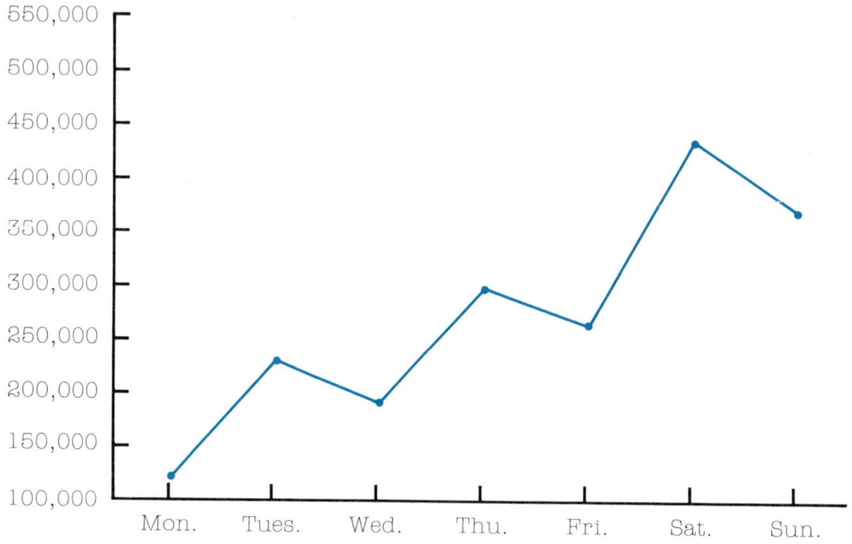

Figure 11-6 Line graphs make it easy to discern a pattern.

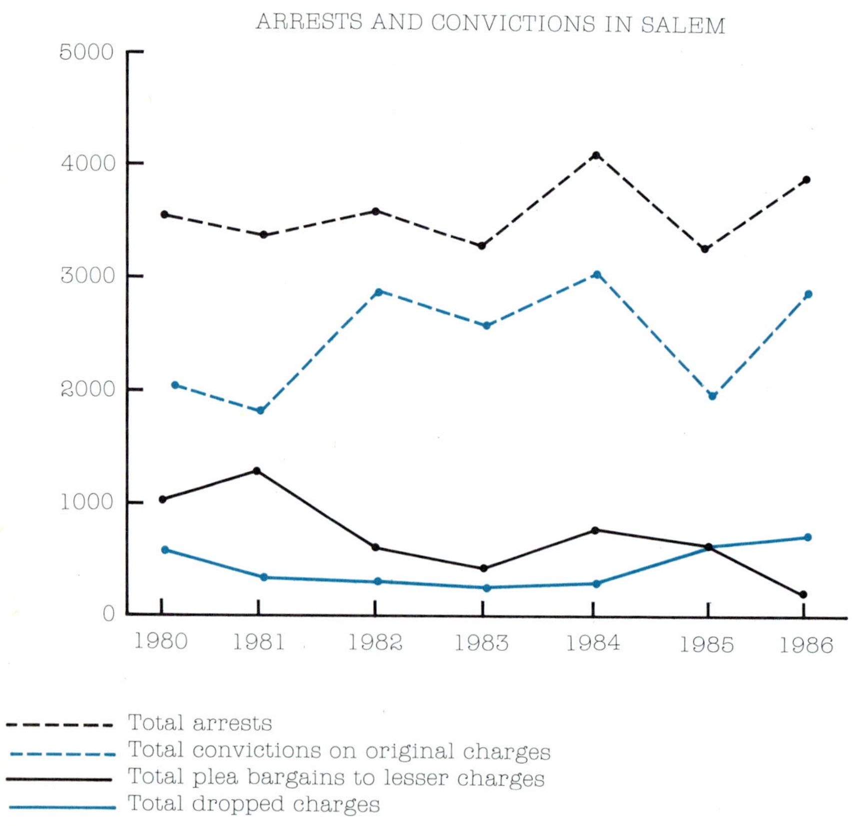

Figure 11-7 On a line graph, we can plot values for several factors.

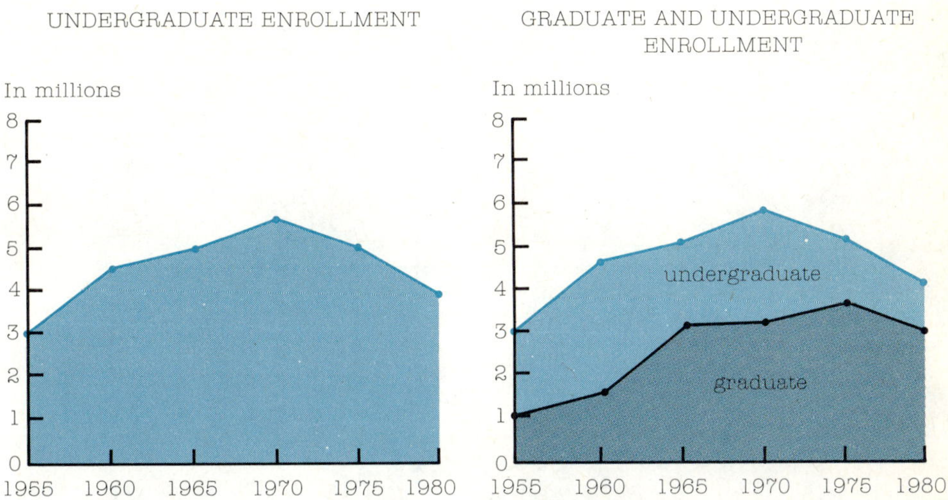

Figure 11-8 Shaded variations of line graphs, like the surface graph (left) and the strata graph (right), can add emphasis and clarity.

area. For clarity, most graphics experts caution against using more than five lines (factors) in a single chart.

As Figure 11-7 illustrates, multiple lines in a graph must be identified clearly. This objective can be accomplished by labeling each line or by varying the nature of the line. One line might be thin and solid, another solid and thick, still another a dashed line, and a fourth dotted. If you are working with color graphics, it is very simple to use lines of different colors to distinguish data. When you use different types of lines, supply a key at the lower left to describe each element.

Two important variations on the line graph, the surface graph and the strata graph, provide dramatic emphasis (see Figure 11-8). The *surface graph* shades the area below a single plotted line to emphasize a particular trend or to reinforce your text. The *strata graph,* or band graph, shades the areas below multiple plotted lines (that do not intersect), making possible quite dramatic comparisons.

PIE CHARTS

Pie charts, or circle charts, are particularly useful when you need to show the proportion of a part to the whole. The circle's circumference represents 100%, or the total numerical value of the data. The pie is divided into segments, each of which represents a precise percentage. The components of this kind of chart are illustrated in Figure 11-9.

A pie chart can tolerate only a limited number of segments before clarity suffers. Most graphics experts recommend that such a chart have no more

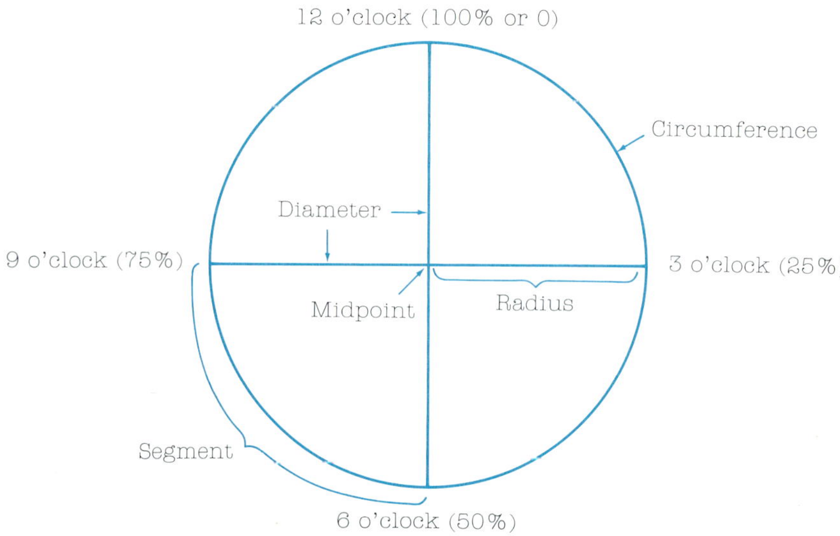

Figure 11-9 A pie chart is used to depict the relationship of a part to the whole.

than eight segments. Unlike the average edible pie, however, this pie rarely has equal-sized slices. Rather, its segments vary according to the numerical relationships among the data. The largest proportion begins at "12 o'clock," followed by a clockwise presentation of the other segments in order of descending percentages. The exception to this rule is that any segment representing composite, miscellaneous information (generally called "other") is placed in the last position regardless of what proportion it represents.

The proportions depicted in a pie chart can have powerful impact in a document. For instance, if we were to turn each year's information in Figure 11-7 into a separate pie chart, we would be struck by how the ratio of convictions to the total number of arrests varies from year to year. Figure 11-10 demonstrates how the information for 1980 would be represented in such a pie chart.

The fact that an individual pie chart would be required to represent each year's pattern demonstrates a shortcoming of the pie chart. It affords only limited opportunity for comparison: part to part. Another weakness of the pie chart is that its segments are not self-explanatory. Labels are necessary to cite both the element and the proportion portrayed. Because labels are readable only on the horizontal plane, a pie chart can become quite cluttered when more than four or five segments are used.

Variations on the traditional circle chart include a ringed structure, which allows comparison of data for more than one time interval, and the raised-

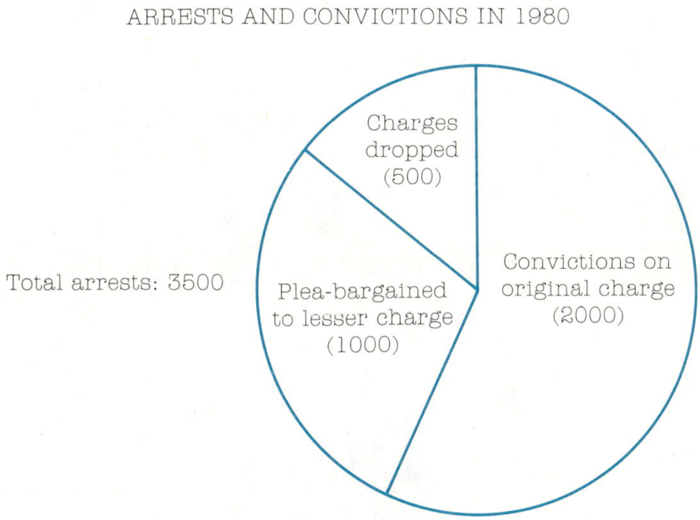

Figure 11-10 Pie charts emphasize proportion, and they allow comparison of parts.

segment pie, which emphasizes a certain segment of the whole. The ringed structure should be restricted to two time periods for the sake of clarity. The raised-segment pie chart can feature only one portion of the pie to be effective. These variations are illustrated in Figure 11-11.

As Figure 11-11 shows, the ringed structure uses shading to distinguish its components further. Note also that this chart requires both a legend describing the time intervals and leaders from the outer segments to their respective labels. The ringed pie chart should be used on a limited basis and only for audiences accustomed to interpreting sophisticated graphic aids.

The raised-segment pie chart, on the other hand, can be used with either technical or nontechnical audiences. It is a valuable tool for emphasis when only one element requires highlighting.

CONVEYING QUALITATIVE INFORMATION

Some technical information falls into the realm of qualitative data rather than quantitative, or numerical, data. Appropriate use of graphic aids can help depict this kind of information in interesting and emphatic ways. Descriptions of procedures can benefit from the use of flowcharts, for instance. Definitions can be clarified by line art and diagrams. Classification

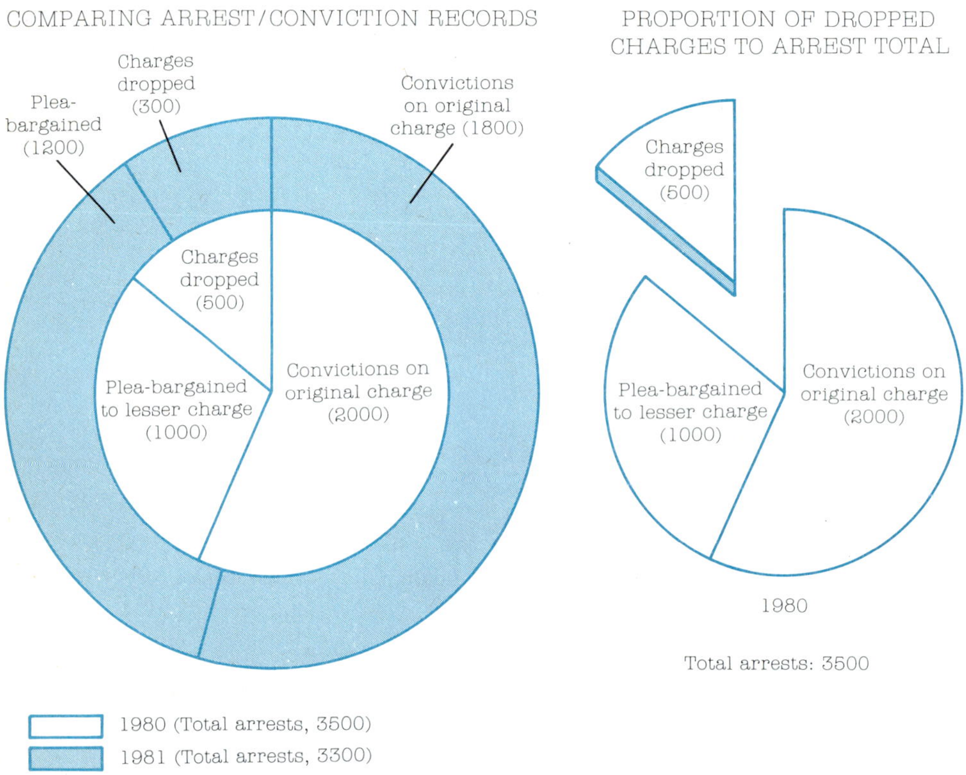

Figure 11-11 The ringed pie chart (left) compares more than one time interval. The raised-segment pie chart (right) distinguishes one segment from the rest of the chart.

systems can be summarized through the prudent use of tables and organization charts. The description of a mechanism can be supported by a high-quality photograph. And so on. Let us briefly survey each of these means of depicting qualitative information graphically.

TABLES

The same reason for using a *table* to display numerical data holds true for qualitative data as well: tables can summarize a lot of information succinctly and clearly. We observe the same guidelines for the layout of columns, rows, and headings, but the nature of the data differs. Words or brief statements, rather than numbers, appear in the columns and rows. Consider the abbreviated newspaper weather report presented in Table 11-2. Cities are listed in alphabetical order as line headings; column headings follow the

WEATHER CONDITIONS

City	Yesterday	Today	Tomorrow
Akron	Pt. Cldy	Rain	Clearing
Albany	Rain	Clearing	Sunny
Albuquerque	Sunny	Sunny	Sunny
Anaheim	Cloudy	Sunny	Pt. Cldy
Atlanta	Sunny	Sunny	Sunny
Bangor	Rain	Rain	Clearing
Boise	Pt. Cldy	Rain	Rain
Boston	Rain	Cloudy	Clearing

Table 11-2 Tables convey qualitative information by using words or phrases in the columns and rows.

chronological sequence of yesterday, today, and tomorrow. To save space, the newspaper uses abbreviations to represent weather conditions familiar to all.

Terse, even cryptic or "telegraphic" language is acceptable in tables of qualitative information; the writer can expand on these elements in the text. Condense these packets of information as much as possible before you build the table.

FLOWCHARTS

Flowcharts trace the sequence of events, steps, or activities that are necessary to perform a given procedure. One of the strongest features of this kind of graphic aid is its ability to demonstrate alternatives and the actions resulting from the choice of each option. Qualitative data depicted in this manner can achieve a degree of coherence and concreteness that is not always possible through narrative alone. Readers can trace the process by which goals are accomplished. Tasks along the way can be pinpointed. Contingencies can be demonstrated with brevity and clarity. Flowcharts are a powerful tool.

The chief components of a flowchart are symbolic shapes, connecting lines, directional arrows, labels, and chart title and number. The symbolic shapes can be quite varied, and often a template (stencil) specific to a technical discipline is available. For instance, chemists can use a template that has figures representing a Bunsen burner, heating coil, test tube, and beaker. Computer scientists use a programming template that includes symbols for manual operations, decisions, input/output, and so on.

When such symbols are used within the confines of the discipline to which it applies, *and* the audience will unquestionably understand them, no

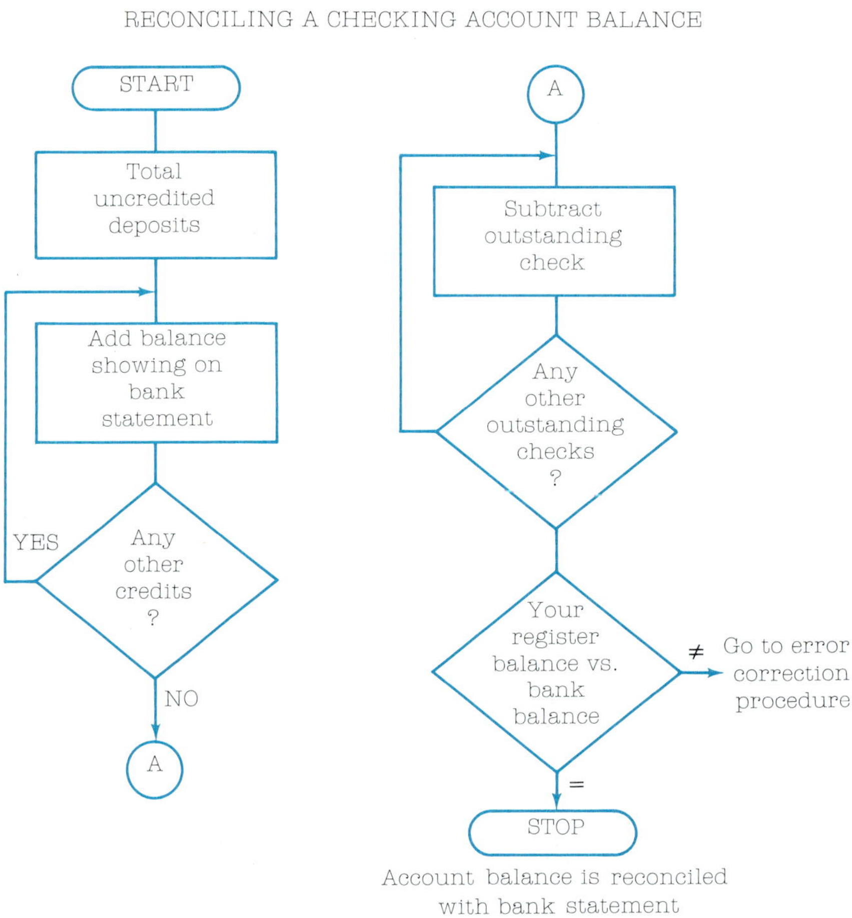

Figure 11-12 A flowchart depicts the steps of a procedure in chronological sequence.

key or legend is required. However, if the audience is likely to be unfamiliar with the symbolic shapes, you must explain them in the narrative just prior to your presentation of the graphic aid. In either case, using a template can standardize the size and form of the flowchart and save the writer time in developing the graphic aid.

A flowchart can be designed in either a vertical or a horizontal format, though the horizontal choice is more common. Figure 11-2 illustrates the use of the across-the-page format to accommodate branching activities. Figure 11-12 illustrates the use of the vertical format to list the steps involved in reconciling a checking account. The vertical format works better

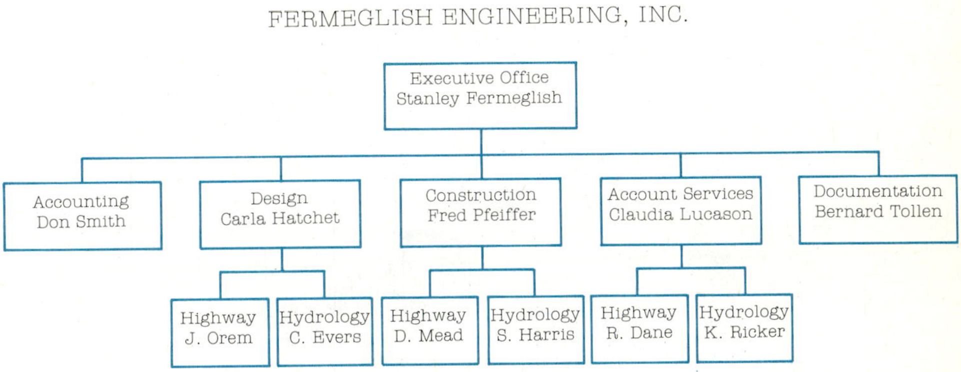

Figure 11-13 Organization charts portray reporting structures.

in this case, because repetition—that is, *looping*—is required at certain steps. The circled letter A indicates continuation.

A flowchart should be reduced to the least possible number of steps through the same kinds of grouping techniques you learned when we discussed writing process descriptions. It is also necessary to determine where decisions must be made before one can proceed to the next step. Be certain to include the appropriate options. If repetition is required, show it as well.

ORGANIZATION CHARTS

Organization charts graphically depict the structure of businesses, agencies, or departments. Using a set of rectangles arranged at several levels, this kind of chart precisely conveys the interrelationships among parts and the structural composition of the whole. Great clarity can be achieved through this visual representation, and actual narrative can be minimized.

Though they are developed in much the same manner, there are three kinds of organization charts. The most common type cites individuals and depicts the hierarchical reporting and responsibility structure in which they operate. The president of a company, for instance, has many vice presidents, each of whom has several people (directors) reporting directly to him or her, who in turn have people reporting to them. The reporting structure is made clear through a series of branches and levels.

A second type of organization chart focuses on the units of an enterprise rather than on individuals. It is possible to state the name of the individual primarily responsible for that department, but the emphasis remains on the entity rather than the person. Figure 11-13 illustrates the use of this approach to describe the inner relationships of departments within a small construction engineering firm.

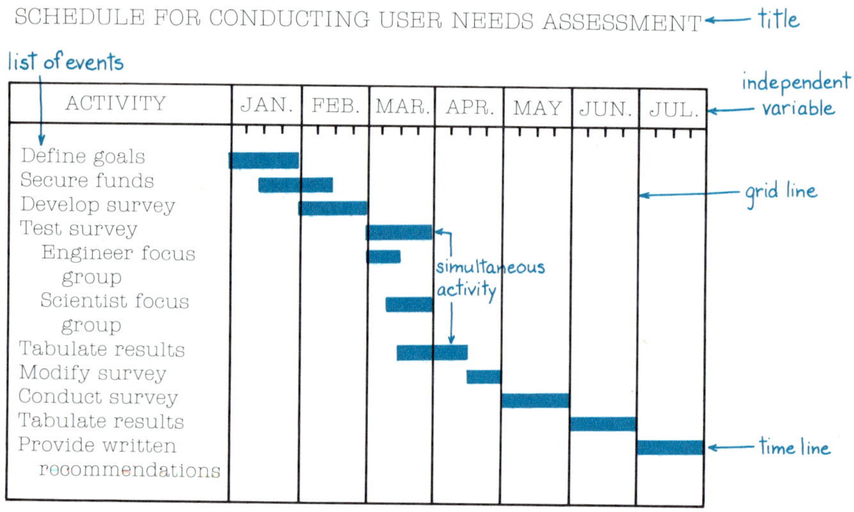

Figure 11-14 A Gantt chart compiles complex scheduling information in a handy table.

The third type of organization chart uses this same tiered pattern to demonstrate functional relationships. This kind of chart is especially useful in defining activities and subactivities in a complex project. Within the documentation department of a major computer manufacturer, for instance, functional levels might include development of marketing literature on one side of the chart and the development of user documentation on the other. Beneath marketing literature, we might find the development of brochures, press releases, sales kits, and advertising. Beneath user documentation, we might find the development of repair guides, product specifications, and user manuals.

TIME CHARTS

A *time chart* shows the time required for carrying out a series of activities, procedures, or tasks. It is especially important to technical documents such as proposals and project reports. This graphic aid doubles as a planning tool through its ability to depict a complex schedule from its starting point to its anticipated completion. During the execution of a project, the time chart is modified as appropriate.

The most commonly used chart, called a *Gantt chart* (named after the engineer who invented it), is illustrated in Figure 11-14. Other names for this device are *time line* and *milestone chart*. A more sophisticated graphic aid in this category, called a *PERT chart* (Program Evaluation Review Technique), reflects relationships and interdependencies between scheduled events. This kind of chart shows the maximum period of time that a firm can take to complete each step in a total process and still finish within the overall time allotted. The development and maintenance of a PERT chart generally requires a computer.

The list of events is arranged sequentially and should be as inclusive as possible. The grid is used to depict time in terms of minutes, hours, days, weeks, months, or years. Finally, the lines, arrows, or symbols depict the time required or elapsed for each activity from start to finish. As you can see in Figure 11-14, it is quite possible to conduct several activities simultaneously, depending on the nature of the project. In fact, few projects have only absolutely linear, discrete events.

To prepare a time chart, list all the activities that will be involved. Then group them into related categories to minimize the number of items. (In Figure 11-14, for instance, we see that two separate focus groups are subordinate to the overall activity of testing the survey.) Arrange these activities or categories as sequentially as possible. Select a brief, action-oriented statement to describe each event, such as "tabulate results." Determine what unit of time to use at the top of the chart. Plot the start date and finish date of each activity. Major activities should be plotted in darker lines than their subordinate events and should encompass the entire period covered by these lesser elements.

LINE ART

Technical documents can benefit from the presence of simplified drawings, or *line art*. Like all the graphic aids we have examined, line art can add clarity to a narrative or stimulate reader interest. No other graphic aid, however, can be such a powerful boon to audience understanding: a picture is worth at least a thousand words.

The style and subject matter of this kind of graphic aid are limited only by the writer's artistic skills and imagination. Almost as varied are the purposes for which line art is used. They include defining or comparing characteristics, representing an abstraction, identifying a structural arrangement, demonstrating stages of development, and portraying a condition.

When these drawings carry labels to identify components or a structural arrangement, they are called *diagrams* or *schematic diagrams*. This particular form of line art is useful in documents that require partition analysis or description of a mechanism.

Figure 11-15 is a single example among the many possibilities that line art offers. If you recognize the need for drawings of this caliber in a document but are not yourself a skilled artist, consult a graphics specialist. However,

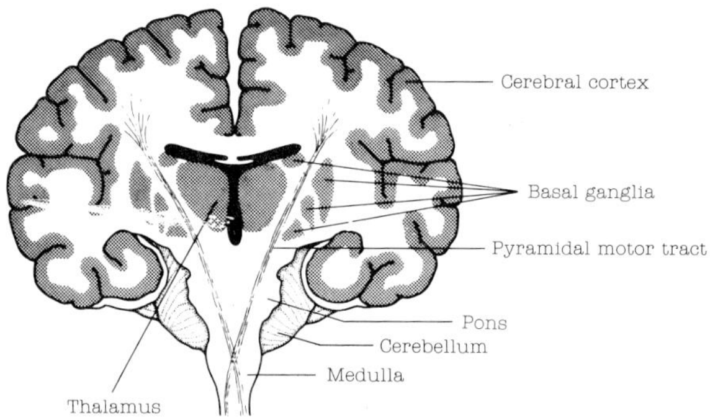

Figure 11-15 An endless variety of styles and techniques makes line art appropriate to a wide range of applications.

plan for the added time and expense that calling on a professional may entail.

Unlike a photograph, which includes the surrounding environment of the subject, line drawings eliminate any unnecessary detail. Thus their use enables the writer to focus on key points and minimize areas of little concern.

Whenever labels are used, they should appear on the horizontal plane (as do labels for pie charts), even if the part itself is oriented vertically or diagonally. Labels should be placed as close as possible to the element in question. If size allows, place labels *within* the element. Otherwise, use a short line (leader) running from the element to the label.

The key axiom of preparing a line drawing is "Keep it simple!" Begin by determining the purpose of the drawing. What will be the best way to feature the structure, components, or characteristics of the subject? Test your answer to this question by producing a rough sketch of the drawing. This kind of drawing is also called a "thumbnail sketch" because of its brevity and conciseness. Experiment with size until you are confident that the drawing will accommodate the elements you need to feature. Are the parts of the drawing in proportion to one another? Do they present a unified look? At this point, does the sketch satisfy the purpose of the narrative? Will it, for instance, enable readers to compare the characteristics of various fungi at a glance, or follow the phases of the moon? Perhaps only a small section of the element is required. Or maybe further magnification would help highlight important details.

Manipulate the sketch until you feel it satisfies your needs. Add more detail only if appropriate. Then lightly add the necessary labels. Right now you are trying to determine positioning and size. Revise as required. Then darken the lines that should appear in the final drawing. You are ready to render the final drawing carefully—or to ask an artist to do so.

PHOTOGRAPHS

Photographs are best used in technical documents when it is important to your audience that the subject be portrayed exactly. Photographs are frequently required to illustrate

equipment to be included on specification sheets or in manuals.

before-and-after studies of medical problems in journals.

site surveys in proposals or geographic studies.

stages of development in primary research reports.

Photographs are also valuable in oral presentations of technical material (most film developed as a photograph can also be produced as a 35-mm slide).

Although photography can be a very exciting medium in which to display technical information, its use in documents is limited because most writers are not expert photographers. Instead, it will be your job to identify the need for photography and to arrange for a professional to take and process the pictures. Considerable planning, time, and funding are required when photography is important.

Before selecting a photographer, be certain the subject you plan to photograph *should* be photographed. Is film the best medium? Or would line art better illustrate your points? If photos are necessary, should they be black and white or color? (Using color entails a much greater production expense.) Should the subject be "shot" in its natural surroundings? Or should it be isolated and photographed against an uncluttered background? Would conducting a photo session on site present any problems?

For instance, imagine taking pictures of a pit mine. The very size of the subject would raise problems of scope. Lighting would be difficult to control. Access to the best vantage points might be restricted. These are the kinds of issues you should try to resolve before calling in professional help. A professional photographer can help address more technical matters, such as the angle of the shot and the most appropriate lens.

Many specialized photographic approaches exist that prove especially valuable in the treatment of technical subjects. For example, the camera can be combined with a high-power microscope to magnify the subject. In microbiological studies, magnification can be raised to such high levels that cellular life can be photographed. Special telescopic camera equipment can record images many light-years distant from the earth. Another alternative,

Figure 11-16 Use a photograph when readers must see exactly how the subject looks.

aerial photography, can provide an unusual perspective on sites. Underwater photography allows us to view worlds untouched by humans. Lennart Nilsson, a renowned medical photographer, has pioneered methods for photographing *within* the body. His work has provided doctors with new methods of studying disease, reproduction, and growth.

Photos do not always require titles, but a figure number and caption are usually appropriate. Also provide a credit line—the name of the professional photographer—in very small type at the lower right-hand edge of the photograph.

Figure 11-16 shows potential customers exactly what a particular computer system's central processing unit looks like. Seeing this photograph will help them determine how much floor space the unit requires.

WEATHER CONDITIONS

City	Yesterday	Today	Tomorrow
Akron	Pt. Cldy	Rain	Clearing
Albany	Rain	Clearing	Sunny
Albuquerque	Sunny	Sunny	Sunny
Anaheim	Cloudy	Sunny	Pt. Cldy
Atlanta	Sunny	Sunny	Sunny
Bangor	Rain	Rain	Clearing
Boise	Pt. Cldy	Rain	Rain
Boston	Rain	Cloudy	Clearing

Table 11-3 Unnecessary detail and lines distract readers.

SOME GENERAL GUIDELINES FOR CREATING AND POSITIONING GRAPHIC AIDS

Here is a list of guidelines to help you make sure your graphic aids are as effective as possible.

1. Keep graphic aids *simple*. Unnecessary detail detracts from the point, just as unnecessary words stand in the way of clarity. Table 11-2 originally looked like Table 11-3, in which meaningless lines and symbols distract readers.
2. Provide *unity* in your graphic aid. The space devoted to a good graphic aid is treated as a single entity, and internal elements look as though they belong together. You can achieve these effects by using connecting lines, even spacing, the same plane and scale throughout the graphic aid, complementary symbols (such as bullets or wedges for subheadings), standardized lettering, and dimensional shading.
3. *Balance* each graphic aid. In a balanced design, if you superimpose four quadrants on the page, elements of the graphic aid are distributed as evenly as possible across each area. This criterion requires careful planning (*not* manipulation of data) so that the space is used wisely and attractively. For instance, in the case of an organization chart wherein there are many more members in one department than another, it might be possible to compensate for the seeming imbalance by using smaller blocks on one side.
4. Provide visual *contrast*. The graphic aid must stand out from its background, and each of its internal elements must be distinguishable from the

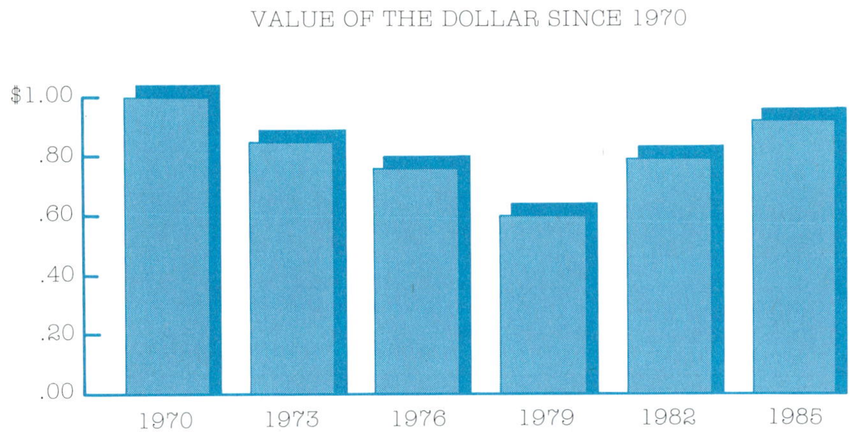

Figure 11-17 Altering the measurement intervals alters (or even obscures) meaning. Compare the impact of this bar graph to that of the one shown in Figure 11-1.

others. You may use lines to separate the text from figures and captions. Use shading, color, and different typefaces to distinguish internal elements.

5. Be sure the graphic aid is *meaningful.* The layout, size, type of graphic aid, symbols, titles, and labels—all must support the point you are making. For instance, return to the bar chart shown in Figure 11-1. The writer wanted to emphasize the sharp drop in the dollar's value from 1970 through 1979 and its encouraging rise in value since 1980. Consider how much less meaningful the chart would have been if the same scale of measurement had been used on a shorter vertical axis and the width of each bar had been enlarged. The chart would have conveyed exactly the same information, but the *meaning* of that information would have been obscured by the minimal difference between columns. Figure 11-17 demonstrates how the chart would look under this second set of (unfortunate) conditions.

6. Place the graphic aid as close as possible to the point where it is discussed, but never *before* it is introduced. Conventions often mandate that figures and tables be placed at the top or bottom of a text page. If the graphic aid (such as a map) is too large to be incorporated into the text, place it in an Appendix and refer readers to it.

7. Always refer to the figure or table in the text; refer to it by its number.

8. Use the most appropriate graphics tools to produce your visual aids. All of the figures and tables reproduced in this chapter (except the photograph) can be created using the following supplies, which are available at office and art supply stores.

white typing paper (*not* erasable)
grid or graph paper
pens of varying tip thicknesses and colors
No. 2 pencils with gum erasers
press-on shading sheets
press-on lettering
templates (circles, flowchart symbols)
compass
protractor
X-acto knife
rubber cement
clear tape
correction fluid

A WORD ABOUT COMPUTER GRAPHICS

With the exception of photographs, all of the figures and tables described in this chapter could also have been developed with the help of computerized graphics systems. In your future jobs, you will almost certainly have an opportunity to work with computers in generating your graphic aids. In fact, you may already be familiar with some of the possibilities. With certain systems, for instance, you can "paint" the pictures you want, using only a touch-sensitive pad and a stylus. And simply supplying the values to the computer with a preprogrammed graphics capability lets you create bar and line charts.

Table 11-4, for instance, was generated by asking the computer system for a report on the latest sales figures for certain geographic regions. When the same data were transported into a graphics program, a bar chart was also produced automatically (Figure 11-18).

As Table 11-4 and Figure 11-18 illustrate, computers can save development time and increase the accuracy of graphic aids. Some computer programs allow users to perform calculations that are then translated into graphic representations. Data can be manipulated easily and changes effected immediately. The innovations of computer science allow the user to rotate the perspective on a subject, change color, and alter dimensions. With a minimal investment of time and energy, the user can try several alternative representations of information on the display screen before producing a hard copy of the best version.

SALES REPORT — SECOND HALF 1985	Chicago	New York	Atlanta	Dallas	San Diego	Seattle
July	$10,000	$16,000	$ 8,000	$10,000	$20,000	$ 8,000
August	13,000	11,000	15,000	12,000	13,000	7,000
September	18,000	13,000	20,000	11,000	17,000	9,000
October	21,000	12,000	16,000	10,000	12,000	8,000
November	22,000	10,000	16,000	11,000	15,000	11,000
December	12,000	8,000	10,000	12,000	13,000	7,000
Totals	$96,000	$70,000	$85,000	$66,000	$90,000	$50,000

Table 11-4 A computer query prompted the computer to build this table automatically.

In the engineering sciences especially, computer graphics can be important on-the-job tools as well as aids in documentation. Initial design work, for instance, can be done more efficiently when development procedures are automated. Engineers can revise existing designs and model alternative designs without the traditional investment of time and effort. The costs of design preparation are reduced drastically when computer-assisted design applications are available.

Even though computer graphics save production time and dollars, they should conform strictly to the standards set forth in this chapter. They should be unified, balanced, clear, and meaningful. You, not the computer, will decide whether its output is good enough.

When computer graphics equipment is at your disposal, the temptation to use visual aids frequently in documents is strong. Exercise discretion. Remember to consider the purpose of the document and the needs of your audience. Apply stringent selection criteria in choosing the right kind of graphic aid for either quantitative or qualitative information, just as you would if you were developing visual aids manually. The overuse of graphics defeats their purpose and distracts the reader from the narrative portion of the document.

CHAPTER SUMMARY

Graphic displays of information can help boost audience interest as they clarify the text. The reasons for using a graphic aid can be as varied as the

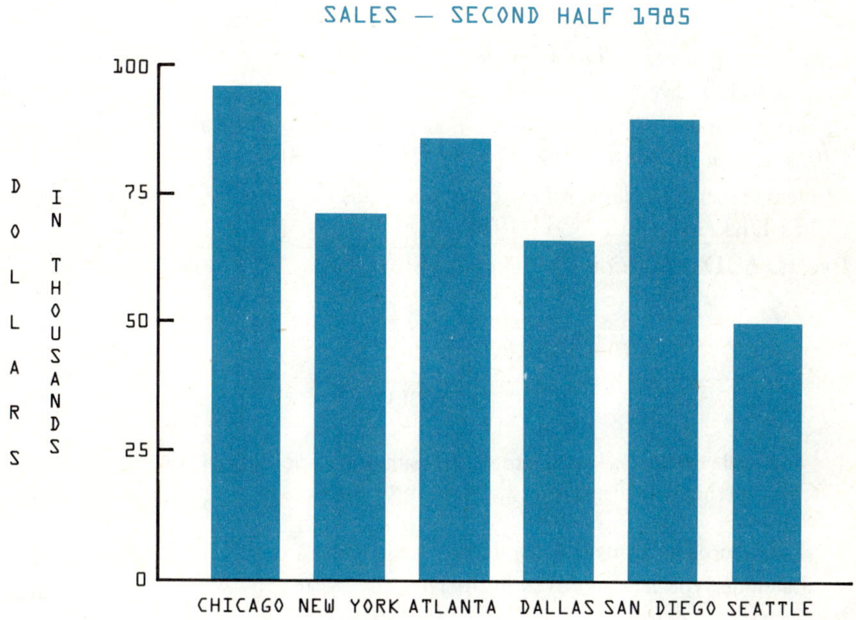

Figure 11-18 This bar graph was built automatically from the information given in Table 11-4.

media available, but the purpose of the document and the needs of the audience should always be considered.

Quantitative information lends itself best to depiction in line graphs, bar graphs, and pie charts, as well as in tables. Qualitative data are best expressed in tables, flowcharts, organization and time charts, line art and diagrams, and photographs. Most of these methods can be implemented manually with computer assistance. The most important advice to remember about the use of visual aids is "Keep it simple."

ADDITIONAL READING

The following short bibliography will help you learn more about graphic aids.

Broeckhuizen, Richard. *Graphic Communications.* Ft. Lauderdale, Fla.: McKnight, 1979.

The Camera. rev. ed. New York: Time-Life Books, 1981.

Feininger, Andreas. *The Complete Photographer*. Englewood Cliffs, N.J.: Prentice-Hall, 1984.

Lefferts, Robert. *Elements of Graphics: How to Prepare Charts and Graphs for Effective Reports*. New York: Harper & Row, 1981.

Meyers, Cecil H. *Handbook of Basic Graphs: A Modern Approach*. Belmont, Calif.: Dickenson Pub. Co., 1970.

Pyeatt, A. D. *Technical Illustration*. 2nd ed. New York: Higgins, 1960.

EXERCISES

1. For each of the following topics, create a situational analysis. Develop two different methods of displaying the same information.

- allocation of your monthly budget
- the grades you have received in the past three years, by subject

2. Return to each of the writing assignments you have completed for this course. Determine where your communication of information would benefit from visual presentation as well. Decide which type of graphic aid to use in each case, and be prepared to defend your choices.

3. Examine a textbook, scholarly journal, popular magazine, or report in your field. Which kinds of graphic aids are used most often? Identify the purpose of each example you encounter. How effective is the example? Would another kind of graphic aid have been more appropriate? Why or why not?

4. From a table of information you locate in a field-specific document, create at least two other graphic aids (just as the Salem arrest and conviction records appear in several forms in this chapter). Explain the circumstances (writing situation) in which each would be preferable to display in a table.

EDITING CRITERIA

The following editing criteria, in conjunction with those concerned with the mechanics of writing (see the end of Part I), should be applied to each of your writing assignments *before* you submit it for workshop analysis or for grading. Your fellow students and your instructor will use these same criteria in their evaluations of your work. Criteria specific to each technique we have discussed appear after the set of general concerns.

GENERAL CONCERNS

1. Is the situational analysis appropriate to the assignment? If not, why not? What changes should be made? How will these changes affect the document?
2. Does the writer maintain a consistent role? If not, identify areas of inconsistency and make corrections.
3. Is the level of detail appropriate to the needs of the audience? If not, what changes are required?
4. Is word choice appropriate to the writing situation? If not, identify and correct the problems.
5. Will the audience understand this document? If not, why not? Identify and correct places where an explanation should be expanded or where additional information is required.
6. Does this document adhere to all format specifications? If not, identify and correct all weak areas.
7. Is the topic being discussed clearly identified at the outset? If not, make necessary changes.
8. Is the arrangement of information appropriate to the writing situation? Why or why not? If not, make any necessary changes.

CLASSIFICATION, DIVISION, AND PARTITION

1. Is movement from general to specific (or vice versa) clear? If not, identify and correct all weak areas.
2. Are the means by which categories are formed clearly identified and valid? If not, identify and correct weak areas.
3. Is only one founding premise applied at each level of complexity? If not, identify and correct weak areas.
4. Does the classification, division, or partition include a sufficient number of categories? If not, what additional categories should be included and why?
5. Is each category distinguished from all others? If not, is the overlap explained and acceptable? If neither is the case, identify and correct all weak areas.

6. Are the same points discussed within each category? If not, are omissions justified? If neither is true, identify and correct all weak areas.

7. Are the points in each category discussed in uniform order? If not, is the variation justified and explained within the document? If neither is true, identify and correct all weak areas.

8. Is each category sufficiently discussed in light of the total writing situation? If not, why not? Identify and correct all weak areas.

9. In a partition system, if parts have been "exploded," are they reunited as a whole toward the end of the document? If not, is such a tactic justified? If neither is true, identify and correct all weak areas.

10. If only one category of the several that are possible is the focal point of the document, is ample justification provided in the document? If not, identify and correct all weak areas.

DEFINITION

1. Is a clear three-part definition statement presented at the outset of the document? If not, identify and correct all weak areas.

2. Within the three-part definition statement, is the term being defined clearly stated? If not, state it clearly.

3. Within the three-part definition statement, is the general category into which this term fits clearly stated? Is it accurate? If either of these is untrue, identify and correct all weak areas.

4. Within the three-part definition statement, are distinguishing traits clearly stated? Do they provide an ordering discipline for the discussion? If either of these is untrue, identify and correct all weak areas.

5. Is an appropriate combination of definition techniques used (such as etymology, institutional, and exemplification)? Why or why not? Identify and correct all weak areas.

6. On the basis of this document, would the audience understand the term and concept well enough to apply it in its proper context? Why or why not? Identify and correct all weak areas.

DESCRIBING PROCESSES, PROCEDURES, AND ACTIVITIES

1. Is the process, procedure, or activity clearly stated at the outset of the document? If not, identify and correct all weak areas.

2. Is a three-part definition statement included at the outset of the document? If not, is its omission justified? Why or why not? Identify and correct all weak areas.

3. Within the three-part definition statement, is the name of the process, procedure, or activity clearly stated? If not, identify and correct all weak areas.

4. Within the three-part definition statement, is the category of process, procedure, or activity stated (such as manual, natural, or hydraulic)? Is it accurate? If either of these is untrue, identify and correct all weak areas.

5. Within the three-part definition statement, are phases of action clearly identified? Do they provide an ordering discipline for discussion? If either of these is untrue, identify and correct all weak areas.

6. In the case of the description of a process, is sufficient causal analysis conducted? Why or why not? Is the causal analysis accurate? If either of these is untrue, identify and correct all weak areas.

7. In the case of the description of a procedure, are reasons for actions supplied to suit the needs of the audience? When contingencies or alternatives are appropriate, are they supplied? Is an evaluation presented in the conclusion? If any of these is untrue, identify and correct all weak areas.

8. In the case of the description of an activity (set of instructions), is the overall mood imperative? Are single action verbs used to begin each command? Does each command follow the pattern "Do this to that with that"? Are all materials and equipment listed and all required warnings and cautions included? Is the title succinct and accurate? Is the sequence of actions accurate? If any of these is untrue, identify and correct all weak areas.

9. In the case of the description of a process, is the audience able to understand the process fully and appreciate its significance on the basis of this document? If not, identify and correct all weak areas.

10. In the case of the description of a procedure, are audience members able to understand the procedure fully and would they be able to supervise its execution? If not, identify and correct all weak areas.

11. In the case of an activity description, is the audience able to perform the activity solely on the basis of the information provided? If not, identify and correct all weak areas.

DESCRIBING MECHANISMS

1. Is the specific mechanism identified clearly at the outset of the document? If not, identify and correct all weak areas.

2. Is a three-part definition statement supplied at the outset of the document? If not, is its omission justified? If neither of these is true, identify and correct all weak areas.

3. In the three-part definition statement, is a function-specific name of the mechanism provided? If not, identify and correct all weak areas.

4. In the three-part definition statement, is an accurate general category of mechanism provided? If not, identify and correct all weak areas.

5. In the three-part definition statement, are functional phases of the activity presented? Do they provide an ordering discipline for discussion. If either of these is untrue, identify and correct all weak areas.

6. Are relationships between parts clearly defined and accurate? If not, identify and correct all weak areas.

7. If an evaluation of the mechanism is provided, does it enable readers to appreciate the functional and physical attributes of the mechanism? Why or why not? If not, identify and correct all weak areas.

8. On the basis of this document, can the audience understand the workings of the mechanism? Why or why not? Identify and correct all weak areas.

ARGUMENTATION

1. If argumentation techniques are used, are they warranted? Why or why not? Identify and correct all weak areas.

2. Is the proposition that is being argued clearly identified at the outset of the document? If not, identify and correct all weak areas.

3. Are all related issues answered to in the document? If not, are omissions justified? If neither is true, identify and correct weak areas.

4. If it deals with a proposition of fact, does this document take a generally deductive approach to presenting evidence? If not, identify and correct all weak areas.

5. If it deals with a proposition of policy, does this document take an inductive approach to presenting evidence? If not, identify and correct all weak areas.

6. Is the evidence adequate to convince readers that the proposition is probably correct? Why or why not? If not, identify and correct all weak areas.

7. Are all arguments conducted logically? If not, identify and correct all weak areas.

GRAPHIC AIDS

1. Are graphic aids used each time there is a need? If not, identify and correct all weak areas.

2. Are the graphic aids that do appear in the document justified in light of the situational analysis? If not, identify and correct all weak areas.

3. Is each graphic aid presented in the best possible medium in light of the situational analysis? If not, identify and correct all weak areas.

4. Are all graphic aids clear? Are they unified? Are they balanced? Are they meaningful? If any of these is untrue, identify and correct all weak areas.

5. Are graphic aids integrated into the text in that the writer refers to them and discusses them sufficiently? If not, identify and correct all weak areas.

PART III

Applications

CHAPTER 12

Letters and Memos

CHAPTER PLAN

Through letters and memos, writers disseminate up-to-date information, ask or answer important questions, make complaints or adjustments, forward suggestions and proposals, report on findings, present plans, explore possibilities, document events, establish dialogue, record transactions, summarize meetings, and make announcements. Effective correspondence can mean the difference between approval or rejection of your ideas and action or inaction on your proposals.

To help you develop strong correspondence skills, this chapter will

- examine the conventions and formats associated with letters and memos.
- explain how to get and maintain reader interest.
- introduce the "You" attitude.
- explore the various purposes and styles of different types of letters and memos.

CONVENTIONS OF LETTERS AND MEMOS

Most people in business positions regularly use letters and memos to communicate. In general, letters communicate information to or from people *outside* the organization, while memos communicate information to people *within* the organization. Although the telephone also plays an important role in business communications, many people do not consider a proposal, complaint, transaction, or message to be official unless it is written down. Many organizations, in fact, require their employees to maintain files of letters and memos that serve as permanent records.

LETTERS

Letters and memos differ in format as well as audience. Although particular organizations may have a preferred format, the three most popular letter formats are block, indented, and modified block.

In the block style (Figure 12-1), all lines begin flush with the left margin, including the return address, date, inside address, salutation, and complimentary close. In the indented style (Figure 12-3), all paragraphs begin five spaces to the right of the left margin. Subsequent lines within the paragraph are flush left. The return address begins at the center of the top of the letter, as does the date. The inside address is flush left, as is the salutation. The complimentary close and typed signature line are aligned with the return address. Modified block style (Figure 12-4) is a combination of elements from both other format styles: paragraphs begin flush left, as in the case of block style; however, the other elements (date, complimentary close, typed signature line) follow the format for the indented style.

Letters should always be typewritten or produced on a word processor. They should be single spaced and typed or printed on 8½″ by 11″ white bond (nonerasable) paper. Side, top, and bottom margins should be at least one inch. In addition, all letters should contain the following standard elements:

- letterhead or return address
- date
- inside address
- salutation
- body
- complimentary close
- signature

Letters may also include other, optional items below the signature line, such as the identifying initials of the author and typist, an enclosure or attachment statement, and a list of people receiving copies (see Figure 12-1).

1. *Letterhead or return address:* The letterhead (on preprinted or official stationery) or the typed return address contains the name of the organization (if appropriate), the full street or post office box address, and the city, state, and zip code. (A zip code directory is available at every post office.) In general, do not use abbreviations for the name of the city or state or for words like *street, road,* and *boulevard.* If, however, the recipient of your letter is a close associate and the tone is friendly or casual, it is permissible to use standard abbreviations.

2. *Date:* The date appears four lines below the return address. It can be written in either of two ways, but always spell out the full name of the month:

December 30, 1986
30 December 1986

3. *Inside address:* The inside address appears four lines below the date and should be the same as the address used on the envelope. It cites the name of the person to whom the letter is written, his or her title, the full name of the organization for which this person works, the full street or post office box address, city, state, and zip code. If you are not addressing a specific person, as in the case of making a complaint or returning merchandise to a manufacturer, use the logical department name (e.g., "Complaint Department"). If you do not know the gender of the person to whom you are writing, write the name without a Mr., Mrs., Miss, or Ms.

4. *Salutation:* Convention requires that letters addressed to individuals begin with a greeting statement, such as "Dear Mr. Smith," followed by a colon (:). Again, if you do not know the gender of the reader, use just the name ("Dear R. J. Smith:"). If you are writing to a department, you can use the department name ("Personnel Department:") without the word *Dear,* or you can omit the salutation entirely and simply begin the body of the letter. It is no longer accepted practice to use the salutations "Dear Sir," "Dear Sirs," or "Gentlemen." These greetings imply that only males receive letters. Although it is still acceptable to use the greeting "To Whom It May Concern," it is preferable to use an individual's name. Place the salutation two lines below the inside address.

5. *Body:* The body of your letter should begin two lines below the salutation. Letters should be as brief as possible. In fact, try to restrict most of your letters to a single page, if possible. To achieve this brevity, begin

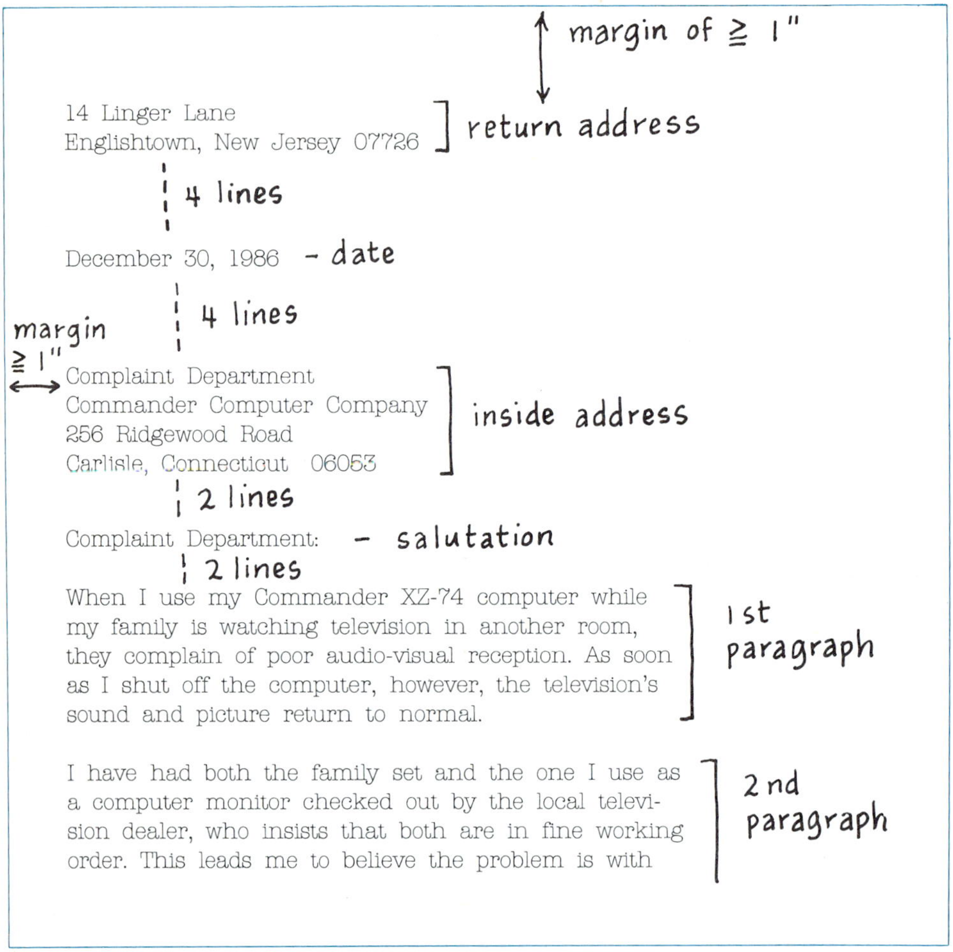
14 Linger Lane
Englishtown, New Jersey 07726

December 30, 1986

Complaint Department
Commander Computer Company
256 Ridgewood Road
Carlisle, Connecticut 06053

Complaint Department:

When I use my Commander XZ-74 computer while my family is watching television in another room, they complain of poor audio-visual reception. As soon as I shut off the computer, however, the television's sound and picture return to normal.

I have had both the family set and the one I use as a computer monitor checked out by the local television dealer, who insists that both are in fine working order. This leads me to believe the problem is with

paragraphs with a strong topic sentence followed by three or four sentences of expansion, specific data, or questions. Closing paragraphs should be no more than two or three short sentences. Skip a line between each paragraph.

6. *Complimentary close:* Two lines below the closing paragraph is a word or phrase called a complimentary close. Some traditional endings are "Sincerely" or "Yours truly," which are suitable for most situations; "Cordially" or "Regards," used with friends or close business associates; and "Respectfully," which is appropriate only when the recipient of the letter is in a position that commands formal respect (such as the CEO of a corporation, a political figure, a religious leader, a distinguished professor, or an eminent figure in a particular field).

7. *Signature:* Sign your letters legibly in black ink. In most cases you should use your full name, but for close associates just your first name

your company's device. Since the machine is still under full warranty for parts and service (but no dealer is within a fifty mile radius of my home), I am returning it to you postpaid with this letter. — 2nd paragraph

Although my request may create a scheduling problem for your service department, I must have my repaired computer back within a week. Since I depend on a computer to conduct my business, I simply cannot afford to go longer than that without it. Thank you for your immediate attention to this problem. — closing paragraph

2 lines

Sincerely, - complimentary close

4 lines

William J. West - signature

William J. West - signature line

2 lines

WJW:sd - typist identification

2 lines

Enclosure: Commander XZ-74 computer, serial no.
143389765 - enclosure notice

2 lines

cc: M. Dettman, W. Manning - list of copies

Figure 12-1 The block style is the most widely used letter format.

might suffice. Four lines below the complimentary close, type your full name. Beneath that, type your job title (do not skip a line between your typed name and your job title).

8. *Identification:* Once you have achieved a certain status on the job, you may no longer have to type your own letters; instead, a secretary or clerk may produce them in final form for you. When this is the case, your typist should indicate his or her role in creating the final document two lines below your typed name and title. Your three initials should be capitalized and appear first, followed by a colon and the first and last initials of the typist in lower case. If you are unable to sign the letter, the typist can sign for you and then add his or her initials just beside or below the signature.

9. *Enclosures or attachments:* If you are enclosing additional materials with your letter, indicate this in the lower left margin two lines beneath the

SCANTEST Interoffice Correspondence

standard company form

Date: April 16, 1987 - date

Subject: Cable signal-amplification problems -- Cluster Builders ← subject

To: Clark Milstead - addressee **From:** Peter Nathan — writer

Organization: Field Service
Location: CA 12
Mail Station: T62
Phone: 341-5553

internal address of writer

cc: Jerry Collins
Sam Franklin
Elinor Weston

people receiving copies

Cluster Builders in Pacifica has been having signal-amplification problems that are hindering their networking capabilities. It sounds like the problem could be the cable itself rather than the links. Since Cluster Builders is one of our controlled-release sites (and

typist's initials or your signature. If materials are stapled or clipped to the letter, use the word *attachment*. If the materials are simply enclosed, use the word *enclosure* or *encl*. You may choose to indicate the number of enclosures, or to describe them, by using one of the following forms:

Enclosure
Enclosures (4)
Encl: Résumé

10. *Copies:* Frequently, copies of letters are sent to individuals other than the person addressed directly. When this is the case, the last line of the letter

their satisfaction affects our success in this new marketplace), I think it would be appropriate if you investigated personally--you have the most expertise in signal amplification.

The system is not down; however, it is running in a severely degraded state. Plan on attaching signal boosters to the line as soon as you locate the problem areas. I'll be able to get you some new cable from Santa Teresa as soon as I hear from you. We'll have to supply the new cable free of charge, of course.

I told Mark Johnson of Cluster Builders that you would arrive on Monday. A copy of my letter to him is enclosed. His telephone number is (614) 555-6789. I'll let you take it from here. Call me if you need any help.

Pete　　signature

Encl.: Letter to Mark Johnson　　enclosure notice

Figure 12-2 Memos are used to communicate information internally.

(flush left and two lines below the previous line) carries the abbreviation *cc* followed by a colon and an alphabetized list of the people receiving copies. It is unethical to provide a copy of a letter to a third party without informing the letter's addressee.

MEMOS

Like letters, memos can be used to convey ideas, make complaints or suggestions, present informal reports, record agreements, and inform others of events. Most companies have memo stationery with a preprinted heading separated from the body of the document by lines. This heading usually supplies or provides room for the following information:

- name of company
- date
- identification of subject being discussed
- name of addressee
- name of writer
- internal address of writer
- people receiving copies (arranged alphabetically)

Figure 12-2 shows a typical memo. Note that unlike the letter, it is not necessary to give the address of the person receiving the memo. Nor are a salutation or a typed signature line needed (although the memo *is* signed). Enclosures, however, are indicated at the bottom just as in letters.

GETTING AND MAINTAINING READER INTEREST

Knowing how correspondence generally is read in the work environment will help you prepare effective letters and memos. Most readers leaf through their in baskets to determine which documents are important. They often glance at only the first paragraph of each document, skip to the first sentences of the body paragraphs, and perhaps read the last paragraph completely. Therefore, if it is important to you to have your messages read, it is up to you to capture and maintain your readers' attention.

One of the best ways to attract attention is to make your first paragraph short and informative. Note in Figure 12-1, for instance, how William West briefly describes the problem he is having with the Commander XZ-74 computer. There are no preambles or wordy introductions—just a succinct statement of facts.

If appropriate, use the first paragraph to refer to previous communication. However, that reference should be in a subordinate position rather than being the focal point of the paragraph. Consider the difference in emphasis between the following two examples:

> Replying to yours of May 10, we wish to state that our research staff has been working for a long time on the problem that you mentioned and has finally succeeded in solving it.

> Our research staff has successfully solved the problem of corrosion you asked about in your letter of May 10.

The second example gets directly to the point and subordinates the reference to the reader's letter.

Subsequent paragraphs should develop the ideas presented in the opening remarks. Specific instances, actions, or events should be presented clearly, but because business readers are rushed, these paragraphs should also be short and direct. Supply background information only as it enables the reader to understand the points you are making. (Note how Mr. West, for example, briefly describes the method he used to determine that the computer, not the television equipment, is one source of the problem.) Use action verbs rather than the passive voice. If possible, try to restrict your correspondence to a single page. If the document will be longer than that, provide a summary statement at the outset so that the reader need only read this abstract if he or she is pressed for time. Or you may wish to present your most pressing conclusion or recommendation in the first paragraph. Here is an example of a summary at the outset:

After speaking with Laura Dodge, vice president of manufacturing, I must acquaint you with the following requirements:

1. All research and development budgets for individual departments must be approved on a per-project basis.
2. All training-related travel must be eliminated for the balance of the fiscal year.
3. Employees will not be allowed to carry over their unused vacation days from one year to the next.

Here is an example of a conclusion at the outset:

Severe budget cuts must be implemented for the balance of this year in order to reduce the overage we will otherwise experience.

Additionally, in informal reports or proposals, you might wish to use headings to carry your reader from section to section. If the preceding summary example were expanded, it might use the following headings above each major point of discussion:

Research and Development Budgets

In previous years, research and development dollars have been allocated in lump sums to individual design departments. Unfortunately, with a serious budget overage facing us the remainder of this fiscal year, we must remove *all* unspent R & D dollars from department operating plans. Effective immediately, only approved projects will receive funding. Use form A-90 to solicit approval on a per-project basis. . . .

Travel Budgets

Effective immediately, travel related to training courses must stop. . . .

Vacation Budgets

A new policy preventing employees from carrying over unused vacation days into the next fiscal year goes into effect immediately. We are forced to implement. . . .

The closing paragraph of a memo or letter should always be positive, whether it is directing the reader to act, thanking him or her, making a request, or presenting a suggestion. If the reader is directed to act, the requirement should be stated clearly, rather than merely implied, and in a polite manner. Note how, for instance, Mr. West gives the computer manufacturer a deadline for repairing the equipment. However, rather than making an angry demand, he supplies a credible reason for his request.

THE "YOU" ATTITUDE

As you have already learned, knowing as much about your audience as possible is important to the effectiveness of any document. In the case of correspondence, this is especially true because you are usually communicating with a specific individual or group. In the case of a letter, that audience is outside your organization; in the case of a memo, your reader belongs to the same group that you belong to.

Unlike most of the technical writing you have done thus far, correspondence speaks directly to the reader through the use of conversational language and a personal, courteous tone. This combination of qualities is known as the "You" attitude—an approach that displays a concern about the individual's needs and interests. The most distinctive characteristic of this writing style is its use of personal pronouns.

For instance, in Figure 12-1, William West uses a conversational, personal tone even though he is addressing a complaint department rather than a single person. We see such phrases as "your company," "I am returning it to you postpaid," and "thank you for your immediate attention." In the memo from Peter Nathan (Figure 12-2), we see such phrases as "you investigated it personally," "get you some new cable from Santa Teresa as soon as I hear from you," "you would arrive on Monday." Notice that the writer in both of these cases also uses the first-person pronoun, "I," another trademark of this conversational tone.

In addition to personal pronouns, the "You" attitude requires that your message be directed to your reader's greatest concerns or interests. In his letter to the complaint department (Figure 12-1), for example, William West correctly determined that his reader's most important concern would be the complaint itself—poor audio-visual reception on a television set. He then provides the proof that the computer is the cause of the problem and concludes his letter with a call to action.

In the memo (Figure 12-2), Peter Nathan explains the problem to Clark Milstead and suggests its cause. In addition, he tries to motivate him by explaining the importance of the site and by complimenting his competency (thus helping to ensure his cooperation).

Vocabulary and level of detail should be geared to the educational level and experience of your audience. Since he is addressing a lower-level staff member in a complaint department, West avoids any technical description of his problem. In contrast, Nathan is fully aware of Milstead's expertise and does not hesitate to use applicable terminology.

A writer also reflects the "You" attitude by using a professional, courteous tone. Rather than conveying his annoyance, William West employs an objective tone that demonstrates his concern. Peter Nathan speaks casually yet respectfully to his associate, offering assistance as needed, although the urgency of the problem is not overlooked. When writing correspondence, think of yourself as a representative of your organization. You must reflect both sincerity and competence. If you handle the situation well, readers will think well of your company. If you are rude or abrupt, readers will not be satisfied. As we look at different types of correspondence, you will be able to see the importance of the "You" attitude more clearly.

THE MANY PURPOSES OF CORRESPONDENCE

As mentioned earlier, the reasons for writing a letter or memo are as diverse as business situations themselves. The most common types of correspondence we will examine are letters on memos of inquiry or response, complaint or adjustment, announcement, report, proposal, transmittal, sales, and application. We will discuss transmittal letters in Chapter 14 because they are a component of the formal report. Sales letters will be covered in Chapter 16, where we discuss how to promote products and ideas. Finally, we will cover application letters in Chapter 18, where we focus on finding a job in a technical field.

LETTERS AND MEMOS OF INQUIRY OR RESPONSE

As their name suggests, letters or memos of inquiry pose a question or make a specific request. They are short, direct, and polite. Questions or requests can be posed simply:

> Can you tell me when the construction on McClellan Road will be completed? I have to commute on this road and am wondering how much longer I will be inconvenienced.
>
> Please send me a copy of your annual report to consider as I develop my investment portfolio.

Or they can be more technical:

> How have the new performance figures for the Commander XZ-74 computer been calculated? My own system doesn't even approach those levels.

Address the inquiry letter or memo to the person who is best able to provide the information required. If you are writing a letter and do not know the name of the right person, find out by phoning the company and asking. When an inquiry is sent to the wrong person or organization, it is likely that the response will be a long time coming, if it comes at all.

The inquiry letter or memo should be particularly brief. After all, you are asking a favor, and you certainly would not want to take more of the reader's (favor grantor's) time than is appropriate. The minimal information to be included is:

- an explanation of who you are (if your reader does not know you).
- the request itself.
- an explanation of how you will use the requested information.

The response to such a letter or memo should be helpful and courteous even when the request is denied.

Figure 12-3 illustrates the inquiry letter. (It also illustrates the indented letter format.) In Figure 12-4, you can see how the response typically is handled. (This example illustrates the modified block letter format.)

LETTERS AND MEMOS OF COMPLAINT OR ADJUSTMENT

If you have not already been one, someday you might be an unhappy customer, a disgruntled employee, an irate citizen, or a dissatisfied user. To get results, you might be inclined to write a letter of complaint. Worse yet, you might be the person who receives complaint letters from such individ-

886 Brittany Way
Tempe, Arizona 85037

January 15, 1987

Mr. Joseph McDaniel
Arizona Solar Energy Commission
Capitol Tower, Room 502
1700 West Washington
Phoenix, Arizona 85007

Dear Mr. McDaniel:

I am a construction engineering student at Arizona State University and am writing a report on house plans suited to passive solar technologies. I understand from Steven Kline, a local architect, that he has recently written a report for your agency on just that subject. He recommended I request a copy directly from you, since he did not feel he was authorized to distribute the report.

Would you please send me a copy of this report? From Mr. Kline's description of it, the report will prove particularly helpful to me. I would be more than happy to send you a review draft of my report before its final submission to ensure that I have used the information correctly.

Thank you for your help. I look forward to receiving the report as soon as possible. I am enclosing a check for $5.00 to cover postage and any charge there might be.

Yours truly,

Carol D. Hill

Carol D. Hill

Encl.

Figure 12-3 The inquiry letter is polite, direct, and brief.

ARIZONA SOLAR ENERGY COMMISSION
Capitol Tower, Room 502
1700 West Washington
Phoenix, Arizona 85007

January 19, 1987

Ms. Carol D. Hill
886 Brittany Way
Tempe, Arizona 85037

Dear Ms. Hill:

Here is the report Steven Kline wrote for our agency, entitled "The Feasibility of Incorporating Solar Energy and Conservation Technologies in Local Home Designs and Construction." I hope it will be useful to you. May we recommend that you also consult the fine solar collection in the Science and Engineering Library at ASU. It has one of the largest holdings on this topic in the country.

I am also returning the check you sent. The report is yours free of charge. Although it is not necessary for you to send us a review draft of your report, we would very much like to have a copy of the final report for our own collection.

Best of luck in writing your report. If you have any questions about the enclosed report, please feel free to call.

Sincerely,

Joseph McDaniel
Solar Engineer
(602) 555-7986

JRM:fb

Enclosures (2)

Figure 12-4 The response to a letter of inquiry should let the reader know right away whether the request is granted.

uals. Whether you are the writer or the reader, the complaint scenario requires clear-headedness, diplomacy, and professionalism if problems are to be resolved.

Despite a natural tendency to rant and rave, it is best to simmer down *before* composing a letter or memo of complaint. Recipients of letters containing harsh language, hyperbole, or threats are far more likely to take their time in responding. No one likes to deal with an unpleasant situation. On the other hand, recipients of letters with a clear, rational analysis of the problem are more willing to explore the problem thoroughly and provide a solution in a timely manner.

To write a letter or memo of complaint, follow these steps:

1. Use a serious, professional tone rather than an angry, hostile one.
2. Describe the problem clearly, in concrete, verifiable terms.
3. Explain how long the problem has been occurring.
4. Cite any remedies you might have already taken to resolve the situation.
5. State *exactly* what you expect the person reading the letter or memo to do.
6. Provide a reasonable time frame in which the reader of the letter or memo can act.

Figure 12-1 demonstrates a mature means of expressing a complaint. West explains the problem directly and concretely, briefly outlining how he has come to the conclusion that the problem lies with the computer equipment. Despite his politeness, it is quite clear that he expects the company to repair the faulty system—and to do so in a very few days.

The response to a complaint letter or memo must re-establish good will and build confidence. Figure 12-5, an answer to West's letter of complaint, demonstrates how these objectives can be accomplished.

To write an effective adjustment letter or memo, follow these simple steps:

1. Cite the status of the problem resolution.
2. Explain the reason for the problem.
3. Thank the reader for bringing the problem to your attention.

If the problem is not the fault of the manufacturer, that should be explained carefully. If possible in such cases, you should suggest what the cause might be and what actions the complainant might take to resolve the issue.

It is also possible that solving the problem will take considerable time. Rather than let the complainant sit and stew, write a letter or memo explaining the status of the complaint and provide an estimate of when the problem is likely to be resolved. Figure 12-6 illustrates this tactic.

COMMANDER COMPUTER COMPANY

January 12, 1987

Mr. William J. West
14 Linger Lane
Englishtown, New Jersey 07726

Dear Mr. West:

Your Commander XZ-74 computer has been repaired and is being returned to you with this letter. We regret any inconvenience the absence of your computer may have caused you, but we are confident the problem will not recur. To help compensate you for your problem, we are including a discount coupon worth 25% off any Commander software package you like.

The problem with your system was in the cord connection between the keyboard and the TV you use as a monitor. Because the cord was unnecessarily long, the television signal coming from the monitor was able to leak out and interfere with the other television the family was watching. By shortening the cord, we were able to eliminate the problem.

We thank you for bringing this matter to our attention. The product developers have been alerted to the problem and are making modifications for future product releases.

Sincerely,

Patrick F. Marlowe

Patrick F. Marlowe
Manager, Customer Relations

PFM:rl

Enclosures (2)

256 Ridgewood Road / Carlisle, Connecticut 06053 / (203) 555-1492 / Telex: ComCo

Figure 12-5 The adjustment letter must meet the reader's needs.

Bondurant Construction Company
81 East Putnam Avenue
Smithville, Illinois 99412

February 18, 1987

Mrs. Marjorie Hollingsworth
42 Bluebird Lane
Meadow Park, Illinois 99403

Dear Mrs. Hollingsworth:

We regret that the construction of our new medical facility across the street from your home is causing excessive dust levels in your house and yard. The project will continue until at least May of this year, and this condition may persist until that time.

In such a large construction site, it is nearly impossible to prevent the distribution of dust to neighboring areas. However, we have instructed the construction crew to wet down the area on a more frequent basis than they have been doing. Although this will help the problem, it will not eliminate it.

It is also likely that the problem will be minimized as the work moves from the preliminary stages of ground breaking and foundation laying to the less uprooting stages of work within the building's frame. Work of that nature will likely begin during the next two weeks.

Again, we are sorry for any discomfort the construction is causing you. We will make every effort to limit the dust. Thank you for your patience.

Sincerely,

Robert O'Malley

Robert O'Malley
Construction Supervisor

Figure 12-6 Even when the situation cannot be resolved quickly, the complainant deserves a prompt reply.

SCANTEST Interoffice Correspondence

Date: April 22, 1987

Subject: Cable signal-amplification problems -- Cluster Builders

To: Peter Nathan

From: Clark Milstead
Organization: Field Service
Location: CA 12
Mail Station: T62
Phone: 341-2653

cc: Jerry Collins
Sam Franklin
Elinor Weston

I just returned from Cluster Builders in Pacifica and wanted to update you on the situation there. The system was running in a highly degraded mode by the time I arrived on Monday, and many files had actually been damaged. Mark Johnson was quite cordial

LETTERS AND MEMOS OF ANNOUNCEMENT

One of the simplest letters or memos to write is one in which a specific announcement is made to readers. Occasions requiring such correspondence include the appointment of an individual to a new post, the introduction of a new product or service, and the announcement of a meeting. In these straightforward messages, tell your readers the announcement in the very first sentence. Follow with any appropriate particulars that qualify the announcement or further explain it. Here are excerpts from memos announcing various events.

to me, but I could tell that the Cluster people are very angry. Can't say I blame them, either.

The problem really is our fault. The cable was defective, just as you suspected. However, the signal variance had already caused the file damage because our surge protectors were not working correctly either.

The Santa Teresa site is shipping the proper cable to Pacifica, and I will supervise its installation tomorrow. The surge-protection problem, on the other hand, is not going to be solved by a new cable. That is going to take more development work, and I request that you ask Jerry Collins to go down to Pacifica with me tomorrow to check out the problem.

Given the delicate nature of this controlled-release site, I am sure you will agree with me. Please call me today if you have any questions.

Clark

Figure 12-7 Informal reports can be presented in correspondence.

I am pleased to announce the appointment of Mary R. Miller to the position of vice president of the Fabrication Technologies Division, effective immediately. Mary has been with Stellar Corporation for the past fifteen years, holding such posts as director of Fabrication Technologies, chief design engineer, and international delegate to the User Association.

Effective fourth quarter 1987, Product Management announces the availability of MAGIC 8 fourth-generation software, offering systems development capabilities to the spectrum of users throughout an organization. . . . Traditional programming languages such as COBOL, FORTRAN, and

Bondurant Construction Company
81 East Putnam Avenue
Smithville, Illinois 99412

December 12, 1986

Mr. Bernard Klemsky
44320 West Bertram Drive
Shaker Heights, Illinois 99458

Dear Mr. Klemsky:

We appreciate the opportunity to provide you with our bid for the construction of a carport attached to your home. As I mentioned when I inspected your home, the sandy loam upon which the carport would be constructed requires special techniques to avoid excessive settling of the concrete slab. Additionally, the current slope of the house's roof will require retrofitting before we can attach the carport roof to it. I am sure you would agree with us that the carport must be well engineered to withstand the severe winters we experience in this area.

Plan of Action

I. Preparation: Grading of lot to balance area upon which carport will sit. Use of wire mesh underpinning prior to laying of concrete slab. Retrofitting of roof to strengthen its arch and gutter system.

II. Construction: Laying of concrete slab. Erection of masonry side and back walls. Extension of roof and its gutter system.

III. Completion: Roof tiling. Wall painting. Cleanup.

Figure 12-8 The informal proposal lends itself well to letters and memos.

Klemsky - 2 - December 12, 1986

Schedule

Start Date: December 19 (dependent on weather)

Construction: December 23: laying of slab
December 27: erection of masonry walls
December 30: extension of roof and gutter system

Completion: January 2: roof tiling
January 5: wall painting
January 6: cleanup

Qualifications

Bondurant Construction Company has been in business for over 25 years. Our construction crews are all licensed, experienced professionals who take great pride in their work. We recommend you speak with your neighbor, Mr. Steven Smith, at 44100 West Bertram Drive. We recently completed a covered patio and attached greenhouse for him.

Cost Estimate

Materials:	$1800.00
Labor:	1100.00
	$2900.00

Terms: Due upon completion of project. If paid in full within 45 days, no interest will be applied. If you wish to pay in four equal monthly installments, you can do so at a 15.5% interest rate. All overdue payments are subject to a 21% penalty interest charge.

Conclusions and Recommendations

Because of the severe winters in the area, we believe that the property value of your home will be increased significantly when

Figure 12-8 (cont.)

Klemsky - 3 - December 12, 1986

you provide protection for an automobile. Hence, the cost of this construction can be built into the selling price of your home should you choose to place it on the market, as you had mentioned doing.

We recommend you begin the work by the start date quoted above, or winter weather may prevent construction until the spring of 1987.

Thank you once again for your consideration. We are confident we can build you an attractive, durable carport in a cost-effective, timely manner.

Sincerely,

Henry Trimble

Henry Trimble
President

Figure 12-8 (cont.)

BASIC lack the functionality and flexibility to satisfy today's groundswell of information requests. As a result, data processing professionals are overwhelmed, and knowledge workers are frustrated. The requirement exists for a higher-level language usable by both the DP and end user community alike. MAGIC 8 is built to serve such a need. By design, the core product is easily learnable and can improve productivity throughout the ranks of an organization.

Please plan to attend a meeting about field technical support to be held on Wednesday, August 26, 1987, in the Roadrunner Conference Room at 1 PM.

The attached agenda will be followed; please come prepared for your portion of the program. Thank you.

REPORT LETTERS AND MEMOS

At times, informal reports are presented in letters or memos. The subjects of such correspondence can include meeting results, trip results, investigation of problems, call follow-up, and situation analysis.

Informal reports should be held to one or two pages and should employ the "You" attitude. As mentioned earlier, it may help to use headings to organize the information and guide the reader through the document.

The opening paragraph of a report letter or memo introduces the subject and the purpose of the report, while additional paragraphs provide supporting detail in descending order of importance to the reader (most important to least important). The closing paragraph should present the most important conclusions and recommendations.

Figure 12-7 is the trip report filed by Clark Milstead (see Figure 12-2) after analyzing the situation at Cluster Builders. Here we can see a conversational tone coupled with a brief yet highly informative presentation.

PROPOSAL LETTERS AND MEMOS

In situations where a brief proposal is in order, a letter or memo may be appropriate. This brief version of a proposal provides the same kind of information you would expect in a formal proposal (see Chapter 13) but in much abbreviated form and with a "You" attitude. When preparing an informal proposal, include the following units of information:

- statement of the problem
- background information
- statement of objectives
- possible solutions
- plan of action
- qualifications
- facilities and resources
- schedule
- costs
- conclusions and recommendations

Letter proposals are common in bid situations for small or short-lived jobs. For instance, a home owner may ask local contractors for written bids for the construction of a carport. The letters that come in might look very much like Figure 12-8.

CHAPTER SUMMARY

Letters and memos are distinct from other kinds of technical writing because they use direct address, reflect the writer's personality, and serve as good-will vehicles for the organization represented. Although letters are used for external communication while memos are used for internal communication, both generally are succinct, casual, and informative. Each has its own formats and conventions, as well.

Because the "You" attitude requires you to demonstrate concern for your reader and understanding of his or her needs, it is especially important to conduct a careful situational analysis before writing a letter or memo. You must have accurate information if the tone, organization, and level of detail of your letter or memo are to be effective.

Letters and memos have a variety of uses, including inquiry, response, complaint, adjustment, announcement, report, and proposal.

EXERCISES

1. List three problems you might be called upon to solve with a letter in your job. Consider why these problems would arise and possible solutions. Cite the characteristics of the people who might pose each problem. What is most important to them in the solution? Consider your own role in solving the problem. Will you be immediately responsible, or will you be delegating the responsibility to someone else? What kind of personal attitude will you need to portray to the reader? What will be the role of your employer and co-workers in solving the problem? What image must you project for the organization?
2. Using the same three problems, consider how to convey information about your actions concerning each problem to your supervisor. List any action this person will be required to take. If co-workers must be contacted to help you solve the problem, list their duties.

WRITING ASSIGNMENTS

1. Select one of the three problems used in Exercise 1. Write a one-page letter to the troubled reader that will solve the problem and present your organization in a favorable light.

2. Using the same problem, create a memo to either a co-worker or a supervisor in which you initiate problem-solving activity. Use the lists you developed in Exercise 2 to help you.

NOTE: Editing criteria for a range of letters and memos appear at the end of Part III. Apply them to your assignments before you submit them to your instructor.

CHAPTER 13

Proposals

CHAPTER PLAN

The proposal is frequently used in business, academic, and scientific environments to present solutions to problems or to apply for contracts or funding. To help you learn to write effective proposals, this chapter will

- explain the importance of proposals.
- explore the structural components of proposals.

THE IMPORTANCE OF PROPOSALS

A proposal presents a specific solution to an existing problem. Most proposals are solicited, that is, an individual or organization with a problem asks other individuals or organizations to present their suggestions for solving that problem. Ideally, the person or organization offering the best solution will be awarded a contract, granted research funds, or given the go-ahead to implement the proposed solution. Thus, proposals must compete with each other, and the stakes are often high. That is why proposal writing is a key skill.

Many organizations, such as research laboratories, construction firms, developers, engineering and architectural firms, defense contractors, and educational institutions, rely on proposals to help them acquire new projects. If one of these organizations loses a contract because of an ineffective proposal, its financial status can be seriously affected. In contrast, when a contract is awarded because of a solid proposal, it can be the foundation for hundreds of jobs and long-term growth. For example, in 1983 an aircraft manufacturer won a contract to build 200 attack helicopters for the Department of Defense (at a cost of $2 million each). Because of this contract, the manufacturer was able to build a new facility and hire over 800 people. This manufacturer, a small-time vendor in comparison with its competition, won the contract because it had submitted the best proposal—one in which the solution, the suggested implementation, and *the way in which these ideas were presented* were realistic, logical, and effective.

In other situations, a good proposal can make the difference between a laboratory acquiring or being denied funds from a government agency to initiate a study of the possible relationships between nutrition and aging. A good proposal can help ensure that a civil engineering firm is awarded a contract to build a road for a municipality. On the basis of a good proposal, a large manufacturer may choose one company's mainframe computer to be the heart of its information network rather than the computers made by other companies.

In few other forms of communication can success be measured so dramatically. Ideally, if a proposal meets all the specifications provided by the initiator, is well written, and provides creative solutions, it will win out over the competition. Because proposals can be pivotal points in the economic status of an organization, a tremendous burden is placed upon those who write them. Therefore, it is in your best interest to master this form of communication.

THE COMPONENTS OF A PROPOSAL

In most cases, an organization or agency solicits proposals by issuing a request for proposal (RFP), which usually spells out detailed content and format specifications. In turn, companies or individuals competing for the chance to implement a solution develop a proposal that conforms to the RFP—often working on it for months. If the proposal does not conform precisely to these specifications, it may be eliminated from the competition. By standardizing the format, decision makers are free to concentrate on the solutions being suggested. The proposal in these cases is analogous to a controlled experiment in which as many of the variables as possible are made consistent.

There is such a thing, however, as an unsolicited proposal—one that is initiated by the someone who notices a problem and wishes to suggest a solution to it. The audience may not be aware of the problem, nor is it anticipating the proposal. In this second case, the success of the proposal may enhance or alter the problem solver's reputation or career as much as it may help the recipient.

Such a proposal does not necessarily have to conform to a set of specifications. Nonetheless, the writer of this type of proposal is wise to consider the same categories of information that are requested for solicited proposals. Let us examine the key components of both the solicited and the unsolicited proposal:

- front matter
- statement of the problem
- background statement
- statement of objectives
- possible solutions
- plan of action
- qualifications
- facilities and resources
- schedule
- costs
- conclusions and recommendations
- preliminary bibliography
- appendix

FRONT MATTER

Whether solicited or unsolicited, most proposals include various front matter elements, such as a title page and a table of contents. The title page lists the title of the proposal, which should be as clear and descriptive as possible, the author, and the submission date. A security statement like "For internal use only" may also be included.

The table of contents makes the information in the proposal accessible to your readers. Major divisions and subdivisions are listed along with the page numbers on which they begin. If your proposal is more than four pages in length, it's usually a good idea to include a table of contents.

We will take a closer look at front matter elements in Chapter 14.

STATEMENT OF THE PROBLEM

In the solicited proposal, the reader knows what the problem is. It may be to build a particular kind of bridge, design a building, provide adequate computing facilities, improve an educational program, conduct a survey of consumer attitudes, or eliminate pollution in a particular waterway. Why then must the writer state the problem again? The answer is simple: readers must feel confident that the bidding organization *understands* the problem's scope and ramifications. A statement of the problem is *not* merely a reiteration of the words appearing on an RFP. Instead, it is an *analysis* of what the proposal writer understands is the problem. It is standard operating procedure for the firm drafting the proposal to investigate the problem for itself to confirm its details and to offer new insights into it.

Consider these statements about water seepage on a small dam. The first appears in the RFP. The second is written by a local hydraulic engineering firm that has investigated the problem and is bidding on the repair project.

Statement of the Problem

The Carter Dam five miles due east of Smithville has shown signs of water seepage that could potentially breach the pervious walls and present a danger to the families living downstream. This problem was first noticed on December 12, 1986. Since this first indication, the site has been monitored daily for increased seepage. Slight piping has occurred since that time. Hence, the water level has been reduced to stabilize the situation. The lower water level is not desirable, however, so measures must be taken to prevent the seepage.

Statement of the Problem

Through on-site testing and evaluation, this firm confirms that water is seeping through Carter Dam at a rate that allows piping to occur. The piping has remained in its beginning stage since the reservoir level of the dam has been maintained at a relatively low level. The presence of potentially dam-breaching piping is evidence that the material internal to the dam is highly pervious. If the existing seepage problem is allowed to progress to an advanced state of piping and high water velocity at the downstream toe (causing excessive uplift forces), the result could eventually be a break through the dam or overturn of the entire structure. Clearly there exists a threat to downstream families and their property until the seepage is controlled.

Note that each writer uses a clear topic sentence to indicate the nature of the problem. The engineering firm, however, uses the solicitor's statement as a starting point for its own preliminary research. Realistically, the firm could not be expected to bid on a reconstruction project—nor would it want to—without investigating the problem itself. Hence, the engineering firm's statement of the problem demonstrates that its staff has in fact taken the initiative to do some research of its own.

In the unsolicited proposal, the writer is not only describing a problem but also bringing this problem to the attention of the reader for, perhaps, the first time. Here is a statement of the problem from an employee's proposal to the vice president of his company. The company is currently undergoing a quality-improvement process that the employee questions. The employee proposes a means of changing the program to eliminate the problem, thereby improving the program's effectiveness.

Statement of the Problem

As the person in charge of the program, you well know that the company is undertaking an exciting quality-improvement process that involves every one of us. That program is founded on the belief that quality means conformance to requirements. This is a valid equation, but it needs further definition if we are to *apply* it to our day-to-day activities. Specifically, unless we devise a way of developing *good* requirements *that can be measured,* we will not necessarily improve quality. If we start with a weak requirement and conform to it 100%, we will wind up with a weak result. . . .

Note that the employee demonstrates a desire to contribute to the success of the company. He points out a logical fallacy that could have disastrous results but does not antagonize the person responsible for the program. The writer takes care not to invalidate the formula but to enhance it so that it is more applicable.

BACKGROUND STATEMENT

Whether solicited or not, a proposal will probably be read by a variety of people, ranging from the technical experts who helped write the RFP specifications (if the proposal was solicited) to the members of a budget committee. The proposal must be informative to both kinds of audiences, yet each type of reader requires different levels of discussion. From what we have learned about audience analysis and meeting reader needs, we know that such a disparity can create a communication problem. A good proposal solves this problem by providing a background section early in the document. This portion of the proposal can be used to introduce key terms and concepts for nontechnical readers and to demonstrate to the expert that you (and members of your organization) have researched the project seriously. The background section also provides a forum in which you can demonstrate your sensitivity to a problem (most effective in the unsolicited proposal).

In a proposal requesting funds for researching the effects of contraceptives on the nutritional balance of their users, the writer provides the following background information:

Background Statement

The Food and Drug Administration approved the first hormonal oral contraceptive in 1960. Since that time, the production of these drugs and their varieties has increased dramatically. Now more than 10 million women use oral contraceptives on a regular basis. Recent evidence published in the *New England Journal of Medicine* (vol. 23, issue 12) indicates that a high proportion of these women suffer from nutrient deficiencies. At the same time, dangerously high blood levels (toxicities) of other nutrients are suspected. In fact, no effective oral contraceptive has been developed that does not have metabolic effects. Many of these reactions have negative nutritional implications. Blood sugar, lipids (fats), and protein, as well as amino acid levels, are all affected. Blood levels of many, if not all, of the vitamins and minerals are affected. For example, very deficient blood levels of folic acid have been reported. Also reported are high vitamin A levels that border on toxicity. . . .

The information supplied in this background statement demonstrates the writer's familiarity with the problem and is therefore best placed after the statement of the problem. This serves as a good rule of thumb in ordering the elements of a proposal if the RFP does not specify a particular arrangement. In addition to explaining technical terms (e.g., *lipids*), this background statement surveys existing information to help build the writer's case for continued research in this field.

In the following background statement from an unsolicited proposal, the emphasis is on demonstrating the writer's sensitivity to the reader's problem. Ultimately, this writer will propose a set of software packages to solve the data processing manager's backlog problems.

Background Statement

Since the mid-seventies, the number of end users in any given organization has grown significantly. Data processing professionals know this better than anyone else because they are responsible for keeping these inexperienced computer users satisfied—a complex and difficult task, at best.

Virtually every department within a large organization depends upon the data processing center to some degree. This dependence can sometimes be a source of frustration for both end users and DP staff members. The end users may not appreciate the complexity of their requests and their needs are not always clearly defined. They don't always understand that they must wait for their reports to be developed. On the other hand, to meet user needs, the applications programmer might have to create numerous iterations of a program before the users are satisfied with a report's format and data. This process takes time.

The data processing staff wants to serve its organization's nontechnical users as best as it can. But, often it isn't possible to do so without interrupting other, equally important work.*

The writer of this background section focuses on the reader's "hot button"—the issue that most bothers him or her. An adequate solution to this problem would indeed be of interest; therefore, the reader's attention is now captured, and this unsolicited proposal will most likely be read.

STATEMENT OF OBJECTIVES

One of the key sections of your proposal is the statement of objectives. It indicates to your reader what you hope to accomplish if the proposal is accepted and also serves as a measuring device when methods, schedules, costs, and conclusions are presented later in the document. Often, objectives can be stated in a list whose items are arranged in an order that helps support the proposed actions. Objectives should be described as concretely as possible. For instance, rather than stating "decrease traffic congestion" as a goal in a highway engineering plan, it would be more effective to give a specific percentage of decrease, areas in which the traffic flow would be minimized, and a timetable.

Consider this excerpt from the statement of objectives in a Land Use Plan (a form of proposal).

Management Planning Objectives

By April 1987, we hope to satisfy the following objectives:

1. Air Quality
   - meet or exceed federal and state air-quality standards

*"What Is a Solution Center?" (Waltham, MA: Honeywell Information Systems, Inc., 1983).

2. Water Quality
   - meet or exceed federal and state water-quality standards
   - provide for on-site water use by park visitors; minimize impact on downstream capacities and volume
3. Soil
   - minimize impact on soil stability and productivity
4. Fire
   - protect the public, improvements, and natural resources from the adverse effects of fire
5. Vegetation
   - provide a variety of, and a diversity within, vegetative communities
   - protect threatened or endangered plant species
6. Wildlife and Fish
   - maintain key terrestrial and aquatic wildlife habitats
   - balance wildlife populations with their habitats' carrying capacity
   - fill vacant ecological niches, giving emphasis to indigenous, unique, endangered, and threatened species
   - protect threatened or endangered animal species

These objectives include a standard of measurement ("meet or exceed federal and state air-quality standards"), and they are very descriptive ("minimize impact on soil stability and productivity" rather than "minimize impact on soil"). Note that a deadline for meeting the objectives is also provided. The *means* by which the objectives will be satisfied are not mentioned; that information will be covered in subsequent sections of the proposal.

POSSIBLE SOLUTIONS

Whether this section is called "Possible Solutions" or has another label, such as "Survey of Techniques," "Survey of Methodologies," "Alternatives," or "Options," its role in the proposal document is twofold:

- to demonstrate to readers that you (or your organization) have considered a variety of alternatives for solving the problem
- to focus on the specific solution to be recommended in the proposal

In this section, objectively survey several realistic alternatives that may or may not have occurred to your readers. Base these alternatives on research and/or experience, being certain to document sources of information in standard footnote or bibliographic style. (See Chapter 14.)

Then focus on the particular solution you wish to recommend. Such focusing will involve eliminating the other alternative(s) and recommending the chosen method. Factors to consider in both disregarding and recommending alternatives include time, personnel, effectiveness, cost, design flaws (strengths), and efficiency. This focusing process prepares readers for the detailed discussion that comprises the remainder of the proposal.

To see how a good possible solutions section should be organized, let us return to the proposal for eliminating water seepage at Carter Dam (Figure 13-1). The writer briefly presents four construction options and then recommends the first because it is the most time and cost-effective. In this possible solutions section, the writer discusses three basic points for each category: effectiveness, construction complexity, and cost. The information is presented consistently in this order so that the reader may follow the information easily. The suggestions given in the concluding paragraph rely solely on the information provided in the preceding discussions—no new data are introduced. Finally, the section ends by focusing on a single option, laying the groundwork for continued discussion of that option in the balance of the proposal.

PLAN OF ACTION

A plan of action details the steps necessary to implement the proposed solution to the problem. Most often this section follows the possible solutions portion of the proposal and relies heavily on process description techniques. (See Chapter 8.) Points of discussion include:

- stages of development
- materials to be used
- potential difficulties (and solutions to these obstacles)

At this point, the proposal tells readers *exactly* what can be expected if you are given the contract. In fact, if the proposal is accepted, it is entirely likely that the resulting contract will include your plan of action nearly word for word. Depending on the requirements stated in the RFP, the plan of action should be direct and specific. If the RFP does not state precise requirements, try to remain as flexible as possible (since you will be held accountable for this plan).

The following plan of action is part of the unsolicited proposal cited earlier in this chapter in which the writer calls for creating measurable requirements to improve quality. The writer is quite specific, yet not inflexible.

POSSIBLE SOLUTIONS

According to our research, several repair options are possible. Each differs in effectiveness, construction complexity, and cost.[1] These are: 1. Criteria

- cement grout curtain [2] 2. Options
- internal concrete cut-off wall
- gunite blanket
- earth and rock-fill berm

Cement Grout Curtain[3] 3. Option 1

The cement grout curtain can achieve the degree of water tightness most suitable to Carter Dam's foundation conditions.[4] It could be constructed by a combination of 4. Effectiveness split spacing and closure methods, both relatively easy procedures.[5] This method would create a permanent wall 5. Complexity that conforms to the shape of the dam's interior and subsurface. We estimate that a cement grout curtain could be constructed for the entire half-mile of the dam to a depth of 25 feet below the base of the dam for a cost of approximately $750,000.[6] 6. Cost

Internal Cut-Off Wall[7] 7. Option 2

An internal cut-off wall is virtually impermeable.[8] The 8. Effectiveness strength and stability of this wall ensures that no excessive lateral forces can be exerted on the dam. This wall is constructed by means of a sophisticated multistage procedure involving elaborate excavation, pile-driving, and measurement techniques.[9] The results would be a wall 9. Complexity extending far below the surface of the dam at a cost of approximately $4.5 million.[10] 10. Cost

Gunite Blanket[11] 11. Option 3

A gunite blanket placed at the upstream face of Carter Dam would be relatively impervious, allowing only a reasonable amount of seepage to pass through the dam walls.[12] The foundation composition of the dam makes this option unattractive, however, because water that is not allowed to pass through the dam's face will instead pipe through the sand and gravel of the base. The result could be serious erosion and ultimate dam failure. If the construction of such a blanket were undertaken, it would be a simple matter of lowering the volume of the dam to expose the upstream face and then applying several sprayed layers of gunite.[13] The cost of such a solution would be approximately $400,000.[14]

12. Effectiveness
13. Complexity
14. Cost

Earth and Rock-fill Berm[15]

15. Option 4

An earth and rock-fill berm could be constructed to force seepage further downstream from the toe. It would not, however, prevent the seepage problem.[16] Construction would require amassing earth and rock fill (with bulldozers) and then compacting this pile to design specifications.[17] A 40' x 25' berm is recommended. Our examination of the site indicates that all materials for the berm could be obtained on-site. Therefore, the cost of the berm construction would be approximately $40,000.[18]

16. Effectiveness
17. Complexity
18. Cost

Sampson Engineering realizes that both time and cost must be considerations in addition to all parties' concern with effectiveness. For that reason, we suggest that a cement grout curtain be chosen as the best solution to the piping problem at Carter Dam. This option would effectively prevent seepage, is sufficiently simple to be constructed within the six-month time limit proposed in the request for proposal, and is within the budget limitations cited. Certainly, it would be ideal to construct an internal cut-off wall. However, the prohibitive cost and elaborate procedures prevent this option from being seriously considered by our firm. [19]

19. Focus on preferred alternative

Figure 13-1 The possible solutions section is used to survey alternatives and recommend a specific one.

Plan of Action

The training of employees to develop measurable requirements can be implemented simply and inexpensively. The training exercise should be incorporated into the existing quality-improvement program, where it can enhance the success of the program while it takes advantage of the resources already in place. In this way, employees are already gathered for discussions, meeting rooms have been arranged, and furnishings and equipment are available.

It is proposed that the discussion of measurable requirements constitute an actual class and that it be conducted immediately following the employees' completion of the introductory sessions of the Quality College.

The topics to be covered in this class would include:

- communicating in writing
- considering your audience
- considering complexity
- considering role
- using precise vocabulary
- identifying the topic
- stating the task
- stating the purpose
- qualifying information
- presenting constraints
- defining success
- following up

The class would require the following materials:

- overhead transparencies to guide discussion
- individual workbooks that provide examples and exercises
- final test

It is projected that the class will require four hours of contact time and can be conducted in one of three ways: (1) as a single unit, (2) as two units of two hours each, or (3) as four units of one hour each. In this writer's opinion, less than four hours of exposure to the materials will prove ineffective. After successful completion of this class, employees should continue to apply the concepts learned to the remainder of the Quality College.

The reader of the plan of action weighs the methods proposed to determine if they will yield the promised results. If it seems likely that they will do so, the proposal is that much closer to being accepted and implemented. Therefore, the importance of this section cannot be overemphasized. It proves to readers that you have planned carefully and helps instill confidence that you can get the job done. The effect this section can have on your readers does not depend on length, but rather on demonstrated competence. The preceding plan of action for modifying the Quality College is short but packed with specific information. The Sampson Engineering firm's plan of action to fix Carter Dam would be very long but should be no less well organized. That proposal would have to pinpoint several phases of implementation, such as design and measurement, lowering water level, positioning techniques, split spacing, hole closure, testing, clean-up activities, and raising water level. In each of these subsections, as in any document employing grouping techniques, the writer is expected to conduct a parallel discussion of method, materials, and potential difficulties.

QUALIFICATIONS

The qualifications section of the proposal presents the individual's or organization's credentials. It spells out the experience, education, training, and knowledge that make the proposal writer qualified to do the job effectively. Some of the points that might be included in this section are:

- years of experience (or years in business)
- similar projects (give names)
- variety of projects (with references)
- special recognition (awards, citations, news coverage)
- specialized education and training
- professional memberships
- related activities (for example, consultation provided to charities or government agencies)

In qualifications sections that focus on an individual rather than an organization, the discussion might have a biographical flavor, such as the following statement by a system designer who wishes to be considered for a special networking design project.

Qualifications

I have worked in data processing since the late 1950s, with a career spanning an unusually wide range of areas, including programming, system design, system engineering, and marketing. I have worked on application system development, operating system design and implementation, data-base management, and networking system software design. Since 1974 my work in distributed systems has included working with users and prospective users of XYZ computers (both large-scale and minis), defining system requirements, coordinating implementation, and working with system designers developing distributed products. I have lectured widely in the United States, Europe, and Japan and am currently a member of the speakers' bureau of the Association for Computing Machinery. I am the author of many technical articles, six specific to distributed systems.

Another alternative is to include an actual résumé. If this is done, the résumé should appear in the appendix of the proposal and merely be summarized in the body of the document. (See Chapter 18 for a discussion of résumés.)

If, however, the qualifications of an organization are being presented, they should appear in the qualifications section in brief yet comprehensive form, and a copy of the company's brochure or annual report should be placed in the appendix. Toward the end of the qualifications section, mention that the brochure or report is attached. Here is a sample of a qualifications section for an organization. Note its promotional rather than entirely objective tone.

Qualifications

Maxwell Rigging Company of Gloucester, Massachusetts, has been in business since 1945, when it was founded by Robert Maxwell, the famous yachtsman and traveler. At that time, the company served as the primary subcontractor for several local yacht-building enterprises, including Smithson Doubleday, the oldest yacht-building company in the United States. In recent years, as the demand for yachts has declined, Maxwell Rigging has expanded its business to accommodate rigging machinery for smaller vessels as well. In fact, two ships have recently been completed that will be entered in the next America's Cup races.

Maxwell Rigging prides itself on quality of workmanship. Recognizing that those who use our equipment will be betting their lives on the safety and security of our lines, the company ensures that rigging undergoes extensive testing both under laboratory conditions and on board ship before it is approved for use. In the forty years Maxwell Rigging has been in business, no systems have failed under stressful weather conditions.

The employees of Maxwell Rigging are all accomplished seafarers, carpenters, and engineers. Each is expected to pass a rigid entrance examination that tests book knowledge and hands-on experience. Most employees have been with the company for years. They are proud of their work, and Maxwell Rigging is proud of them.

For a further description of our qualifications, please refer to the informative brochure we have included at the end of this proposal.

FACILITIES AND RESOURCES

The facilities and resources section of the document usually comes next. It lists the physical facilities and equipment that you and/or your organization have to support the implementation of the proposal. Once again, it is possible to refer to the annual report or brochure included in the appendix, but it is preferable to present as well a paragraph or two of discussion within the body of the proposal.

Facilities include all permanent structures, such as office buildings, warehouses, classrooms, test sites, and laboratories. *Resources* include audiovisual equipment, bulldozers and excavators, trucks, tillers, assembly lines, computers, dollars, and personnel (number and type). The following discussion is excerpted from a grant proposal in which a school for emotionally disturbed children describes its facilities and resources.

Facilities and Resources

Our school is located in central Seattle, where it is easily accessible by bus or automobile. Its campus covers 11 acres of playing fields, dormitories, classrooms, and assembly areas (such as the cafeteria, gymnasium, and auditorium). The total number of buildings, including the maintenance warehouse, is 14.

The four small classroom buildings are clustered at the center of the campus to reinforce the educational emphasis of the school. These buildings are decorated in a less clinical way than that found in public schools, in the hope that the educational environment will seem more inviting to these troubled youngsters. For instance, upholstered chairs with side tables are used instead of traditional school desks. Class size is limited to ten students so that individual attention can be paid to each student.

Each of our ten faculty members is a certified counselor or special education expert with the state of Washington (which has one of the strictest qualifying codes in the country). Each has at least six years of experience with emotionally disturbed students. Two of our members have their Ed.D.s in special education. Two others have Ph.D.s in related fields. Four other members have their M.S. degrees in education. . . .

SCHEDULE

As its name suggests, this section provides readers with your anticipated project or activity schedule. Generally, schedules are categorized into phases of action with corresponding approximate dates. When deadlines are important to a project, this section can be a pivotal component of the proposal. For instance, if it is essential that a project be completed within a certain fiscal year in order to be funded, the proposer must demonstrate that the

Activity	Start Date	Completion
Site inspection, surveying, measurement	4/3/86	4/10/86
Site preparation	4/15/86	4/30/86
Foundation excavation	5/1/86	5/12/86
Pipe laying	5/14/86	5/28/86
Cabling	5/14/86	5/28/86

Table 13-1 A schedule can be presented in table form.

organization can fulfill its obligation to finish by that time. If, however, the project is long term, precise dates may not be significant factors in the awarding of a contract. In either case, dates should be as realistic as possible. When projections are overly ambitious, an organization may find itself having to make costly adjustments in personnel and overtime allotments.

Categories of activities should be arranged in sequential order with clear headings for each. If certain activities are simultaneous, columns might be used to demonstrate their parallel nature. As Table 13-1 (a building contractor's schedule) demonstrates, it is possible to present this information in the form of a table. (See Chapter 11 for more information on preparing tables.)

COSTS

Most proposals must discuss the cost of implementing the activities or projects described. The cost section is especially important in bid proposals, where contracts are to be awarded, and in grant proposals, where funds are to be secured. This section should be as detailed as possible, and because readers will hold you to the approximated costs, your statement should be very carefully researched and realistic.

The cost statement can take the form of a table, or it can be a description of expenses. Both types of statements are illustrated in the following examples. If a variety of expenses are to be incurred, the table format is preferable.

Costs

The following advertising costs are based on the 1985 *Consumer Advertising Index*. This table identifies ad subject, publication, unit size and cost, number of insertions, and total costs. Subtotals are shown per product type. Publications reserve the right to alter their advertising space costs.

SPARTAN PERFUME FOR MEN

Publication	*Unit Size*	*Unit Cost*	*No. of Insertions*	*Total*
Full Line:				
Business Week	2 p.	$ 52K	5	$260K
Sunday supplements	⅛ p.	2.5K	8	20K
				$280K
Spray Cologne:				
GQ	2 p.	$ 16K	6	$ 96K
Man's World	1 p.	9K	10	90K
California Dreamer	1 p.	4K	4	16K
Celebrated Man	2 p.	4K	3	12K
				$214K
Soaps/Deodorant:				
Penthouse	2 p.	$ 27K	3	$ 81K
Playboy	2 p.	34K	2	68K
				$149K
		SPARTAN Products Total:		$643K

Costs

The costs of the indirect-gain thermal storage walls are based on four structural components. If the south-facing wall is used as a solar collector, the costs for the Hamilton Building are estimated at $15,000 for the collector itself, $1,400 for double glazing on windows within the south wall, $4,200 for the storage medium (bulk rocks), and $8,350 for the heat distribution mechanism (copper tubing). Personnel costs are estimated at $65 per hour for 80 hours for a total of $5,200. Total costs are estimated at $34,150. Cost figures may change depending on availability of materials at the time this project is scheduled to begin (when construction on the Hamilton Building has been completed). This firm has no control over the building construction timetable.

In both examples, the cost section relies on the detailed discussion of materials and methods presented in the plan of action. Nonetheless, each writer mentions again the type of material to refresh the reader's memory and make the cost statement clearer.

In the first sample, costs are actual quotes; a generally accepted pricing guide is cited in the introductory remarks. In the second case, costs are merely estimates. To make this clear to the reader, the writer is careful to present a qualifying statement ("if the south-facing wall is used . . .") and to repeat the word *estimated* several times in the pricing discussion. Note, too, that each writer indicates that costs are subject to change and the possible reasons that changes might occur. In both cases, changes would be outside the control of the proposer. The changes would affect all other proposers as well and, therefore, would not be a factor in the reader's decision-making process.

CONCLUSIONS AND RECOMMENDATIONS

The conclusions and recommendations section of a proposal is a synthesis of the key points, proposed methods, schedule information, and costs. In this section you have a last chance to convince readers of the merits of your proposal. Topics discussed and the order of information should reinforce the action you wish readers to take. In a situation where one method of construction is proposed over another, for instance, the writer would be wise to state again the preferred method, briefly indicate why it is the best alternative, and discuss the expected benefits of the proposed actions.

Generally, headings are not used within the concluding remarks. Instead, rely on strong, logical transitions to guide readers through this section. An assertive tone is appropriate here to help demonstrate your abilities. Finally, no matter how long the proposal itself may be, try to restrict your concluding remarks to a single page.

The following conclusions and recommendations section comes from an engineering firm's proposal to install a solar heating system in a small office building. It is part of the same proposal from which the preceding cost statement was taken.

Conclusions and Recommendations

Milton and Associates, P.C., proposes to build a passive, indirect-gain solar heating system for the Hamilton Building being constructed on the southwest corner of Dunlap and 19th avenues. We have selected this indirect-gain method as superior to all other solar-collection methods because it fits comfortably within the $40,000 budget limit set by the owners of the Hamilton Building, and because it would be the most efficient method within the architectural design already being implemented for this building.

The solar-collection design proposed uses thermal storage walls to provide for heat collection, storage, and distribution. The design takes advantage of the thermal characteristics of the wall components to provide a constant heat source. The proposed glazing techniques would create a heat trap as well as provide an attractive look to the southern wall of the building. The architect's blueprints, shown to our firm, indicate wall structures of sufficient mass to store and distribute the heat effectively over a period of time without causing undue temperature variations. Given that the Hamilton Building is in a Sunbelt city, this method of heat collection and distribution is enhanced by an average of 300 sunny days per year. If this method is implemented, no other heating mechanisms would be required for this building. The projected cost savings over conventional electric or natural gas heating systems would pay for the solar-collection system within two years.

Our firm proposes that construction of the passive, indirect-gain solar heating system begin as soon as the southern wall of the Hamilton Building is completed. As indicated in our discussion of schedule, it would take only two work weeks to complete the project. Total estimated expenditures are $34,150.

Use your concluding remarks to present your strongest argument for being awarded the contract, project, or funds. This streamlined version of the whole proposal is vital to the success of the document for several reasons. First, the hurried reader may look at *only* this section of the proposal before deciding to read the document. Hence, the more convincing the arguments found in these few paragraphs, the better the chance that the entire proposal will be read and considered. Second, other readers may use this section to refresh their memory before making a final decision. Don't waste words here. Present every significant point; delete all minor points. Third, this section may be distributed at a meeting or to a committee as a self-sufficient statement. Those readers might never see the entire proposal. In short, this section must be able to substitute for the proposal as a whole.

The best way to be certain that all key points have been covered is to review each section of the proposal. Not all sections may need to be summarized or alluded to in the conclusion, but be sure that all your major points appear once again. Your major points need not necessarily follow the order used in the proposal, but they should be arranged to support the proposed solution, action, or project.

omit

PRELIMINARY BIBLIOGRAPHY

Proposals often end with a preliminary bibliography or footnotes to indicate that initial research of the project, activity, or solution has been undertaken (see page 328). These citations also afford readers an opportunity to explore issues first hand, should they wish to. See Chapter 14 for a more complete discussion of the bibliography.

APPENDIX

If you include a résumé or annual report, or any additional supporting data, in your proposal, you will need to do so in an appendix. Be sure to allude to the information found in the appendix in the proposal itself, so readers will know where to look for it. See Chapter 14 for a more complete discussion of the appendix.

PUTTING IT ALL TOGETHER

Thus far in this chapter, we have examined portions of proposals but have not looked at an entire document. Figure 13-2 presents a complete brief proposal written by a researcher who wishes to investigate the feasibility of recharging a city's groundwater supply with treated effluent and rainwater. The writer is *not* proposing to initiate a groundwater recharging project; rather, she is asking her supervisor for the go-ahead to *study* the possibility of such a project. In effect, she is proposing to develop a research report. Graduate students often write similar proposals before embarking on a thesis or dissertation project, and researchers may write proposals like this before beginning an involved study.

CHAPTER SUMMARY

Proposal writing is a key skill in business, academic, and scientific environments. Through proposals, funds are allocated, projects are initiated, and research is conducted.

Although proposals vary according to the specifications of the person or organization soliciting the document, certain elements are standard, including statement of the problem, background information, statement of objectives, possible solutions, plan of action, qualifications, facilities and resources, schedule, costs, conclusions and recommendations, preliminary bibliography, and appendix. The order of information and level of detail depends on the specifications spelled out by the organization requesting the proposal (in solicited proposals) or according to the needs of the audience (in unsolicited proposals).

Proposal to Research the Feasibility
of Recharging Groundwater Supplies
with Treated Effluent and Rainwater

Prepared by:

Renatta Carlson
Chief Hydrologist
Bureau of Water Quality Control

February 20, 1987

FOR INTERNAL DISTRIBUTION ONLY

Figure 13-2 A sample proposal.

TABLE OF CONTENTS

- ii -

Figure 13-2 (cont.)

STATEMENT OF THE PROBLEM

Tucson, Arizona, is the largest city in the world to rely solely on groundwater for its water supply. Tucson's groundwater supply is naturally recharged by the Santa Cruz and Avra Valley watersheds. These bodies also supply water for Pima and Pinal County irrigation projects. With increased usage by farmers and urban dwellers, the water supply is decreasing at a rate faster than it is being replenished. Current estimates indicate that 21 million acre-feet (a.f.) of water are available. Each year, 250,000 a.f. are withdrawn while only 95,000 a.f. are replenished through natural sources. With no change in this pattern, it is estimated that only a 135-year supply of water remains. However, because Tucson's growth rate is virtually compounding each year, this estimate is far too conservative. It seems more likely that the remaining groundwater reserve will last less than 80 years.

Needless to say, alternative sources of water must be located. Currently, the planned Central Arizona Project (CAP) offers a temporary solution to the problem. However, it is likely that no water will be realized from this project until at least 1997. Measures must be taken to ensure that our water supply can be replenished prior to that time.

BACKGROUND INFORMATION

Tucson's sewage-treatment plant has been in operation since 1962. Recent expansion has increased its capacity twofold. Presently 20% of the treated effluent (waste water) is used for crop irrigation of nonfood sources, while 80% is discharged into the Santa Cruz River. The treated solids are used for fertilizer and fuel. Any rainfall Tucson receives (approximately 13 inches annually) is allowed to percolate into the soil to recharge the groundwater system naturally.

A preliminary literature search indicates that prior attempts to recharge groundwater systems artificially in locations other than Tucson have been relatively successful. It is important for our agency to study these methods, in particular those using treated sewage effluent and trapped rainwater, to determine if they would be a valid means of replenishing Tucson's groundwater table.

- 1 -

Figure 13-2 (cont.)

The 1972 Water Pollution Control Act established a national program to restore and maintain the chemical, physical, and biological integrity of the nation's waters. The act establishes a national goal to eliminate pollutant discharges by 1987. This program to clean up water pollution includes an expanded system of federal grants to plan and construct publicly owned sewage-treatment plants and a permit system to restrict pollution discharges. If the proposed study of artificially recharging the groundwater proves that it is feasible to do so for the Tucson area, the project would undoubtedly fall within the jurisdiction of this act. Funding, therefore, would likely not be a significant problem for the local government.

OBJECTIVES

As chief hydrologist for the Bureau of Water Quality Control in the Tucson area, I propose that my department be funded to study this issue in depth. Through a thorough search of existing literature and a minimal amount of testing, I am certain we will be able to determine the worthiness of a recharging project of this nature. If a valid means of using sewage effluent and rainwater is found, it could mean more years of groundwater usage for farmers and urban dwellers in this valley. It may help to minimize expected water rate increases as well as reduce this area's dependency on outside sources of water.

POSSIBLE SOLUTIONS

Currently, one major source of water supply for the Tucson area is being considered to augment the natural groundwater table supplies, the Central Arizona Project. This large canal system would route water from the Colorado River basin through Arizona. It would dispense water to key storage basins located throughout the state, including the Tucson water basin in the Avra Valley. Construction of this project was to have begun in earnest in 1986 in the Tucson area, but to date, no serious construction has been undertaken (except in the northern portions of the state) because of budgetary debates in Congress. This

- 2 -

Figure 13-2 (cont.)

project has become a political football, with proponents labeled "anti-environment" and opposers designated "isolationist."

At this point, the cost of the canal system has more than trebled since its inception, with no end of inflationary trends in sight. Its key proponent, Congressman Robert Hill, is no longer able to fight for the issue because of serious health problems. Hence, Arizona is left with a lone voice, Congressman Dana Edel, to provide direction and force action on this important issue. Realistically, it is entirely possible that CAP will never be wholly implemented. Even if it is, it will provide only a partial solution to Tucson's growing water supply depletion problem.

A more immediate solution to the problem may be to make better use of existing resources, specifically, sewage effluent and rainwater. Our advanced sewage-treatment plants may indeed have the capability to purify waste water properly so that it can be used to recharge Tucson's dwindling groundwater supply. Currently, the sewage-treatment plants that serve metropolitan Tucson discharge more than 37 million gallons of treated effluent per day into the Santa Cruz River. By the year 2000, effluent discharges are expected to exceed 67 million gallons per day. Our current aggressive programs help to ensure the purity of that reintroduced water. However, nothing is being done now to make that water a part of the groundwater recharging system. It is being lost, in great part, as the Santa Cruz flows to the sea.

Another important solution to the problem would be the construction of a new reservoir to store rainwater for reintroduction via percolation into the groundwater system. These last two methods of reclaiming resources seem attractive and warrant further investigation.

PLAN OF ACTION

Several cities and valleys have been investigating the potential use of treated sewage effluent and have begun serious experimentation in this area. These cities include Albuquerque, New Mexico; Laughton, Texas; and Yuma, Arizona. Scientific laboratory research has been conducted as well as by the Universities of Texas, Arizona, and Iowa.

I propose that my department be given the time to review this literature and to travel to and consult with hydrologists at these key sites. It

- 3 -

Figure 13-2 (cont.)

would also be advisable to conduct several laboratory experiments within our own facilities to corroborate the evidence uncovered. Such a study would result in a comprehensive feasibility report for use by the Bureau of Water Quality Control.

QUALIFICATIONS

My ability to direct such a project is attested to by my title, chief hydrologist. The budget committee should be aware of a recent article about my experimentation with water-purification techniques published in the Water Science Journal last July. My findings (that federal reclamation standards are inadequate) have been very well received among my peers and have sparked considerable debate over current, less effective purification practices used in some treatment plants. A copy of my résumé and this article are attached to this proposal.

FACILITIES AND RESOURCES

The facilities used for research will consist primarily of the holdings of our own department as well as those available in the University of Arizona's water sciences collection. We will test the primary research of other cities and universities in the laboratories of the Bureau of Water Quality Control. Prior to preliminary research, it is difficult to estimate whether our laboratory is adequately stocked to perform all required experiments. However, past experience has shown that our laboratories are sufficiently advanced and well equipped to accommodate most modern testing procedures.

Three workers, including myself, would comprise the research team. Our salaries are already budgeted even though workhour expenses are shown below in the Costs section of this proposal.

SCHEDULE

A total of five months would be required to conduct adequate primary and secondary research. That time would be distributed in the following way.

- 4 -

Figure 13-2 (cont.)

PHASE I: Two months are required to study existing literature in the University of Arizona's water sciences collection. This phase will begin as soon as project approval is received. If, after this initial research is conducted, groundwater recharging seems feasible, the research team will move on to Phase II. If the solution does not seem feasible, the project can terminate at this point.

PHASE II: One month is required to meet with principal researchers in Albuquerque, Laughton, and Yuma. If after this phase of study recharging still seems feasible, researchers will move on to the third and final phase. If the solution does not seem feasible, the project can terminate at this point.

PHASE III: Two additional months are required to conduct corroborating research of key experiments performed in Albuquerque, Laughton, Yuma, and the aforementioned universities, and to apply these results to equipment and facilities that could be used in the Tucson area. At this point, researchers will be able to evaluate the feasibility of recharging Tucson's groundwater supply via treated sewage effluent and captured rainwater.

COSTS

The following figures are estimates of workhour, travel, and testing expenses for the five-month project. If, however, only a portion of the project is conducted before recharging is deemed not feasible, the costs will be correspondingly lower.

Workhours (3 people/5 months)	=	$45,000
Travel	=	8,000
Testing	=	16,000
TOTAL	=	$69,000

- 5 -

Figure 13-2 (cont.)

CONCLUSIONS AND RECOMMENDATIONS

From initial research of groundwater recharging projects, and from full knowledge of Tucson's dwindling resources, we believe that further study of such alternatives is warranted. We recommend that funding be granted within the next few weeks so that our department can begin study in earnest. We feel confident that such research will improve our overall water treatment program as well as offer a cost-effective solution to diminishing water resources.

PRELIMINARY BIBLIOGRAPHY

Berkman, Christopher C., and Harold W. Gislon. Groundwater Replenishment Theories. Encino, CA: Dickenson Publishing, 1983.

Boone, David L., and Louise E. Kuretzi. "Treatment Plant Technological Advances." Hydrology, 49 (January 1984), 62–66.

Bovee, Courtland. Contemporary Water Recapture Methodologies. New York: John Wiley and Sons, 1983.

Morris, Henry M. "Applied Sewage Effluent Treatment Techniques in the Albuquerque Water Quality Control Program." Water Engineering, 64 (February 1982), 79–90.

Pride, William C. "Results of Rainwater Recapture Tests in Yuma, Arizona." Water Engineering, 65 (March 1982), 44–52.

Ulanoff, Maurice K., and Robert J. Stone. "Progress Report on Laughton, Texas, Effluent Treatment Applications." Hydrology, 45 (January 1983), 16–25.

- 6 -

Figure 13-2 (cont.)

EXERCISES

Do one of these exercises before writing your own proposal.

1. Find some government bid proposals in your library and look them over. Do the proposals received measure up to the specifications put forth by the government?
2. Evaluate the proposal concerning the recharging of groundwater supplies using the editing criteria at the end of Part III.
3. Examine your local newspaper for articles dealing with proposals being considered by the state or city government. What components of the formal proposal does the article list? Develop a proposal outline from the information cited in the article.
4. Contact a local engineering firm. Request to see a recently accepted proposal. Speak with the proposal writer to determine what he or she feels are the strengths and weaknesses of the document. Ask how many months went into preparation of the proposal. How many technical contacts did the writer require in order to develop the document?

WRITING ASSIGNMENT

Using an upcoming research project in your major field as your topic, write a proposal that explains how you will conduct this research project. Your audience is the instructor for the course in which this project will be developed. Your competition is your entire class—imagine that your instructor will accept only the two or three best project proposals.

Your proposal should accomplish the following objectives:

- win favorable attention from your reader immediately
- get you started on researching your project
- help you focus your project
- demonstrate to your instructor that you have a solid foundation on which to begin your research project
- Convince your instructor that you should be one of the students who receives the go-ahead on the project proposal

NOTE: Apply the editing criteria at the end of Part III to your assignment before you submit it to your instructor.

CHAPTER

14

Reports

CHAPTER PLAN

Reports, the most commonly written type of technical document, can communicate a range of vital information: project status and outcome, experiment results, conclusions drawn from analyses, policy recommendations, decisions made at meetings and conferences, reviews of literature, and surveys of conditions. The more effectively you write your reports, the more seriously your information will be taken. To help you write effective reports, this chapter will

- examine the importance of reports in technical environments.
- discuss the difference between formal and informal reports.
- discuss the different kinds of reports.
- outline the report writing process.
- describe the various components of formal reports.

THE IMPORTANCE OF TECHNICAL REPORTS

A report is an objective description of something that is happening, has happened, or may happen. Different circumstances, audiences, purposes, and events call for different kinds of reports. For instance, if you are an agronomist (a specialist in field crop production) writing to an audience of product developers in a chemical manufacturing company, you might present a report on the effects of three herbicides on *Plantago ovata* (a hemplike grass). Such a report would focus on experimentation methods and results. If you are an agronomy major in a university graduate program, you might write a report about *Plantago ovata* for your weed science professor, focusing on key researchers working with this crop, the characteristics of the crop under both natural and laboratory conditions, and the feasibility of planting it in the southwestern desert of the United States. Your report in this case would be a survey of existing literature. Or perhaps you are an agronomist with the U.S. Department of Agriculture addressing a group of Arizona farmers who have planted *Plantago ovata* as an alternate crop. Your report might explain the reasons for crop failure after the farmers used a recommended herbicide. Such a report would take an investigatory approach to the problem, concentrating on causes, effects, and recommendations for future preventive measures.

A technical report usually records information about something that has happened, draws conclusions about the event, and offers recommendations for further action. As a technical staff person, the verifiable data and expert advice you provide in reports will serve as the basis of important decisions in your organization. Ideally, your reports will be the catalysts for new policies and actions. An ability to develop clearly written, logical reports is crucial to any person serious about advancement in corporate, industrial, technical, and scientific environments. It doesn't matter how exceptional your ideas are—unless you are able to communicate them effectively in a competitive context, little notice will be given to them.

INFORMAL REPORTS

Much of the writing done in industry and technical fields falls into the informal report category. Although the objectives and preparatory steps that apply to formal reports apply to informal reports as well, the components of the informal report are less extensive, the structure more flexible.

Informal reports are usually brief (less than 10 pages) and often take the form of a letter or memo (see Figures 12-2 and 12-7). Sometimes, as in the case of accident, trip, or expense reports, these documents are prepared on a standard, preprinted form. Frequently, letter and memo reports make use of the "You" attitude (see pages 284–285). Your underlying purpose in these informal reports is to provide a *succinct* statement of the event or facts at hand. Details are held to a minimum.

FORMAL REPORTS

Formal reports are usually lengthy (10 or more pages) and often adhere to a precise format and well-defined structure. They are designed to build a case carefully and to argue its conclusions and recommendations. The tone of a formal report is objective; there is no evidence of the "You" attitude here.

Formal reports serve as permanent records of key decisions made within an organization. They are often widely reviewed by an audience with divergent interests. Frequently, a formal report is written by a group of people, each bringing his or her special expertise to the project. The formal report can take many months to develop. In fact, the formal report may be issued in several stages: preliminary, progress, and final.

KINDS OF REPORTS

Both formal and informal reports can be categorized into five basic types, based on the purpose of the document and the needs of the audience: updates, analyses, experiment reports, investigation reports, and literature reviews.

UPDATES

The simplest, most direct report is one that updates its readers. Very often this report is informal and takes the form of a memo addressed to an internal audience. The reader is generally acquainted with the topic and focus of the document.

A *status report* updates readers on a regular basis (daily, weekly, or monthly). It usually concentrates on a number of projects under development and details their latest accomplishments, schedule changes, deadlines, or resources (budget remaining, personnel requirements, or materials on hand). Documents of this nature often lay the foundation for broad planning and scheduling.

A *progress,* or *interim, report* relays any and all developments on a specific project that have taken place since the last progress report was issued or since the project began. Both progress and status reports cite measurable activities as the substance of the document and rarely include conclusions or recommendations, although discussions of the best way to proceed are sometimes presented.

A *trip report* is as much a justification of a trip as it is a listing of all that happened during the trip. Usually filed immediately following the trip, this report states the objectives of the trip, the people with whom the traveler met, descriptions of events and meetings, details of what was accomplished, and items that need to be acted on.

A *call report* records the business conducted during a visit (for example, a sales call) or a telephone conversation. It briefly lists topics of discussion, decisions, and items that need to be acted on.

ANALYSES

Analyses are generally undertaken prior to the execution of a particular action. Their function is to determine the factors that will be involved in that course of action.

Feasibility studies are the most common form of analytical report. These studies help readers determine the effects of various proposals. A feasibility study might, for instance, examine a number of public transportation systems that could be implemented in a city. After analyzing population growth patterns, economic conditions, and accessibility, this report might recommend that a specific transportation system be implemented.

More often than not, a feasibility study is commissioned by a key decision maker or decision-making committee. In these cases readers need to know their options but must rely on the writer's expertise to help them determine the best one. Frequently, these reports are developed by a consulting firm, which can offer an unbiased, objective view of the problem and the solution. (Throughout this chapter, we will be looking at portions of a report written by a consulting firm for a state agency that discusses the feasibility of incorporating solar energy systems in home designs.)

A *forecast* is another type of analytical report, often used in the business world. This kind of report examines past history, current trends, and foreseeable directions; it outlines anticipated changes in technology or in the marketplace and examines the impact of those changes on a specific issue. From comprehensive forecasts decision makers can determine for themselves the best actions to take.

Modeling reports, or "what if" analyses, describe hypothetical situations. This type of report allows readers to examine the effects of different variables on existing conditions without actually undergoing the risks involved in implementing these steps. Modeling reports provide readers with a set of possible scenarios, thereby enabling them to anticipate future requirements.

These reports are usually based on computerized models in which data can be manipulated and recalculated easily. Models can also take physical form—usually in miniature scale—as in the case of design alternatives. The resulting written report, which is based on the model, requires keen interpretive skills because you are usually dealing with a range of possibilities and interrelationships.

EXPERIMENT REPORTS

Reports that discuss experiments cite all of the elements involved in the experimental effort: objectives, controls, variables, equipment, materials, schedules, procedures, and measurement criteria. Experiment reports are written in chronological order with careful attention paid to every development. Conclusions and recommendations for action based on the results of the experiment are also included. Objectivity and precision are paramount in this kind of report, because the reader may wish to replicate the experiment from the information supplied. The report on *Plantago ovata* on pages 483–505 is primarily an experiment report providing an examination of the plant's tolerance of three herbicides.

INVESTIGATION REPORTS

Investigation reports are written after a particular event has occurred or a condition has developed. These reports reconstruct the variables or events that led to the event or condition in order to determine a source of the problem. The outcome of an investigation report is usually a recommended course of preventive action.

In early 1984, for example, a six-state energy blackout occurred in the western United States. Several power companies worked together to uncover the reasons behind the incident. The report they produced described the source of the problem (insufficient load-leveling equipment) and recommended to all power companies in the area several specific steps that would prevent future occurrences.

LITERATURE REVIEWS

Literature reviews survey existing literature, expert opinions, and experiment results in order to yield a compendium of the available knowledge on a certain topic. The information in literature reviews usually forms the basis of specific conclusions and recommendations. Grouping and definition skills are particularly important to these kinds of reports.

THE REPORT WRITING PROCESS

By now, you should be familiar with the four-phase technical writing process of analyzing the situation, gathering and interpreting information, planning and arranging your materials, and writing and revising the document. Formal and informal reports rely on this process, too. Because of the amount and complexity of information and the extended time frame in which it is developed, the formal report, especially, must adhere strictly to this process.

ANALYZING THE SITUATION

When developing the situational analysis, pay particular attention to the needs of your audience.

- Who are your readers? Are they a relatively homogeneous group of individuals who share the same degree of expertise, or are you facing a mixed audience?
- What are your readers' expectations?
- How familiar are your readers with the issues you will be covering in the report?
- How will your readers apply the information in the report?

Many formal reports are addressed to readers with a variety of backgrounds whereas informal reports receive more limited and homogeneous distribution. In the case of the formal report, some of your readers may be more familiar with the subject than you; some may be your peers; still others may know nothing about your topic. When facing an audience with mixed backgrounds you must try to inform the less expert readers without patronizing the trained specialists. Conversely, you must address the needs of the specialists without losing the novices. A general rule of thumb is to direct your discussion to the segment of your audience that will be most influential in the acceptance of your conclusions and recommendations.

When we look at the components of a typical formal report, you will see that certain sections are designed to meet the needs of a lay audience (the background statement, the executive summary, and the glossary), while other sections can include detailed and sophisticated information (e.g., the methods and results sections). Before you begin to write your report, therefore, one of your first steps should be to prepare a written analysis of the mixed needs of your audience. (Review Chapter 2, if necessary.)

A clear idea of your purpose is also extremely important. Is a report really necessary, and if so, what do you expect it to accomplish? What kind of report will it be—an update, an analysis, an experiment report, an investigation report, or a literature review? Answering two questions about your audience—"What are my readers' expectations?" and "How will they use this information?"—will also help you determine a clear purpose. Write down that purpose as succinctly and concretely as possible.

Knowing your audience and your purpose will help you determine your role as a writer as well as the most effective level of detail, method of exposition, arrangement of information, types of examples, vocabulary, and emphasis. Since most organizations provide their employees with a set of format specifications, you must also plan for these requirements at the outset. Anticipate the information your format specifications will require before you begin your research.

Be certain to allow yourself sufficient time to produce a high-quality document in a cost-effective manner. If you are part of a team that is producing the report, be sure the responsibilities of each writer are clear. It may even be advisable to create a chart to keep track of activities to avoid duplication of effort or the wasting of valuable time.

GATHERING INFORMATION

With a clear statement of audience, purpose, and role in hand, you are ready to begin researching or experimenting. Be as comprehensive as possible: it is easier to discard data at the writing stage than it is to collect it. List your resource materials accurately: you will be required to make complete citations in your report when you paraphrase or quote others.

As you gather information, interpret it. Group information according to logical relationships (refer to Chapter 6, if necessary), weed out overlapping and redundant data, highlight conflicting points (and attempt to resolve conflicts before proceeding), and determine what gaps in your data need to be filled in. You may find yourself gathering information according to the requirements of format specifications. Thus, even as you do your initial reading, you may be able to pigeonhole information into such categories as "statement of method," "background," or "scope."

PLANNING AND ARRANGING

Begin planning the report by developing a preliminary outline. Consider at this point the expository techniques that will be most appropriate for presenting your information. Which terms and concepts require definition? Are you going to describe procedures or equipment? If so, incorporate the appropriate descriptive techniques into your outline. If your conclusions and recommendations require argumentation, arrange your evidence to persuade your readers to accept what you say. Revise the outline to emphasize your main points and to eliminate redundancy.

Anticipate the graphic aids your report will require. Which information should be displayed in a table, line graph, drawing, map, time line, organization chart, or pie chart? Consider the examples you could use to clarify points. Which sources will you quote directly and which will you paraphrase? Prepare the citation information so that it can be incorporated, with quotations and paraphrased material, into the rough draft.

WRITING AND REVISING

Now you are ready to write a rough draft of the report. Working from your outline and notes, write each report section. Do not worry that the order may not be the one used in the final document. Right now, your goal is to produce as complete a rough draft as possible and to flesh out your discussion.

Once the rough draft is complete, you need to distance yourself from the report. You've probably been working on it for days, if not weeks. Work on something else. Or give yourself a little reward—you might go to the movies to divert yourself. Put the report aside for at least a few hours or, better yet, a whole day, if your deadline schedule permits it.

Then, with a refreshed perspective, apply the editing criteria for reports (pages 477–478) to your draft. Rework it until you feel that every criterion has been satisfied.

Carefully type and proofread the final copy. Typographical errors can create misunderstandings *and* unfavorable impressions. Adhere to all required specifications for report components (cover, title page, letter of transmittal, and so on). If specific action on the reader's part is necessary, indicate this clearly in the letter of transmittal.

REPORT COMPONENTS

Although an informal report is generally brief enough to be read from beginning to end, a good formal report is designed *not* to be read in its entirety. Rather, it is designed to be read by a person who must extract information quickly and easily. Thus, information must be presented in brief, self-sufficient sections so that various readers with differing interests can benefit from the report.

Let's take a look at the individual report components that have evolved to facilitate quick and easy access to information. Certain components are likely to be present in most reports. Formal reports might include many of the elements discussed next as individual sections of the document. Informal reports might address some of the same elements without delineating them as separate sections. The amount of space dedicated to any one of these sections depends on the needs of the audience and your purpose in writing the report. In a research report, for instance, the methods section, which is usually a subsection of the introduction, may be elevated to a major section.

FRONT MATTER

The term *front matter* refers to the sections at the beginning of a formal report that prepare the way for meaningful discussion. (An informal report generally does not have front matter.) These sections also make the information in the report easier for the reader to find. Although a report might not include all of the following elements, a formal report is likely to include most of them. The order used for these elements will depend both on specifications and on the most logical way of developing any specific topic.

Letter of Transmittal

A letter of transmittal is used when the audience for a report is either a known individual or a specific group of persons. It is attached as a separate piece of paper to the cover of the report. It explains that the report is attached, gives its full title, and outlines certain elements of interest to the reader. It may explain who authorized the report or when it was requested. It may cite problems encountered during the development of the report, or it may call for specific action on the reader's part. In addition, it may contain special acknowledgments to people who have helped. If the audience of the report is not known, the same purpose can be served by a general preface that follows the title page. Figure 14-1 shows a typical letter of transmittal.

Cover

A cover or jacket is usually required for a lengthy formal report. Covers are made of heavier stock than standard paper. They protect the report, enable it to stand upright on a bookshelf, and add an air of professionalism. The cover is secured to the report by clamps, rings, or prongs. A label, centered on the cover, should include the following information:

- the full title of the report
- the author's name and affiliation
- the date of submission

DESERT SUN CONSULTANTS, P.C.
5815 North 25th Drive
Phoenix, Arizona 85066
(602) 555-5392

December 29, 1986

James F. Warnock, Executive Director
Arizona Solar Energy Commission
Capitol Tower, Room 502
1700 West Linden Street
Phoenix, Arizona 85007

Dear Mr. Warnock:

I submit the enclosed report, entitled "The Feasibility of Incorporating Solar Energy and Conservation Technologies in Local Home Designs and Construction,"[1] under the terms of the contract (no. 478-19) between my consultation firm and your agency. It has been developed according to specifications we agreed upon on October 7, 1986.[2]

1. Title of enclosed report

2. Authorization

The status of the report appears only in the case of updates, where *preliminary, progress,* or *final* should be indicated. Figure 14-2 shows a sample label.

The label should be white, and the typeface should match that used in the report and the letter of transmittal. The binder should be a conservative color and a standard size (to hold 8½″ × 11″ paper).

Title Page

Many formal reports include a title page in addition to the cover. Like the cover, this page includes the full title of the report, the author's name and affiliation, and the submission date. The title page also includes a statement of authorization if the report has been requested rather than initiated by the

The report examines a wide range of active solar technologies, including solar collectors, water heating systems, space heating systems, and storage devices. It examines many passive solar technologies as well, such as Trombe walls, water walls, greenhouses, earth integration, thermal storage devices, and distribution techniques. It also examines a variety of energy conservation technologies that can be built into new homes or added to existing homes. Based on reasonable design and construction costs, availability of materials, and time constraints, several recommendations, which use a combination of these technologies, are included. [3] 3. Summary of report

I am indebted to Donald E. Osborn, projects manager of the Solar Energy Research Facility at the University of Arizona, who shared his facility's research findings and solar energy designs with me.[4] I hope this report will prove satisfactory. 4. Acknowledgment

Sincerely,

Steven Kline
President

Figure 14-1 The letter of transmittal briefly outlines the report for the reader.

The Feasibility of Incorporating
Solar Energy and Conservation
Technologies in Local
Home Designs and Construction

Steven Kline
Desert Sun Consultants, P.C.

December 29, 1986

Figure 14-2 The label identifies the topic and the author of the report.

author. The information is spaced neatly both vertically and horizontally on the title page. Figure 14-3 illustrates this report component.

Abstract

An abstract is a synopsis of the entire report brief enough to be read in only a few minutes. It provides an overview of the key elements of the report, enabling readers to digest important information without having to read the report in its entirety. In fact, an abstract enables a reader to determine if the entire report is *worth* reading; hence, a well-written abstract helps interest readers in your report.

There are two basic types of abstracts: indicative and informative. The *indicative abstract,* as its name suggests, indicates the kind of information included in the report. It is a description rather than an exposition. The *informative abstract* actually condenses the information in the report into a brief factual statement.

An indicative abstract provides a synopsis similar to the type used in the letter of transmittal in Figure 14-1. It is the type of synopsis you'll find in annotated bibliographies and card catalogues. Here is a sample of an indicative abstract. Note that it outlines what is covered in the report but does not attempt to provide a lot of information.

The purpose of this report is to discuss the feasibility of incorporating solar energy and conservation technologies into local home designs and construction.[1] Primary and secondary research sources were studied,[2] and careful consideration was given to the following criteria: required materials, ease of installation, required design modifications, efficiency, reliability, durability, and initial and operating costs.[3] The report concludes that a combination of active and passive solar technologies can and should be used by local builders.[4] Several of these hybrid systems are recommended, and designs for their implementation are included.[5]

1. Purpose
2. Method
3. Description of discussion
4. Key conclusion
5. Recommendations

The informative abstract, on the other hand, is intended to provide readers with sufficient information to render the reading of the entire document optional. It provides in condensed form all of the essential information contained in the report.

The informative abstract includes a brief discussion of the purpose, method, and scope of the document. It discusses the key sections of the text in the order followed by the report. Important conclusions and recom-

THE FEASIBILITY OF INCORPORATING
SOLAR ENERGY AND CONSERVATION
TECHNOLOGIES IN LOCAL
HOME DESIGNS AND CONSTRUCTION

Steven Kline
Desert Sun Consultants, P.C.

Authorized by James F. Warnock, Executive Director
Arizona Solar Energy Commission

December 29, 1986

Figure 14-3 The title page gives the title, the author's name and affiliation, the name of the person requesting or authorizing the report, and the date.

mendations are also included. In addition, an informative abstract may cite the special features of the report to help interest the reader in reading the entire document.

Brevity is the key; abstracts should rarely be more than a single page long, single-spaced. Nevertheless, abstracts should be written in complete sentences, and there should be graceful, logical transitions between sentences. Do not use abbreviations unless the reader will understand them without referring to the complete document. In fact, all information should make sense without any reliance on references to the report itself.

The following is an informative abstract of a 50-page report. The annotations in the margin point out the various report components that are condensed.

Recent legislation requires companies to distribute management positions equitably among their male and female personnel.[1] [2] Taking a broad-based view of adult education,[3] this report examines the feasibility of establishing specialized training programs to help women in the construction industry attain management positions.[4]	1. Introduction 2. Problem 3. Scope 4. Purpose
Traditionally, the construction industry has been dominated by males. In the past 10 years, the proportion of women in construction has risen from 3% to 12%; yet the proportion of women in management positions is only 1%. Research demonstrates that women in construction management positions exhibit the same skills as their male counterparts and remain in their jobs as long. Nonetheless, women are not hired for management positions, nor are they able to progress as quickly to management responsibilities.[5]	5. Discussion
This discrepancy exists because women have not received adequate training in the specialized business methods of the construction industry. Special education programs for women in industry are now being offered by the federal government as well as by construction engineering departments of major universities. Additionally, the National Association of Women in Construction has designed several programs to move women into management positions.[6]	6. Conclusions
Organizations can bring the skills of their male and female personnel into parity, thereby meeting federal guidelines, by encouraging women to attend specialized training programs or by establishing an in-house program.[7]	7. Recommendations
If an in-house program is established, it should fit the specific needs of the company, assess and select motivated	

women for participation, and provide incentives for participation.*

An abstract should be written only after the entire report has been completed. Here is a set of instructions for writing effective abstracts.

1. Photocopy your report so that you can mark it up as necessary.
2. Read it cover to cover without making any notations. As you read, imagine you are the audience.
3. Several hours later, reread the entire report. This time use a highlighting pen or underlining to pinpoint the most important information. Be certain to include information from each of the major sections.
4. Paraphrase each of the highlighted sections, limiting the total length to one page of single-space type.
5. Arrange these paraphrased statements in logical order (as close to that appearing in the report as possible).
6. Review the statements you have selected. Do they convey the essence of your report? If not, fill in the gaps. Is there redundancy? If so, select only the essential points. Be sure to maintain the third-person voice throughout.
7. Supply graceful transitions between sentences and paragraphs. Avoid a telegraphic style.
8. Several hours later, read the abstract to determine if it is self-sufficient. Provide explanations of terms or concepts that are not clear without reference to the report.

An *executive summary* is a specialized form of abstract addressing the needs of nontechnical readers—usually senior management. An executive summary focuses only on the conclusions and recommendations made in the report. These are arranged in order of priority to the reader. If the preceding abstract were written as an executive summary, it might read like this:

To comply with federal guidelines requiring organizations to distribute managerial positions equitably between male and female employees, our company must improve women's familiarity with construction business practices.[1]

Female employees should be encouraged to participate in a specialized in-house training program. This training program should fit the specific needs of the company. We should assess and select motivated women for participation and provide incentives for them.[2]

1. Primary conclusion
2. Recommendations

*J. W. Fleury, "Educational Programs for Women in the Construction Industry" (Phoenix Chapter of the National Association of Women in Construction: December 5, 1979).

TABLE OF CONTENTS

Acknowledgments

An acknowledgment is a statement of credit given to an individual or a group of individuals who have assisted the author of the report in a significant way. Such a statement names the person credited, the contribution the person has made, and the effect of this contribution. An acknowledgment is placed on a separate sheet of paper and follows the abstract of the report. See the appendix ("Tolerance of *Plantago ovata* to Three Herbicides") for a good example of an acknowledgment.

Table of Contents

The table of contents, which appears after the abstract and before the introduction, is a sequential listing of the topics contained in a report. The ii

Figure 14-4 The table of contents directs readers to various parts of the report.

heads used within the report body are given, with their page numbers, so that readers can find information in the report quickly and easily. (The report must be in its final form, including the pagination, before the table of contents can be completed.) Depending on the complexity of the report, several levels of subordination may appear in the table of contents, although you should not go beyond three levels. (See Chapter 4 for an explanation of levels of subordination in a decimal outline.)

Although several variations are possible, the format illustrated in Figure 14-4 follows an acceptable standard that you can use in the absence of specifications provided by your own organization. When more than one level of subordination is used in the table, the primary head appears in all capital letters. Subordinate heads are indented five spaces for each level of subordination and appear in upper- and lower-case letters. For readability, a series of periods is often used between the head and the page number.

Lists of Tables and Figures

The tables and figures that appear in a formal report should be listed in the front matter so readers can find them easily. Each list should appear on a separate page, and each table and figure should be identified by title and page number. Here are two brief samples taken from the feasibility report on solar energy technologies:

List of Tables

List of Figures

INTRODUCTION

The introduction of a report, like that of any document, sets the stage for further discussion. The information supplied in the introduction helps readers understand the points you make in the discussion and conclusion sections. Although all of the subsections listed here may not be required in a single report, each is valuable under given circumstances. When you are not required by specifications to include certain subsections, choose only the most appropriate ones.

Statement of the Problem

The statement of the problem is an overview of the situation being examined in the report. It defines the central problems or questions. Here is a sample problem statement taken from the feasibility report on solar energy.

1.1. Statement of the Problem

[1] During the past five years, electricity rates in the Phoenix area have more than tripled. This increase has occurred because the Arizona Public Service (APS) is passing on to consumers a percentage of the construction costs of its Palo Verde nuclear power plant. The plant, now scheduled to begin operation in January 1987 and to be fully operational by mid-1989, is already almost two years behind its original schedule. Safety, design, and engineering problems have increased the original cost estimate of the project by more than 400%. Each new expense has translated into a rate increase for consumers, who are already paying an average of $200 each month for electricity. Once the plant is totally on-line, consumers will pay an even greater percentage of the overall construction costs.

1. Overview: first part of the problem

[2] Added to this situation is the rate at which the population of Phoenix is growing. The huge demand for new housing allows builders to construct houses that simply meet current energy needs. Builders are clearly not anticipating the higher costs and lessened availability of energy that homeowners will have to face less than five years from now.

2. Overview: second part of the problem

[3] In light of these two problems—expensive energy and energy-inefficient homes—consumers are becoming increasingly interested in alternative forms of energy, such as solar energy, and in alternative design and construction techniques that optimize energy use.

3. Overview of solution

Notice that this problem statement makes sense without any reliance on other sections of the report. It also sets the stage for the statement of purpose that follows.

Statement of Purpose

This subsection of the introduction acts as a kind of contract between you and your readers. Here you explain what readers can expect to gain or understand by reading your report. Generally, the purpose statement is restricted to two or three concise sentences (and frequently can be taken verbatim from your situational analysis). Here is a sample:

> The purpose of this report is to recommend the best alternatives to the subsidized housing program in Cincinnati through an examination of the current program.

A second example (taken from the solar energy feasibility report) is a bit lengthier.

1.2. Statement of Purpose

The purpose of this report is to determine the feasibility of incorporating solar energy and conservation technologies in new homes in the Phoenix area. Through an examination of various active and passive solar methodologies, we hope to recommend a set of design and construction suggestions that local housing developers can implement in a cost-effective, timely manner.

The purpose statement usually follows the problem statement in the introductory section of a formal report. If these two sections are very brief, individual headings may not be necessary. Sometimes these two categories of information can be handled in the same paragraph.

1.1. Problem and Purpose

All health maintenance organizations (HMOs) in Atlanta currently provide pharmacists with a standard price list for medications; HMOs will pay only those prices indicated on the list.[1] In some cases, a pharmacist's services and drugs cost less than the allotted amount, but more often than not, the pharmacist loses money when compensated according to these prices.[2] The purpose of this report is to examine the possibility of establishing service charges that close the gap between HMO allocations and the actual costs incurred by pharmacists.[3]

1. Background
2. Problem
3. Purpose

Statement of Scope

The statement of scope details *how* the purpose will be accomplished. It outlines the criteria that will be applied, the data that will or will not be discussed, and any constraints that will be placed on the report's conclusions or recommendations. This statement is particularly important in feasibility studies, where it sets the ground rules for the argumentation that follows.

The following sample of a statement of scope explains both what information will be found in the report and the reason for excluding a category the reader might otherwise expect to see.

1.3. Statement of Scope

This report reviews the most current literature on bacterial nitrogen fixation in nearly all Gramineae species.[1] Bacterial nitrogen fixation in rice, although important, has been omitted because the cultural practices and the crop physiology associated with this grass are unique.[2]

1. Report restricted to literature review
2. Reason for omitting one topic

A statement of scope might also focus on the ways in which each topic is structured. In the feasibility study of solar energy technologies, the statement of scope looks like this:

1.3. Statement of Scope

This report examines active and passive solar technologies as well as energy conservation methods that could be used by local builders of new homes.[1] For each technology, the following information is presented:

- mechanism description
- statement of materials required
- description of installation
- list of necessary design modifications
- statement of efficiency
- statement of reliability and durability
- statement of start-up and operating costs

[2]

A brief survey of conventional technologies (e.g., electricity and natural gas) is included so that a meaningful comparison can be made.[3] A description of Arizona's solar conditions is also included. The summary of facts integrates the data into categories based on overall cost-effectiveness.[4] The conclusions focus on the feasibility of implementing any or all of these design technologies.[5] The recommendations cite specific actions this firm would like to see executed by both the Arizona Solar Energy Commission and local builders.[6]

1. Categories
2. Points of discussion within each category
3. Comparison
4. Focus of summary of facts
5. Focus of conclusions
6. Focus of recommendations

Statement of Method

The statement of method describes how the information in the report was obtained. Depending on the type of report, this statement may be quite simple or quite elaborate. It may in fact be the focal point of the report, as in the case of research or experiment reports, where the results would be suspect if the methods were unconventional. In literature review reports, however, the method statement may be very brief. Let's look at two examples and the kind of information each conveys.

1.4. Method of Acquiring Data

This report was researched in the Solar Collection of Arizona State University's Science and Engineering Library and at the University of Arizona's Environmental Research Laboratory in Tucson. Current articles and designs were the primary sources of information. Interviews with architects and solar engineers also provided some information.

In this first example, we are told where the information for the report was uncovered, the nature of those sources, and how some supplemental information was obtained. In reports that are based almost entirely on secondary research, no more elaborate statement of method may be necessary.

However, whenever the readers of a report are likely to scrutinize the results of research, the method statement should be more comprehensive. Consider this statement from a report that assessed the information processing requirements of a police department.

1.4. Data Sources

In developing a description of the information processing requirements of the Des Moines police department, we used a number of data sources. A primary source was our firsthand study of the Des Moines police department's current information processing system, which included interviews with users about the system's operation and effectiveness.[1] Another major source was a review of the pertinent literature, particularly reports written by United States and Iowa law enforcement agencies, bureaus, and committees; papers and books on police operations and data handling; and reports on information system studies, system designs, and operational results in various state and local police jurisdictions throughout the country.[2] A third source was the information about data handling in police departments presented in meetings held for grantees and

1. Primary research

2. Secondary research

contractors by the Iowa Governor's Committee on Law Enforcement.[3] A fourth source was the information collected during visits to the police departments of Omaha and Lincoln, Nebraska, to review their current information systems and their future plans.[4] A fifth source was the technical approaches developed in a number of previous CPTL projects about information processing.[5] A complete list of data sources is given in Appendix VI.

From this set of source materials, a picture of the information processing needs of a police department in a medium-sized city like Des Moines has been developed.[6]

3. Secondary research
4. Primary research
5. Secondary research
6. Restatement of purpose

In this method statement, the writer lists each of the major sources of information used in the report. The aim is to demonstrate that no important source was overlooked and that the results reflect the most current thinking on the subject.

Background Statement

A background statement provides general information that helps prepare the reader for the more detailed discussions to come. Its length and substance vary according to the complexity, audience, and purpose of the report. For instance, an expert audience may require only a very brief statement that presents some new developments of which they may not be aware. A nonexpert or mixed audience, on the other hand, may need an extended explanation of key factors, trends, terms, or procedures. Consider a report about brain-function testing written for a hospital funding committee composed of medical experts and nontechnical investors. The report writer might include in the background statement a brief mechanism description of the brain to help explain such functional components as the brainstem, the cerebellum, the limbic system, the cerebrum, and the cortex. Having provided this foundation, the writer can be more confident that the entire committee will understand the research outlined in the discussion sections. Here is a portion of such a background statement:

1.5. Background

The human brain is made up of many parts that work together to govern our thoughts, actions, and reactions.[1] One part of the brain, which dates back to our earliest days on earth (over 500 million years ago), is called the *brainstem.* The brainstem is made up of three subparts—the midbrain, pons, and medulla—and determines our general level of alertness. This "ancient" brain is also responsible for such basic body functions as the regulation of breathing

1. Overview statement
2. First category

and heart rate. Attached to the brainstem is the *cerebellum*, the portion of the brain that coordinates muscle movement. This report will point to strong evidence that the cerebellum acts as a storage facility for simple learned responses.

3. Second category

Also near the brainstem is a group of cellular structures collectively called the *limbic system*. Two key parts of this system are the hypothalamus and the pituitary gland. Electrical and chemical messages from this system maintain body temperature; blood pressure; blood sugar levels; balance; eating, drinking, sleeping, and waking urges. . . .

4. Third category

If the majority of the audience is relatively knowledgeable about the topic of a report, they need only skim the background statement, but the nonexpert readers will benefit from it greatly. There may be times when the nonexpert reader needs even more detailed discussion than you can provide in a brief introduction. When this is the case, include the information in an appendix and refer readers to it or list several key texts they can consult for more information.

Generally, however, the background statement simply needs to provide enough information so that discussion can take place. Let us look at another typical background statement, taken from the feasibility report on solar energy technologies. It presents definitions of the key terms the writer uses in the discussion section.

1.5. Background Statement

A solar energy device is a series of mechanisms designed to collect radiation from the sun and transform it into heating or cooling energy or into electrical or mechanical power.[1] These results can be accomplished by either active or passive means.[2]

1. Definition
2. Description of categories

Active systems use some mechanical subsystems, such as fans or pumps, to transfer solar energy from collection points to storage facilities. These subsystems also usually involve roof-mounted collectors, fluid-transfer components, storage tanks, and precise control devices.

3. First category

Passive systems, on the other hand, do not use mechanical subsystems; instead, the design and structure of the building fosters the collection and storage of solar energy. Distribution of energy occurs through convection, conduction, and radiation. Passive solar systems require special designs based on careful consideration of local climates.

4. Second category

An example of an active system is a solar hot-water heater with a pump attachment. An example of a passive system is an attached solar greenhouse used for space heating.

5. Examples

In practice, passive systems are often modified into *hybrid* systems, in which some mechanical assistance—blowers, thermal dampers, switches, or controls, for instance—is used to augment heat flow.[6]

6. Qualification

DISCUSSION

The discussion section of the report is the core of the document. This section carries out the promises made in the statement of purpose.

This portion of the report is generally the longest. Hence, the use of clear headings for specific points of information is very important to your readers. Information within a subsection should not depend on readers' familiarity with any other subsections of the discussion, nor can you rely on their having read the introduction, since reports are designed to be skimmed rather than read in their entirety. The following sample taken from the discussion portion of the feasibility report on solar energy technologies demonstrates this self-sufficiency.

2.3. Passive Solar Technologies

Passive solar buildings are designed to use the natural movement of heat and air (i.e., radiation, conduction, and convection) to maintain comfortable temperatures within the buildings.[1] The passive solar technologies we will examine include Trombe walls, water walls, solar greenhouses, earth-integrated structures, thermal storage devices, and distribution techniques.[2]

1. Basic overview of subject
2. Points covered in this section

2.3.1. The Trombe Wall[3]

3. First point

A high-mass wall with an external glaze that traps solar radiation at the surface of the wall is often referred to as a Trombe wall after the French solar engineer who popularized the idea.[4] During the day, the blackened external surface of the wall becomes increasingly warm. The heat builds up between the glazing and the wall and gradually moves inward, warming the cooler regions of the wall. Within a period of several hours, the inner surface of the wall is warmed, and heat is passed to the room itself. The

4. Definition
5. Mechanism description

time required for this process can be calculated according to the thickness and composition of the storage wall. Typically, the process takes 8 to 12 hours, which means that heat is delivered to the interior areas in the evening hours, when it is needed most.

Concrete masonry units with cores of mortar or grout provide sufficient mass to allow adequate heat build-up during the day.[6] Standard-sized double-glazed units can be mounted in wooden frames; joints can be sealed against heat loss; and diversion frames can help prevent rainwater penetration.[7]

6. Materials

7. Installation

The content and organization of your discussion will vary according to your purpose and the needs of your audience. The feasibility report about solar energy technologies uses a division structure (see Chapter 6) to organize the technologies into three categories: active, passive, and conservation. Within each category, the writer conducts an identical discussion of points with elements arranged in graded order of importance, since readers will be making decisions about the advisability of each technology.

In a report on mass spectroscopic analysis of pesticide quantities in human tissue, the writer's purpose was to disclose the results of an experiment using new equipment. The writer included a mechanism description (Chapter 9) as part of the discussion to explain how the equipment measures pesticide retention rates. Using a procedure description (Chapter 8), the writer outlined the steps taken to test the equipment and to set up the experiments. This discussion followed a simple-to-complex arrangement so that readers were introduced gradually to more complicated information. This method of ordering information was appropriate because the readers, knowledgeable about traditional pesticide testing procedures, were unfamiliar with the particular mechanism and techniques being described in the report.

In a report on the various transportation options available to the city of Concord, New Hampshire, the writer organized possible solutions into categories according to the mode of transportation. Within each category, there was a parallel discussion of points. The table of contents reflects the overall plan:

The report about the information processing requirements of a police department used still another arrangement. It began with an examination of the current system, then described the recommended system. The final major point of discussion was the implementation of the recommended system. The first two points were discussed as a comparison; the third point was primarily a procedure description.

The best method of organizing the discussion section is the one that best supports your conclusions and recommendations. A general-to-specific (deductive) order may be most appropriate when it is important to create a context and then focus on a particular level of investigation. When a writer is faced with busy readers whose attention must be captured quickly, a descending order of importance works well. On the other hand, it may be effective to build to a climax of crucial points and, if so, an ascending order of points may be valuable. Whatever method of organization you choose, be sure to back up your points with sufficient information to support your conclusions and recommendations.

Results

Primary research, such as an exploration, an investigation, an experiment, or a statistical survey, always yields results of some sort. A report based on primary research must describe those results. Generally, a subsection of the discussion, the results section, is dedicated to this description. This subsection is a *non*interpretative compilation of findings. Interpretations and conclusions are reserved for a separate major section.

The results subsection will form part of the foundation for the conclusions and recommendations sections of your report; however, it is important to note that your results may not always be conclusive. Sometimes the only conclusion that can be drawn is that more research is required.

To develop an effective results section, return to the questions you hoped your experimentation and research would answer. Determine if your research has in fact turned up verifiable answers. If so, list them under headings that reflect the nature of your findings.

Depending on the complexity of the primary research, the results section will vary in scope and length. For instance, the following sample from the *Plantago ovata* report presents the results of a single test in a series of extensive tests. For each type of test, the same pattern of presentation is observed, and to help readers compare results, the author refers to tables that compare data.

3.2. Tests with a Single Herbicide

3.2.1. Effects on Seedling Survival[1]

1. First criterion

Paraquat usually reduced survival of plantago "AZ 316" (Table 1) and "AZ 329" (Table 2) seedlings at rates higher than 0.25%.[2]

2. Paraquat

The results with diquat were less consistent. In some tests, rates of diquat higher than 0.125% caused reduction in the survival of seedlings of both plantago varieties. In other tests, rates as high as 0.25% diquat did not reduce seedling survival (Tables 3 and 4).[3]

3. Diquat

Diuron did not greatly reduce the survival of seedlings of either variety at any of the rates tested (Tables 5 and 6).[4]

4. Diuron

3.2.2. Effects on Seedling Height[5]

5. Second criterion

In some tests, treatment with paraquat at rates of 0.25% or higher significantly reduced the height of "AZ 316" and "AZ 329" plantago seedlings. In all tests, paraquat at rates of 0.5% or higher reduced seedling height (Tables 1 and 2).[6]

6. Paraquat

Diquat reduced the height of both varieties of plantago seedlings in most tests at rates higher than 0.125% (Tables 3 and 4). In the first trial of diquat on "AZ 316," all treatments significantly reduced seedling height (Table 3).[7]

7. Diquat

Diuron reduced the height of plantago "AZ 316" seedlings in one trial. This reduction occurred at the highest rate of diuron tested (0.56 kg/ha) (Table 5). Diuron did not significantly reduce the height of "AZ 329" seedlings at any of the rates tested (Table 6).[8]

8. Diuron

Some burning of foliage was observed at all rates of paraquat, diquat, and diuron. Increases in degree of leaf burn with increasing rates of these herbicides contributed to the significant differences in seedling height observed in some tests.[9]

9. Qualification

3.2.3. Effects on Root Development[10]

10. Third criterion

Paraquat did not reduce root development of plantago "AZ 316" seedlings at any of the rates tested (Table 1). Paraquat reduced "AZ 329" root development at all rates in the first trial and at the highest rate only (0.75%) in the second trial (Table 2).[11]

11. Paraquat

Plantago "AZ 316" root development was significantly reduced at two of the four rates of diquat tested (0.25% and 0.5%) in the first trial. Root development was not reduced at the highest rate of diquat (0.75%) in this trial. When this test was repeated (Trial II), no significant reduction of "AZ 316" root development was observed with any of the diquat treatments (Table 3). All diquat treatments reduced root development of plantago "AZ 329" seedlings in the first trial. In the second trial, no significant reduction of "AZ 329" root development was observed at any of the rates of diquat tested (Table 4).[12]

12. Diquat

Treatment with diuron did not reduce root development of either plantago variety at any rate tested (Tables 5 and 6).[13]

13. Diuron

The effectiveness of a results section depends heavily on the primary research: a well-planned experiment, investigation, or survey will yield measurable, reproducible results; a poorly executed project will not.

Summary of Facts

Lengthy reports with no results subsection can be enhanced by a summary of facts subsection. As its name suggests, this subsection distills the key points presented in the body of the report. Readers can use this section as a quick synopsis of the report. You, the writer, can use the section to reinforce the ideas on which your conclusions and recommendations will be based. This section also integrates data so that interrelationships become clearer and more meaningful.

This section should run approximately one-half page per each ten pages of discussion. This kind of summarizing, then, requires you to exercise considerable selectivity. Here is an excerpt from the summary of facts in the solar energy feasibility report. Note that this synopsis supplements the indicative abstract presented earlier, which had described, rather than summarized, the information in the report.

2.8. Summary of Facts

With the increase in electricity costs and the rapid rise in the population of Phoenix, alternative forms of energy technology, specifically, solar technologies, should be explored for inclusion in local home designs.[1] The low deserts of southern Arizona are particularly suited to solar technologies since this area receives more days of sun-

1. Problem

light per year than most other areas of the United States.[2] With this extra sunlight comes an increased demand for cooling technologies and a lessened demand for heating technologies.[3]

2. Climate

3. Cooling demands

One active system, solar water heating, has been readily 4
accepted in certain parts of the area and is available in several housing developments. Active whole-space heating and cooling are still a long way from widespread implementation in the Phoenix area. Active methods require few design or structural modifications before installation. The cost effectiveness of these systems is exceptional, with users reporting reductions in their electricity bills of 60% to 80%.

4. Active technologies

Cooling systems using passive solar methodologies, 5
such as earth integration, concrete or masonry wall construction, or water walls, can also help retain heat during the winter months. These techniques require expensive design and structural modifications but save more money in energy costs over long periods of time than active systems. Other passive solar technologies, such as the Trombe wall and thermal rock-bed storage, are designed for heating purposes; they are effective in outlying areas and higher elevations (such as Flagstaff), but they do not seem advisable in the Phoenix area.

5. Passive technologies

Conservation technologies that can enhance the effec- 6
tiveness of either active or passive solar systems are particularly attractive to housing developers. Relatively inexpensive features such as high-grade insulation, double windows, overhangs on southern exposures, proper orientation of the house to the sun, landscaping, and the use of masonry or concrete over frame can be implemented in current plans for housing developments without many modifications.

6. Conservation technologies

In recent years, several laws related to solar energy have 7
been passed by the Arizona legislature. ARS 41-571 through 575, for instance, created the Arizona Solar Energy Commission, whose purpose is to coordinate federal and regional solar energy programs for Arizona. The commission develops and implements solar energy standards and codes and publishes up-to-date information about solar energy on a periodic basis. Rule R12-16-101 requires that anyone who manufactures, furnishes for installation, or installs a solar energy device must provide the customer with a written warranty statement. . . .

7. Solar energy laws and regulations

Homeowners may claim a tax credit against their Arizona income tax liability when they install a solar energy device. . . .[8]

8. Incentives

Compared to conventional heating and cooling devices, solar technologies require higher start-up costs. However, even without tax credits, these solar devices have a pay-back period of between one and fifteen years (depending on the nature and expense of the device). These estimates of pay-back are based on known trends of electricity rate increase and do not anticipate any catastrophic events such as electricity shortages, power plant down time, or power plant sabotage.

9. Comparison to conventional technologies

CONCLUSIONS

The conclusions section, which follows the discussion, serves as the interpretive synthesis of the document. Based solely on information presented in the introduction and discussion, this section evaluates research, methods, and findings in order to draw a set of comprehensive, valid conclusions.

To develop the conclusions section of a report, return to your purpose and problem statements to ensure that you have met the objectives you set for yourself. Examine all information in the discussion section to determine what, if any, conclusions can be made. To be most effective, arrange points in descending order of importance. To help readers see clearly the basis of your conclusions, you may wish to place after them in parentheses the page or section numbers where the appropriate supporting information appears in the body of the report.

The conclusions section from the feasibility report on solar energy technologies follows. This section is particularly succinct because the summary of facts prepared readers for these inferences.

3. Conclusions

The primary conclusion of this feasibility study is that the incorporation of solar energy and conservation technologies into home designs and construction is not only possible but highly desirable. Minimal modifications of plans for housing developments, such as the reorientation of houses in relation to the sun, can help each home be more energy efficient. Installation and materials costs for conservation systems can be factored into the cost of the new homes. Building schedules, except for the most sophisticated passive solar structures, are not affected by the addition of these systems.

1. Most important conclusion

A combination of technologies will work best in the Arizona climate. Solar hot-water heaters (which are active systems) help reduce electricity bills, can be credited against consumer tax liability, and are highly efficient. Conservation technologies such as high-grade insulation, masonry construction, and a north-south orientation of the house will increase the benefits of any solar system as well as reduce electricity bills. Passive cooling techniques are more expensive but highly desirable in an arid climate. There has been limited public acceptance of the concept of earth integration, however.[3]

2 — 2. Second most important conclusion

3. Qualification

Note that conclusions must be verifiable. Armed only with the information in the report, readers should be able to develop their own data that support the same conclusions. However, sometimes the information you uncover yields conflicting conclusions that you cannot resolve. When this is the case, you should point out this conflict to readers and then deal with it in one of several ways:

1. Discern a difference between cases and cite this variance as an explanation of the discrepancy in information. For instance, a professor studying *Plantago ovata* in its native habitat of India may provide different data on herbicide usage than does a graduate student who studies these plants in a laboratory environment. Factors contributing to the difference in data are the heartiness of the native plant stock and the natural growing conditions compared to the controlled environment.
2. Discount some of the information yielded. For example, in a statistical survey of Republican voters (which yielded valid results), the writer was able to discount data on a candidate's popularity because the survey was conducted prior to the candidate's announcement that he had accepted illegal campaign contributions.
3. Simply state that an unresolved or unaccountable conflict exists. These situations often arise when you are dealing with expert opinions. For instance, one economic adviser may say that it is possible to end deficit spending without cutting back on vital government programs. An equally respected authority may say that it would be impossible to do so. After carefully examining each position, you may have no choice but to report that these conflicting views exist.

RECOMMENDATIONS

A key section of a report is the set of recommendations that the writer offers to the reader. These recommendations should be based solely on information presented in the report. Written with the scope, purpose, and audience of the report in mind, the recommendations section should be a comprehensive list of *feasible* plans, policies, or actions. For instance, in a situation that involves budgetary or scheduling constraints, options that exceed the monetary or time limitations should not be presented.

Recommendations should be listed in descending order of importance. The length of this section varies according to the complexity of the report. It is often possible, however, to present a simple, general statement:

4. Recommendations

This feasibility study makes the following recommendations:[1]

1. Introduction to recommendations

1. Arizona developers should be encouraged by legislative measures and tax incentives to include solar energy technologies in new housing.
2. This report should be made available, free of charge, to all home developers in the Phoenix area.
3. The Arizona Solar Energy Commission should conduct free public seminars on the solar technology options available to homeowners. In this way, home buyers will be more likely to demand that such technology be built into their homes.
4. If a developer or home buyer wishes to make use of one of the designs included into this report, he or she should contact the architect, since each of the designs is copyrighted.

[2]

2. Recommendations given in descending order of importance

Or a more elaborate set of suggestions can be made. The following sample is an excerpt from a lengthy set of recommendations. Note that each recommendation is briefly discussed, rather than merely presented.

4. Recommendations

In spite of the drawbacks of the mass transit system,[1] it is becoming a preferred mode of transportation for an increasing number of Concord residents. It is recommended that alternative funding sources and their impact on transit users and nonusers be further investigated and analyzed. Investigators should depend on active public discussion and participation.[2]

1. Qualification

2. Recommendation for further study

It is also suggested that alternate modes of transportation, such as car pooling, be examined. Their effect on the city's transportation network as well as the public's willingness to use such methods of transportation should be determined. 3

3. Recommendation for further study

In addition to investigating alternate transportation modes, the engineering department should analyze other strategies for reducing the peak-period volume of traffic. An example of a reduction strategy is a staggered work-hour program, or "flex-time." By spreading traffic congestion out over a longer period of time, the capacities of both the transit and the roadway system could be increased. 4

4. Recommendation for further study

BIBLIOGRAPHY

Most reports include a bibliography, which is a list of sources of information that follows a prescribed format. Each technical and scientific discipline has its own format for documenting sources. For instance, the Modern Language Association generally sets bibliographic style for the languages and "soft" sciences such as psychology and social studies, and it revises those stylistic standards almost every year. The Institute of Electrical and Electronic Engineering has its own bibliographic style. The American Association of Physicists uses yet another style, and so on. It is your responsibility as a writer of technical documents to find out which specifications apply to the report you are working on.

The more comprehensive your bibliography, the more helpful it will be to readers who may wish to do some research of their own. Your bibliography will also help to demonstrate that you have done your homework by consulting a variety of important references.

There are several types of bibliographies: preliminary bibliographies, lists of works consulted, lists of works cited, and annotated bibliographies. Each type has a different purpose and effect.

Preliminary Bibliography

As its name suggests, a preliminary bibliography is a list of documents (articles, books, reports, proceedings) that have been consulted during the early stages of a long-term project. It presents a sampling of available sources and generally includes only a few works. The purpose of a preliminary bibliography is to demonstrate that some research in the area has been done by the developer of the report. In later versions of the report, this kind of bibliography will be supplanted by one of the other types of bibliographies we will be discussing.

A preliminary bibliography appears most often in a proposal or a progress report, where its purpose is to show your readers that you are acquainted with several key books and articles on your topic. The bibliography strengthens such a report by enabling readers to look up information on their own. A sample preliminary bibliography appears in Chapter 13 on page 328.

List of Works Consulted

A list of works consulted cites only those books, articles, papers, and reports that were used in the development of your document. The following list of works consulted shows how particular kinds of documents should be cited. Note that the list is alphabetized.

List of Works Consulted

Andersen, Philip. *Economic Stress in the 1980's.* Kalamazoo, MI: Smithson Press, 1979.[1]	1. Book with one author
Branston, Gerald, and Ursula Connors. *Introduction to Systems Cost-Effectiveness.* Seattle: Touchstone Press, 1973.[2]	2. Book with two authors
Charles, Anton, et al. *Modern Economic Modeling.* Lawrence, KS: University of Kansas Press, 1984.[3]	3. Book with three or more authors
Darren, Homer. "Computers and Economics." In *The History of American Economics.* Ed. Randal Redman. Boston: M.I.T. Press, 1982, 679–685.[4]	4. Chapter in a book
Ellery, Albert. "Economic Energies." *Journal of Modern Economics,* 76 (April 1986), 943–946.[5]	5. Article in a journal
Fuller, Terrence. Lecture: "Constructing a Strong Economy." November 6, 1986, at Rice University, Houston, Texas.[6]	6. Lecture
Gant, William, ed. *Dictionary of Economic Terms.* Detroit: Scholastic Publishing, 1978.[7]	7. Book with editor
"Justification of Inflation." *Wall Street Journal.* January 5, 1986, 7–8.[8]	8. Newspaper article without byline
Lamont, Elizabeth. "Fiscal Fumbling." In *The Proceedings*	

of the 4th Annual Convention of Bank Officers, 72–80. March 12–18, 1986, at Las Vegas.[9]

U.S., Congress, Senate, Special Subcommittee on Economic Evaluation, *Foundations of Sound Economic Policies,* 87th Cong., 2d sess., 1962, 229–230.[10]

9. Paper in a bound proceedings
10. U.S. Congressional document

List of Works Cited

A list of works cited presents only those works from which specific ideas and quotations have been taken. This type of list often takes the place of a footnote page in a formal report. The works are arranged alphabetically and numbered sequentially; when a statement or an idea from one of the works on this list is used, the author simply places that number in parentheses at the end of the borrowed section. Here is an example of how those parenthetical references would appear in a document:

The mucilage properties of the seed coat of plantago have made the plant useful in a number of commercial products, industrial processes, and home remedies. The plantago seed husk has been used as a mild laxative (21); in the treatment of diarrhea, especially in the case of dysentery (50, 73); in the treatment of urinary disorders (29); for the preparation of poultices and soothing lotions (21); in the treatment of respiratory ailments such as coughs (29); and in the treatment of varicose veins (50).

Annotated Bibliography

An annotated bibliography is a list of sources that contains short descriptions of each work cited. Any of the bibliographies we have mentioned could be annotated. Here is an excerpt from *Business and Technical Writing: An Annotated Bibliography of Books, 1880–1980.*

Cobb, Charles. *Practical Communication for Technical and Vocational Arts.* Santa Monica, CA: Goodyear, 1978. Index. 204p.

This text-workbook with tear-out pages has 13 chapters covering such topics as "The Basics of Communication," "Work Orders and Estimates," "Other Necessary Forms and Records." All chapters end in a "learning quiz" which reviews the text and projects that require students to write a response, complete a form, or prepare graphics materials. There are six appendices covering grammar, punctuation, and vocabulary. There are many examples of forms for memos, work orders, job records, insurance claims, maintenance requests, etc.*

NOTES

Often, reports will use notes rather than a bibliography to give information about the documents cited. The documentation style followed in your field will determine whether this use of notes is appropriate. Note specifications also vary from field to field; therefore, it is your responsibility to become familiar with the standards applied in your area. There are several types of notes: footnotes, parenthetical author citations, parenthetical numbers keyed to the bibliography, and endnotes.

The first method, footnotes, uses a raised number or letter at the end of the quoted or paraphrased material that is keyed to reference information appearing at the base of the page. In this case, a bibliography at the end of the report is not needed because resource information appears in the document as required. The following excerpt illustrates this technique:

> A high-mass wall with an external glaze that traps solar radiation at the surface of the wall is often referred to as a Trombe wall after the French solar engineer who popularized the idea.[4]
>
> [4]Jeffrey Clark, *Solar Homes Can Be Cool Homes* (Phoenix: Solar Arts Press, 1979), pp. 41–42.

A second kind of note places the author's last name and the copyright year in parentheses immediately after the quoted or paraphrased material. When this method is used, a bibliography is required so that readers can find the full reference by skimming the alphabetical listing. Parenthetical author citations look like this:

> A high-mass wall with an external glaze that traps solar radiation at the surface of the wall is often referred to as a Trombe wall after the French solar engineer who popularized the idea. (Clark, 1979)

A third method is to insert a number in parentheses after the quotation or paraphrase. This number is keyed to the bibliography (which must be numbered sequentially according to its alphabetical arrangement). In this case, the citation looks like this:

*Gerald J. Alred, Diana C. Reep, and Mohan R. Limaye, *Business and Technical Writing: An Annotated Bibliography of Books, 1880–1980* (Metuchen, NJ: The Scarecrow Press, 1981), p. 57.

A high-mass wall with an external glaze that traps solar radiation at the surface of the wall is often referred to as a Trombe wall after the French solar engineer who popularized the idea. (6)

A final way of presenting notes employs a combination of raised numbers at the point of reference and a separate listing called endnotes, which appears after the recommendations section in lieu of a bibliography. This listing is not alphabetical but follows the sequence in which these quotations and paraphrases occur in the report. Endnotes look like this:

[12]Jeffrey Clark, *Solar Homes Can Be Cool Homes* (Phoenix: Solar Arts Press, 1979), pp. 41–42.
[13]William Watt, "Use of Cellulose House Paints to Increase Heat Gain," *Sun-Life* (February 1984), 53–57.

The citations for footnotes and endnotes always list the author's name, the title of the document, the city of publication, the publisher, the publication date, and the appropriate page numbers.

Unlike the bibliography, in footnotes or endnotes the author's first name comes first. If you choose to include a bibliography in your report in addition to the notes, you do not have to cite the publishing company or the city of publication in your footnotes or endnotes. However, if your footnotes or endnotes will be your only reference listing, be sure to include these elements.

Note that when parenthetical citations are used, readers must turn to the bibliography at the end of the document to find the source of the information. You then run the risk of losing your audience's interest as they flip pages. Footnotes, however, provide immediate access to source information.

APPENDIX

Formal reports sometimes require a section of additional information to be included outside the body of the document. This section, called an appendix (singular) or appendices (plural), usually contains information that is too large or bulky to be included in the body or that may be of interest to readers but is not directly related to the topic of the report.

When an appendix serves as a repository for oversized or bulky items such as maps, drawings, charts, or diagrams, these items are usually mounted on heavier stock than that used for the rest of the document. When you are including tables or 8½″ × 11″ figures, you can use the same paper you used in the report itself. When you are including supplementary materials such as articles, interview transcripts, notes, or survey tally sheets, you can also use the same paper stock. In all cases, each appendix should carry its own title and section number. These should be listed in your table of contents with sequential pagination continuing from the last page of the report's body.

CHAPTER SUMMARY

Reports are the most common type of technical document and employ all the writing skills we have studied in this textbook. Generally, they are written according to pre-established guidelines, standards, or specifications. Careful situation analysis, information gathering, planning, writing, and rewriting are required for high-quality reports.

Reports can be formal or informal and can be further categorized according to their nature and purpose. The categories include updates, analyses, experiment reports, investigation reports, and literature reviews. Many formal reports are divided into various components, including the front matter, the introduction, the discussion, conclusions, recommendations, reference citations, and appendices.

EXERCISES

Do the following exercises in sequence:

1. As you research a report you are currently working on for another class, write each piece of information you intend to use on a note card. Arrange these cards into a logical order that will help you create a detailed outline.

2. Write a situational analysis for your report. Consider the following characteristics of your audience:

- formal training in field
- experience with data presented
- age

- gender
- expectations
- need for this information
- budgetary constraints
- intellectual ability
- level of responsibility

Also consider your own role, the format specifications, and the purpose of the document.

3. Develop a detailed decimal outline that follows accepted specifications (either those assigned by your instructor or those outlined in this book) and that meets your audience and purpose requirements. Use phrases that can be used as headings in the report.
4. Examine each topic in your outline to determine its relation to other topics, its importance to your audience or to the conclusions you will be making, the level of detail it deserves, and the kinds of examples that might be necessary. Fill in any gaps.
5. Examine each major topic in your outline to determine whether it yields results, conclusions, or recommendations. Determine the relationship of any results, conclusions, or recommendations to others. Arrange these in order of importance to your audience and purpose.

WRITING ASSIGNMENT

Choose a topic in your major field that will be useful for you to explore. You may choose a subject that you are working on in another class this term; however, do not select a project that you have completed during a previous term. Write a 10- to 15-page report. Rely on very current information for this report and focus the topic sufficiently so that it can be explored in a report of this size.

Imagine the following set of circumstances:

- You are a project leader in an industry or business related to your field.
- Your audience is your immediate supervisor. This individual is as knowledgeable as you are but is not aware of the specifics of your project.
- Your supervisor has called for formal reports from all project leaders. He or she will use these reports to write a comprehensive report to the division head.
- Although it has not been stated directly, the purpose of this report is also to present an accurate picture of your abilities as a project leader. Your supervisor will

probably use this report to aid in his or her decisions regarding your promotion and salary increase.

- The industry in which you work follows the specifications outlined in this chapter. (See "Report Components.") The reader will accept *no other mode of presentation.*
- You must use current and recent data from at least seven reputable sources.

NOTE: Apply the editing criteria at the end of Part III to your assignment before you submit it to your instructor.

C H A P T E R

15

Manuals

CHAPTER PLAN

As technical equipment has become more complex and consumer products more sophisticated, the manual has become a very important document. Users rely on manuals to explain the operation and capabilities of a mechanism. Equipment manufacturers are obligated by law to instruct owners about the mechanism and warn them of potential hazards. To help prepare you to write effective manuals, this chapter will

- explain the purpose of manuals.
- examine the ways in which the technical writing process applies to the development of manuals.
- discuss elements common to most manuals as well as elements specific to operation and service manuals.

APPLYING THE TECHNICAL WRITING PROCESS TO MANUALS

A technical manual is a document designed to explain the features and functions of a mechanism and guide readers through approved operating procedures. Although manuals can focus on subjects other than equipment, such as policies and procedures, we will concentrate here on operation and service manuals—the types of documents technical staff members are most likely to write.

The operation manual describes normal operating procedures and gives a functional description of the mechanism. Its purpose is to familiarize the consumer with his or her latest purchase, instruct users on proper operation of the mechanism, provide maintenance guidelines, and explain how to store the equipment. The audience for the operation manual can range from nontechnical, first-time users to expertly trained and skilled specialists. In all cases, however, this kind of manual relies heavily on mechanism and procedure description techniques to accomplish its objectives.

The service manual describes potential malfunctions and provides problem-solving procedures. The readers of service manuals are usually trained technicians and specialists, such as electricians, plumbers, or mechanics. Service manuals do not rely as heavily on theories of operation or general mechanism description (although some are included) as do operation manuals. Rather, these documents concentrate on trouble-shooting tactics and repair procedures.

Like the other technical documents we have explored thus far, manuals require you to adapt content, language, level of detail, graphic aids, organization, and style to meet the needs of your audience. To understand how these accommodations can be made, let us look at the ways in which the overall technical writing process applies to writing manuals.

THE SITUATIONAL ANALYSIS

Imagine that you are part of a product-development team that is about to announce that product's availability in the marketplace. The organization you and your colleagues represent is legally bound to publish any and all documentation that will ensure correct usage of the product and protect owners and operators from known potential hazards. Therefore, as a member of that team, you may be required to write part (if not all) of an operation or service manual.

Your first step is to prepare for the writing process by analyzing the situation. Your purpose goes beyond protecting your organization from product liability lawsuits. In addition, you must determine if your goal is to inform readers of the device's features and functions, to explain how to assemble or connect the product, to instruct readers about its proper operation, to discuss design elements, to equip users to analyze and repair malfunctions, to refer to all related documentation, or to list individual parts and give ordering information. In more comprehensive single-volume manuals, all of these purposes may be combined. In the case of complex equipment, it is more likely that individual manuals will be developed for each.

A thorough situational analysis also requires you to understand your audience. Consider as many personal characteristics of your intended readers as possible: age, gender, education; technical training, experience, and abilities; interests, activities, and economic status. Have they ever had a similar mechanism (different model, different features)? Why will they be using the manual you are writing, under what conditions, and how often? Will they read the entire document, or will they refer to it section by section on an as-needed basis?

The less skilled your reader, the more general and less technical you must be. On the other hand, manuals about complex, technical products are frequently written for experts and specialists. In these cases your writing job is somewhat easier because you can assume a fair amount of training and knowledge. Terminology can be field-specific; background information can be minimal. For instance, when you address an audience of electricians, there is little need to explain the warning "Determine that machine is grounded before turning it on." However, if you are writing to nontechnical consumers who have just purchased new microwave ovens, you need to tell them exactly how to ensure that the machines are properly grounded. Figure 15-1 illustrates how to accomplish this task. The document guides the reader through the process by the use of both words and diagrams.

When considering your readers and how they will use the manual, put yourself in their place and think about what information they need. Do they need to know how the device works, or do they merely need to know how to operate it correctly? Do they need to know how to assemble, disassemble, and store it? Do they need to know how to prevent damage to the device? Do they need to know the safety precautions? How often should the device receive preventive maintenance, and must it be serviced by an authorized dealer to retain the warranty? Can readers perform minor adjustments on the equipment themselves? Are there limitations on the use of the device? Prepare as comprehensive a list of these concerns as possible.

Next, explore your own role as writer. If you are the only writer on the project, you have the sole responsibility to provide all the required information and to determine its accuracy. Because manuals can be quite long and involved, you will also be affected by deadlines more adversely than if you are a team member. Most organizations, however, delegate the development of manuals to several staff members, which cuts down on the time

GROUNDING INSTRUCTIONS

This appliance must be grounded. In the event of an electrical short circuit, grounding reduces the risk of electric shock by providing an escape wire for the electric current. This appliance is equipped with a cord having a grounding wire with a grounding plug. The plug must be plugged into an outlet that is properly installed and grounded.

WARNING

Improper use of the grounding plug can result in a risk of electric shock.

If it is necessary to use an extension cord, use only a 3-wire extension cord that has a 3-blade grounding plug, and a 3-slot receptacle that will accept the plug on the appliance. The marked rating of the extension cord should be AC 115-120 Volt, 15 Amp. or more.

Consult a qualified electrician or serviceman if the grounding instructions are not completely understood, or if doubt exists as to whether the appliance is properly grounded.

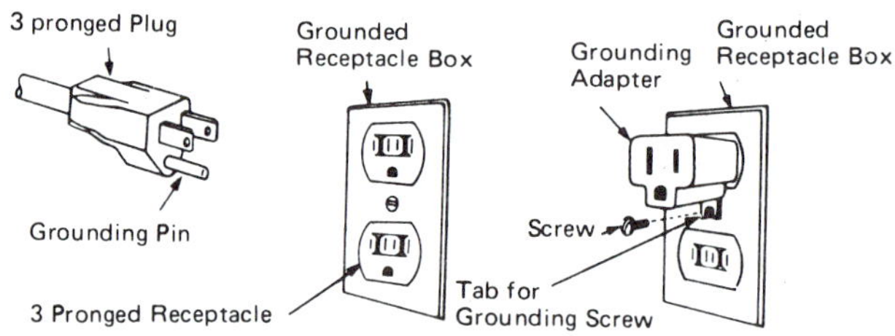

120 V 15A

Figure 15-1 Nontechnical consumers require very precise directions.

involved for any one writer while allowing the writers to bring individual expertise to certain sections of the manual. Without strong supervision, though, team-written manuals can suffer from inconsistent tone and quality, redundancy, and poor organization.

Finally, before you start the actual writing of a manual, it is also important to consider the constraints you face. These may include strict format guidelines, tight deadlines, budget limits, and insufficient personnel.

GATHERING INFORMATION

After analyzing the situation, your next step is to gather information for the manual. Unfortunately, when it comes to writing manuals, few sources of information will be found in the peaceful surrounds of the library. Rather, you will be collecting information—even as it develops—from such widely scattered sources as product or prototype demonstrations, planning and status meetings, interviews with experts, engineering designs and drawings, existing documentation (e.g., preliminary product specification reports), and hands-on experience.

The nature of the information you seek depends, of course, on the purpose and audience of the manual. However, for most operation and service documents you will need at least the following data:

- key terms and concepts
- product specifications
- product functions
- product features
- theory of operation
- normal operating procedures and conditions
- possible abnormalities and contingency plans
- all known potential hazards

PLANNING AND ARRANGING

Assuming that you have met the challenge of securing sufficient information for your manual, you are now ready to prepare an outline. At this point, keep in mind your organization's publication standards as you try to balance audience requirements, the purpose of the manual, and details of the mechanism. Publication standards may stipulate the manual's objectives, audience, content, arrangement, even distribution, or they may be limited to directives about typeface, paper stock, and manual dimensions. If you have some leeway in determining sections, then you bear the responsibility of including all relevant information.

Major sections and subsections of most manuals follow a decimal outline limited to four levels of subordination. (See Chapter 4.) Often, these same headings and numbers will be used in the manual itself, although consumer-oriented manuals may not attach numbers to these levels. Whether the numbers ultimately appear in the document or are restricted to your outline, it is wise to use them at this stage of development to ensure consistency, parallelism, and completeness.

Most manuals follow a general-to-specific arrangement of information; hence, it is a good idea to incorporate this structure into your outline from the start. The outline might begin with a general description of the mechanism, move to a more detailed functional description, and then focus on individual operating components or usage procedures. For the mechanism description you may wish to use either a spatial or a functional order to discuss the interrelationships of components. For the procedure descriptions, you may follow a sequential order or arrange procedures from simplest to most complex.

In addition to planning the text at this point in the development process, it is essential also to plan for the graphic aids you will be using. An effective manual uses as many graphic aids as possible because in most cases (especially in service manuals), the picture or diagram can be as important in explaining a procedure or mechanism as the text. Because you should choose, design, and place graphic aids appropriately, it is important to plan now. Put yourself in your reader's place to determine which elements require the clarity and reinforcement made possible by drawings and photographs. It is not necessary at this point to develop the actual visuals; however, insert a brief descriptive statement into your outline to designate the subject, placement, and type of aid required. If you have not already alerted the personnel who will supply the graphic aids, now is an excellent time to tell them of your plans.

WRITING AND REVISING—STRATEGIES FOR CLARITY

When all of your planning work is done, you are ready to write a first draft of your manual. Without worrying about format specifications just yet, flesh out each major section of your outline so that you have a series of skeletal paragraphs. (See Chapter 5 for a review of this process.) As you further refine these paragraphs, make use of the following strategies for increasing the clarity of your information:

- headings
- paragraph sets
- lists
- examples
- sentence structure
- graphic aids
- warnings and cautions

1. *Headings:* The object of a section heading is to direct readers to the right part of the manual. Therefore, section headings should identify in a word or short phrase the specific topic of each section. Similarly, subsection headings should identify logical divisions of the text. Although these headings should not be repeated within a section, they can be repeated from major section to major section if necessary. In fact, where a parallel consideration of points is required, this repetition is desirable.

When you later move from approved copy to the production stage of the manual development process, be sure that the major section headings ap-

EXPONENTIATION

In a like fashion, you can easily raise a number to a power (this is the same as multiplying a number by itself a specified number of times). The '↑' (Up arrow) signifies exponentiation.

```
PRINT 12 ↑ 5
248832
```

This is the same as typing:

```
PRINT 12 * 12 * 12 * 12 * 12
248832
```

Figure 15-2 Text examples can help clarify instructions.

pear in larger type, bolder print, upper-case letters, and are centered on the page. For subheads, use smaller type, finer print, a mix of upper- and lower-case letters, and position the headings flush left or indented.

2. *Paragraph sets:* Related paragraphs under a subsection can be organized into logical sets and introduced by headings. Figure 15-2 illustrates how this technique can be used to present information clearly.

3. *Lists:* Lists can be used as a visual aid to break up the text and to make the featured information more prominent. Numbered lists can be used effectively in instruction writing to detail each step in a procedure. (See Chapter 8.) Bulleted lists are useful for highlighting features or topics. Frequently, putting information in a list can reduce the amount of text required.

4. *Examples:* As you saw in Chapter 7, examples can be an effective way of making a point. Often, the insertion of an example in the text of a manual demonstrates the point being made and dispels confusion. Computer manufacturers make heavy use of this strategy to ensure that the readers of their manuals understand the commands or data they must input. Figure 15-2 from a Commodore 64 user guide illustrates the benefit of the example.

5. *Sentence structure:* Use the active voice in the narrative portions of your text; use the imperative voice when presenting instructions. Information must be presented simply and clearly. Do not worry about offending people with a simple approach; few readers get upset when information is easy to understand. Use your definition skills (see Chapter 7) for explaining key terms and concepts. Use the procedure description techniques you learned

in Chapter 8 to instruct users on the proper operation of the mechanism. And you can handle the product specifications, functions, and features by using the mechanism description skills you learned in Chapter 9.

6. *Graphic aids:* In Chapter 11 we discussed the importance of reinforcing the written word with graphic aids. Nowhere is this point better taken than in the manual development process. Although you may be faced with budgetary constraints that limit the number of graphics you use, fewer visuals may in the long run result in misused products. In addition, if the visuals that you do include are poorly labeled or hard to understand, your readers will become frustrated.

As you will recall, drawings can be created to focus on exactly what you want to show your readers and are generally less expensive than photographs. Drawings can be cut away to reveal hidden areas of equipment, or they can be "exploded" to emphasize the interconnection of components. However, drawings can cause problems in a manual. For instance, because drawings are abstractions, or because they feature only a portion of a mechanism, readers may have trouble spotting the actual component in context. Technical drawings, although more exact, can confuse the nontechnical reader and should be left to manuals addressing specialists only.

Photographs offer your manual a degree of realism suitable for nontechnical audiences. However, they may be cluttered and may not isolate the important part you are trying to stress. Needless to say, photographs cannot be used to show hidden areas or exploded parts. It is possible to eliminate clutter by using contrasting arrows or outlining, or by airbrushing out the background distractions. No matter what kind of photo or special techniques you use, the mechanism must be pictured accurately and reproduced well if the graphic aid is to be of use. Figure 15-3 shows the use of labeling (for part numbers) and arrows (to depict motion) in a photograph. A caption should identify and explain the photograph.

For the most part, graphic aids should be large and uncluttered, and their meaning should be clear. Refer to their presence by figure number *before* the visual actually appears in the document. To suit the overall order of your manual, you may wish to show the entire mechanism first and then to provide photographs or drawings of important details. Figure 15-4 shows two successive visual aids appearing in a microwave oven operation manual. The labels in the first drawing refer to the second drawing, so the reader has no trouble realizing that the second is an enlarged detail of the first.

7. *Warnings and cautions:* In Chapter 1, you learned that technical writing has an inherent responsibility to provide readers with sufficient and accurate information. Nowhere is this requirement more true than in the case of manuals, which can actually be presented as evidence in liability lawsuits.

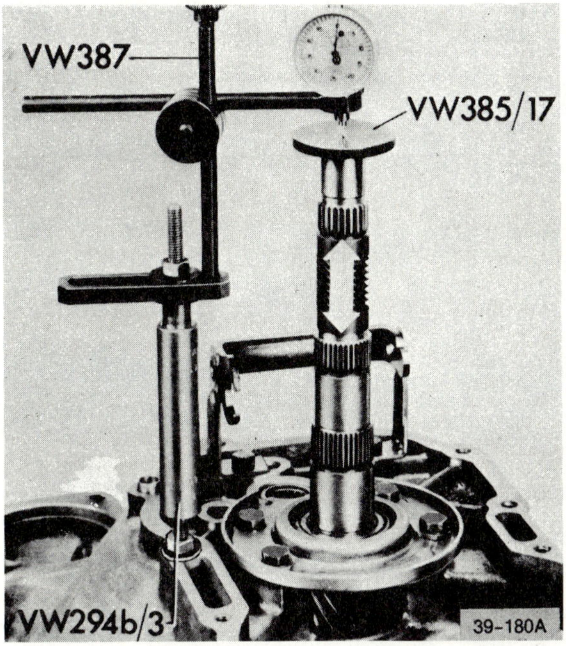

Figure 15-3 Photographs must be clear and uncluttered to be useful to manual readers.

Manufacturers are required by law to provide both proper instruction and proper warning in their manuals. According to the *Journal of Products Liability* 4 (1981), a judgment can be made against a manufacturer if: (1) the product has a defect, (2) the defect is present when the product leaves the control of the manufacturer, (3) injury or damage is incurred, (4) injury or damage has been caused by that defect. Defects can be attributed to manuals themselves if the manufacturer fails to warn readers of all risks or hazards present in the product, if that warning is not adequate, or if adequate and appropriate instructions for use of the product are not supplied.

Warnings (dangers to people) and cautions (dangers to the equipment) must include dangers associated with normal use of the product and those associated with foreseeable misuse. The manufacturer also has a continuing duty to warn of any hazards connected with the product even if the problem is detected after the release and sale of the product. In these cases we find manufacturers issuing product recalls, manual updates, special mailings, and public notices.

As you learned in Chapter 8 when we discussed instruction writing, an

SHARP CAROUSEL®II CONVECTION MICROWAVE OVEN R-8350

Oven Diagram:

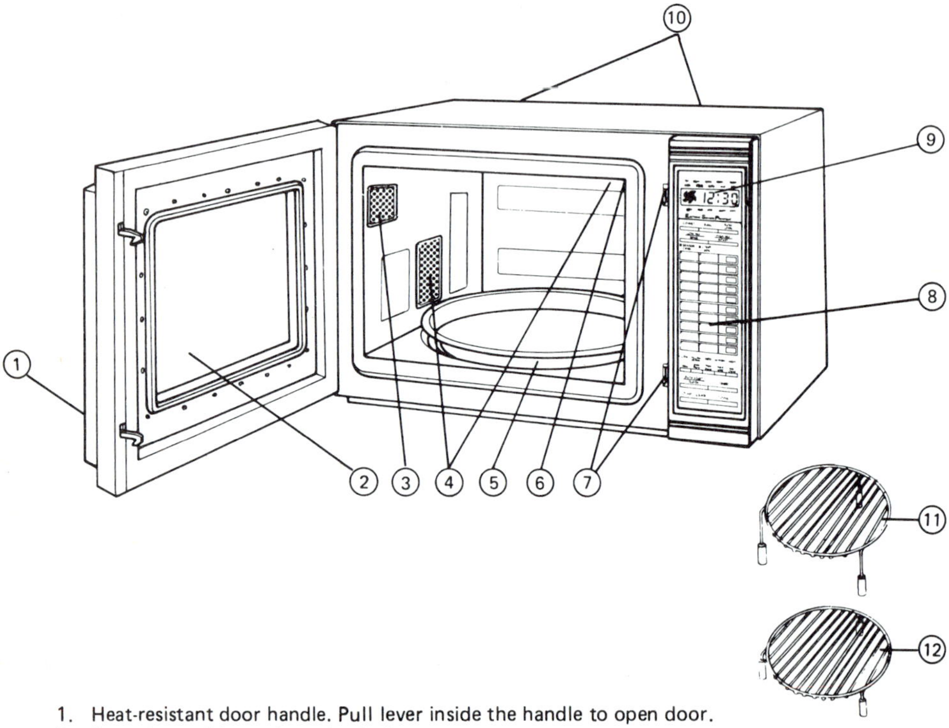

1. Heat-resistant door handle. Pull lever inside the handle to open door.
2. Oven door with see-through window.
3. Oven lamp. The lamp will light when door is opened or during cooking.
4. Convection air openings.
5. Removable turntable.
6. Waveguide cover.
7. Door latches. The oven will not operate unless the door is securely closed.
8. Auto-Touch Control Panel. See page 9 for details.
9. Lighted digital display. See page 9 for details.
10. Ventilation openings.
11. Baking rack.
12. Broiling trivet.

Figure 15-4 The drawing above shows the oven and its major parts; the drawing to the right gives a close-up view of the control panel. The text of the manual parallels this general-to-specific order.

AUTO-TOUCH™ CONTROL PANEL

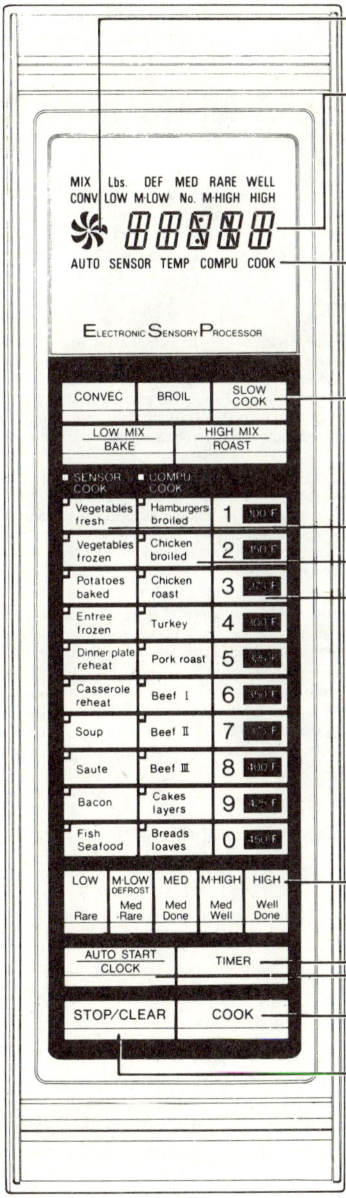

CONVECTION FAN SYMBOL

LIGHTED DIGITAL DISPLAY
Shows readout of cooking time, convection temperature, Sensor or CompuCook setting, cooking instruction or time of day.

INDICATORS
Indicator lights for Convection (CONV), Variable Cooking, Cook, Mix, food doneness (rare, med, well), pounds (Lbs.), number (No.), Sensor, Compu and Auto Start (AUTO).

COOKING FUNCTION PADS
Touch to select Convection, Mix, Broil or Slow Cook settings.

SENSOR COOK PADS
Touch to select Sensor Cook settings.

COMPUCOOK PADS
Touch to select CompuCook settings.

NUMBER AND TEMPERATURE SELECTION (CONVECTION) PADS
Touch to enter cooking times, clock time or convection temperature.

VARIABLE COOKING CONTROL/FOOD DONENESS SETTING PADS
Touch to select microwave power setting. If not touched, HIGH is automatically selected.
Touch to select food doneness setting for CompuCook.
If not touched, Well Done is automatically selected.

TIMER PAD
Touch to set minute timer.

AUTO START/CLOCK PAD
Touch to set clock or auto start time.

COOK PAD
Touch to start oven.

STOP/CLEAR PAD
Touch to erase during programming.
Touch once to stop operation of oven during cooking; touch twice to cancel cooking program.

Please read these warnings and cautions
before proceeding with maintenance and repair work.

WARNING—

• Never work under a lifted vehicle unless it is solidly supported on stands intended for the purpose. Do not support a vehicle on cinder blocks, hollow tiles, or other props that may crumble under continuous load. Do not work under a vehicle that is supported solely by a jack.

• If you are going to work under a vehicle on the ground, make sure that the ground is level. Block the wheels to keep the vehicle from rolling. Disconnect the battery ground strap to prevent others from starting the vehicle while you are under it.

• Never run the engine unless the work area is well ventilated. Carbon monoxide kills.

• Tie long hair behind your head. Do not wear a necktie, scarf, loose clothing, or necklace when you work near machine tools or running engines. If your hair, clothing, or jewelry were to get caught in the machinery, severe injury could result.

• Disconnect the battery ground strap whenever you work on the fuel system or the electrical system. When you work around fuel, do not smoke or work near heaters or other fire hazards. Keep an approved fire extinguisher handy.

• Illuminate your work area adequately but safely. Use a portable safety light for working inside or under the vehicle. Make sure its bulb is enclosed by a wire cage. The hot filament of an accidentally broken bulb can ignite spilled fuel or oil.

• Catch draining fuel, oil, or brake fluid in suitable containers. Do not use food or beverage containers that might mislead someone into drinking from them. Store flammable fluids away from fire hazards. Wipe up spills at once, but do not store the oily rags, which can ignite and burn spontaneously.

• Finger rings should be removed so that they cannot cause electrical shorts, get caught in running machinery, or be crushed by heavy parts.

• Keep sparks, lighted matches, and open flame away from the top of the battery. If hydrogen gas escaping from the cap vents is ignited, it will ignite gas trapped in the cells and cause the battery to explode.

• Always observe good workshop practices. Wear goggles when you operate machine tools or work with battery acid. Gloves or other protective clothing should be worn whenever the job requires it.

CAUTION—

• If you lack the skills, tools and equipment, or a suitable workshop for any procedure described in this Manual, we suggest you leave these repairs to an Authorized VW Dealer or other qualified shop. We especially urge you to consult your Authorized VW Dealer before attempting any repairs on a vehicle still covered by the new-car warranty.

• Before starting a job, make certain that you have all necessary tools and parts on hand. Read all instructions thoroughly; do not attempt shortcuts. Use tools appropriate to the work and use only replacement parts meeting VW specifications. Makeshift tools, parts, and procedures will not make good repairs.

• Use pneumatic and electric tools only to loosen threaded parts and fasteners. Never use these tools to tighten fasteners, especially on light alloy parts.

• Be mindful of the environment and ecology. Before you drain the crankcase, find out the proper way to dispose of the oil. Do not pour oil onto the ground, down a drain, or into a stream, pond, or lake. Consult local ordinances that govern the disposal of wastes.

Figure 15-5 An entire page of warnings and cautions appears at the beginning of the *Volkswagen Rabbit/Jetta Diesel Service Manual.*

adequate warning both identifies the risk and tells readers how to avoid problems. To make warnings and cautions stand out clearly in a manual, use a consistent combination of framing boxes, indentation, color, large and/or bold type, underlining, or italics.

Warnings and cautions that apply to a specific procedure appear within the discussion of that procedure. (See Figure 15-1.) Those that apply to an entire product generally appear in the front of the manual, as in the case of the *Volkswagen Rabbit/Jetta Diesel Service Manual* excerpt shown in Figure 15-5.

WRITING AND REVISING—MEETING SPECIFICATIONS

Most organizations have publications specifications for their external documents (those intended for distribution outside the company). Some organizations even have specifications for their internal documents as well. If you write manuals, it will be your responsibility to become acquainted with those requirements. Do so at this stage of writing to ensure your final document's compliance with them.

In this portion of the chapter we will look at content and format specifications common to both operation and service manuals. We will use a flexible model that allows you to practice incorporating specifications into your documents. However, when you are writing on the job, comply with the more specific guidelines provided by your own organization.

Front Matter

The front matter of most manuals includes the

- cover
- title page
- preface
- warnings
- table of contents
- lists of figures and tables

1. *Cover:* The more often you expect a manual to be used, or the more likely the manual will be used under field conditions, the more durable the cover stock should be. Frequently the cover of a manual is color-coded to identify the product within a related product line. The cover should carry the full title of the manual—a clear, meaningful title, the product line in which it fits, and any required security notices. The cover also usually carries the official logo of the manufacturer.

2. *Title page:* This page gives the full document title (identical to the one on the cover) as well as a descriptive statement of the subject. It carries special instructions about the status of the document or the user. In Figure 15-6, for instance, we are told that this document supersedes a previous edition of this manual. The title page should also present special notices about security clearance or government regulations. Other items listed on the title page include the date the manual was issued (hence, the date of accuracy of the information), the order number, and the manufacturer's logo.

3. *Preface:* The purpose of the preface (which usually appears on the reverse side of the title page) is to assist readers in using the manual. Gener-

DPS 88

SITE PREPARATION MANUAL

SUBJECT

Instructions for Preparing a DPS 88 System Site

SPECIAL INSTRUCTIONS

This manual supersedes DN02-00, dated November 1982. The entire manual has been reorganized and revised; therefore, change bars are not utilized.

The following notice is provided in accordance with the United States Federal Communications Commission's (FCC) regulations.

Warning: This equipment generates, uses, and can radiate radio frequency energy and if not installed and used in accordance with the instruction manual, may cause interference to radio communications. The equipment manufactured after October 1, 1983 has been tested and found to comply with the limits for a Class A computing device pursuant to Subpart J of Part 15 of FCC Rules, which are designed to provide reasonable protection against such interference when operated in a commercial environment. Operation of this equipment in a residential area is likely to cause interference in which case the user at his own expense will be required to take whatever measures may be required to correct the interference.

ORDER NUMBER

DN02-01 March 1985

Honeywell

Figure 15-6 The title page identifies the manual and specifies its intended use.

ally, the preface briefly previews the subjects covered in the manual and mentions the type of reader expected to be using the document. In some cases, the preface cites related documents as well. A legal disclaimer, copyright date, or order number may appear at the bottom of the page.

4. *Warnings:* Warnings and cautions applicable to the product rather than to a specific procedure appear in the front of the manual. We saw in Figure 15-5 that some manuals dedicate an entire page to such information, while others (see Figure 15-6) use the title page or the preface.

5. *Table of contents:* This page lists major section and subsection headings and gives their page numbers. Page numbers are consecutive *within* the section. Hence, they appear as 1-1, 1-2, 2-1, 2-2, and so on. Some manuals provide a separate table of contents for each section to help readers find information more easily.

6. *Lists of figures and tables:* As in the case of the formal report, figures and tables are listed at the front of the document to help readers find needed information. If both figures and tables are used in the manual, separate listings are used for each. Figure numbers run consecutively within each section; table numbers (a separate set from those used for figures) also run consecutively. A descriptive title for each figure or table and its corresponding page number also appear in the listing.

Finally, when preparing the front matter for your manual, consult your organization's publications specifications for such elements as margin width, column width and length, document height and width, paper stock, and typefaces.

Body

The body of the manual will most likely be divided into sections that reflect the requirements of the writing situation. You will want to delineate these sections and subsections clearly so that readers can find them easily. Accessibility of information is especially important for manuals because most readers use it only as a reference on an as-needed basis. Therefore, use tabs, color coding, or marginal bars to divide sections if possible, and use section numbers at the bottom of each page.

Generally, the body of a manual begins with an introduction that explains the purpose and scope of the document. It provides required background information such as definitions and prerequisites. Most manuals also refer readers to related documentation.

Because most manuals follow a general-to-specific arrangement, the next section is usually a general description. In the case of the manual for microwave ovens, the first section after the introduction describes the oven in general terms. The next several sections discuss the mechanism in greater detail, describing individual components such as the control pad and oven compartment. Before discussing particular operating procedures, most manuals present an overview of operation. Then sections treat related activities. Sections will cross-reference one another as required so that readers can skip around in the manual on an as-needed basis. Procedures are usually arranged from the simplest to the most complex.

Back Matter

The back matter provides other important features, including an appendix listing authorized dealerships or parts, an alphabetized index of topics and their respective page numbers, and a glossary of key terms. An index is usually more comprehensive than the table of contents and should be cross-referenced to related topics. Many manufacturers also include a reader-response card (postage paid) at the back of the manual so that readers can ask questions directly or receive updates to the manual.

KEEPING INFORMATION CURRENT

Most manuals for technical products are loose-leaf bound to facilitate updating as new information becomes available. Generally, an addendum is issued to provide insertion and replacement pages to manual users. The addendum has a title page consistent with that of the original document so that readers can immediately associate it with the original. The key difference is the addendum designation, usually a lettering system ("Addendum A," "Addendum D"). A set of special instructions is also included to help readers add and replace pages. Supplemental pages use numbers and letters to maintain the sequence of the original pagination. Hence, pages added to 12-2 would be 12-2a, 12-2b, 12-2c, and so on. Replacement pages generally carry change bars in the margin to designate those areas that have been added or revised. For manuals that do not use loose-leaf binding, supplements supply additional rather than changed information and are issued as separate publications.

The information contained in an addendum or a supplement is incorporated into the manual when the original document undergoes a complete revision. In this printing, the change bars are deleted. When a manual is revised, it carries a revision number in addition to its original order number. As you can see in Figure 15-6, the title page of a revised manual should indicate that the original document has been superseded.

THE OPERATION MANUAL

The typical operation manual for a piece of equipment has multiple purposes, including

- pointing out the features and functions of the mechanism.
- enabling readers to assemble or install the mechanism.
- explaining how the mechanism operates.
- explaining key operating procedures.
- outlining necessary preventive maintenance procedures.
- guarding the reader's safety and protecting the manufacturer's liability.
- ensuring proper storage and/or disassembly of the product.

Each of these purposes usually constitutes a separate section of the manual. In some cases an operation manual will also include sections on service and repair procedures. However, this practice is not recommended in manuals addressed to general consumers because the inclusion of such information may encourage these nontechnical readers to attempt procedures for which they have insufficient expertise.

WRITING FOR THE GENERAL PUBLIC

In the case of operation manuals directed to the general public, such as automotive guides or owner manuals for microwave ovens or video cassette recorders, manufacturers cannot assume any technical expertise on the part of their audience. Such manuals must be easy to understand and easy to use. They must provide carefully developed and tested step-by-step instructions (see Chapter 8), include plenty of helpful graphic aids, and establish a friendly tone. In many instances, the manuals for these products play a critical part in the purchasing decision. If by skimming the manual, the potential user is not impressed by the special features of the product or feels that operation will be difficult, the product may remain on the shelf.

The language of a guide for consumers should be as nontechnical as possible. If it is essential that field-specific terms be used (e.g., *random access memory* or *floppy disk*), provide definitions of these terms in the introduction. Should an acronym be used (*RAM* rather than *random access memory*), give the full term in parentheses the first time the acronym is used *in each section*. If several technical terms must be used, include a glossary in the back matter of the manual. You may wish to code terms appearing in the glossary by placing them in boldface in the text the first time they are used in a section.

Graphic aids in the consumer's manual must be big, clear, and simple. Use captions to help readers understand the meaning of the graphic aid and refer to the figure in the text *before* it appears. Use general drawings and details with cross-referencing to keep readers on track.

The standard manual components should be modified somewhat in documents for consumer audiences. For instance, in the preface, it is especially important to explain the intended use of the manual. Describe the nature of the person the manufacturer expects will be reading the document, and inform readers of the prerequisite skills, tools, experience, and education they must bring to the document.

The introduction of an operation manual for consumers should provide a nontechnical description of the mechanism. Tell readers only what they must know in order to operate the mechanism safely and efficiently. Provide also in this section such information as power requirements, placement of the mechanism, ventilation needs, and warnings, cautions, and use limitations applicable to the whole document.

A detailed mechanism description may or may not be necessary. In the case of a home computer system, for instance, it is wise to discuss individual components and their interrelationships in some detail. Such discussion will better prepare the owner to work well with the system. On the other hand, it may not be necessary to detail the working parts of a video cassette recorder beyond the overview in which you briefly point out their location and function.

Assembly and/or installation is usually the first procedure presented in an operation manual. Be certain to provide clear, step-by-step instructions. Be sure also to give clear warnings about grounding (see Figure 15-1) and testing. In the case of the *Commodore 64 User's Guide,* this section is called "Unpacking and Connecting the Commodore 64" (Figure 15-7). Note that readers are told to check the contents of the carton before beginning. This step serves the dual function of ensuring that readers have all the required components and of familiarizing them with these parts.

Once the consumer has set up the system, he or she is ready to learn to operate it. An operation manual will usually devote several sections to operation procedures. To develop a complete set of operation procedures, list all the activities for which the mechanism might be used, arrange these activities in a simple-to-complex order, then provide a complete discussion of each activity. You may want to circulate this draft of the procedures to technical reviewers to ensure that it is complete and clear.

In the case of the manual for a microwave oven, for instance, the operation procedures include microwave cooking, sensor cooking, convection cooking, mix cooking, broiling, and slow cooking. Each section begins

UNPACKING AND CONNECTING THE COMMODORE 64

The following step-by-step instructions show you how to connect the Commodore 64 to your television set, sound system, or monitor and make sure everything is working properly.

Before attaching anything to the computer, check the contents of the Commodore 64 container. Besides this manual, you should find the following items:

1. Commodore 64
2. Power supply (black box with an AC plug and supply cord)
3. Video cable
4. TV Switchbox (small silver box with short antenna leads).

If any items are missing check back with your dealer immediately for a replacement.

First, let's take a look at the arrangement of the various connections on the computer and how each functions.

SIDE PANEL CONNECTIONS

1. **Power Socket.** The free end of the cable from the power supply is attached here to supply power to the Commodore 64.
2. **Power Switch.** Turns on power to the Commodore 64.
3. **Game Ports.** Each game connector can accept a joystick or game controller paddle, while the lightpen can only be plugged into the game port closest to the front of your computer.

REAR CONNECTIONS

4. **Cartridge Slot.** The rectangular slot to the left accepts program or game cartridges.
5. **Channel Selector.** Use this switch to select which TV channel the computer's picture will be displayed on.
6. **TV Connector.** This connector supplies both the picture and sound to your television set.
7. **Audio & Video Output.** This connector supplies direct audio, which can be connected to a high quality sound system, and a "composite" video signal, which can be fed into a television "monitor."
8. **Serial Port.** You can attach a printer or single disk drive directly to the Commodore 64 through this connector.

Figure 15-7 The first procedure is usually assembly or installation.

with a large heading and marginal bar to make it easy to find. Discussion within each section relies heavily on a table that itemizes the steps, pinpoints the button to be pushed, and displays the mechanism's readout. Narrative is kept to a minimum and is in either the active or imperative voice.

Final procedures usually include cleanup, disassembly, and/or storage. Here it is especially important to safeguard both the consumer and the equipment with clear-cut warnings and cautions. Most operation manuals for general consumers also give routine maintenance and service schedules or check lists to help keep the equipment running correctly.

The back matter may present a summary of operation or use and safety tips. It may also include a reader response card, warranty application, list of available dealers, index, description of other equipment manufactured by the same company, glossary, error messages (in the case of computer manuals), and an order form.

WRITING FOR THE SPECIALIST

The users of high-technology products tend to be well-trained, experienced specialists. No mainframe computer, for instance, will be installed or maintained by a personal computer hobbyist. No spacecraft will be serviced by an amateur mechanic. No laser-ring gyroscope will be operated by a grocery store clerk. Therefore, the writer responsible for developing an operation manual for specialists can speak the language of the field. Explanations can be detailed and complex. Diagrams, charts, and other visuals can be highly technical, although they should still be clear.

Despite the luxury of not having to explain technology to the general consumer, the writer of a manual for specialists must still concentrate on the clear, logical arrangement of information. Information accessibility is even more important in the case of sophisticated equipment because the manual likely is considerably longer and more involved than consumer-oriented documents.

Often, separate manuals are written for related specialized activities. For instance, the *DPS 88 Site Preparation Manual* is one in a series that includes the *DPS 88 Site Planning Template Kit,* the *DPS 88 System Power Sequencing Installation Manual,* and the *Power Conditioning Units Installation and Operation Manual.* Within the series, manuals should be color-coded. At the front of each manual, reference should be made to the others in the series, with an explanation of the function of each.

The sections in an operation manual for specialists may be identical to those in a manual for the general public. However, each section will be more detailed to reflect the needs of the specialist. Undoubtedly, the mechanism description will be carried through to a component and even subparts level, rather than presented merely in overview. Figures 15-8 and 15-9

4.13 GROUNDING SYSTEM

Successful operation of a large scale computer system can be dependent upon proper grounding. Refer to Figure 4-8. If proper grounding techniques are overlooked, ignored, or misunderstood, system performance may be degraded. This may result in costly troubleshooting and maintenance, with the possibility of the system not being operational when needed.

The most important reasons for grounding electronic equipment are:

1. To provide a low-resistance, fault return path to protect personnel and equipment in the event of an ac power fault to ground. The objectives are to keep the voltage potential between the equipment and adjacent structures low enough for personnel safety and to cause the power system protective devices to open before equipment is damaged by a power fault.

2. To provide a low resistance path to earth for lightning current with sufficient shunting effect to keep the potential at the equipment frame within safe limits.

3. To provide a common equipotential plane for the overall computer system installation. The objective here is to keep all of the equipment at the same potential by shorting out equipment potential differences caused by electrical fields of many frequencies.

4. To provide a low resistance path to earth for static electricity which might otherwise build up enough potential to spark, causing electrical noise or system malfunction.

It must be understood that while all ground circuits are eventually connected to earth ground, ground conductors should not be used indiscriminately for various ground functions. Signal grounds must be isolated from other grounds (conduit, circuit breaker load center(s), raised floor supports, etc.) to help prevent interaction between signal path and electrical noise voltages.

4.13.1 EARTH CONNECTION

The most important part of any grounding system is the actual connection to earth. The low resistance path from the system ground point to earth ground is dependent upon the adequacy of this connection, and the design objective should be to reduce this resistance to as near zero as possible. A maximum of 25 ohms is allowed as measured with a ground megger to meet the National Electrical Code.

Earth grounds are usually of two types:

SITE POWER REQUIREMENTS 4-15 DN02-01

Figure 15-8 Text in a manual for specialists is highly detailed. Note the writer's assumption of the reader's familiarity with electrical terms and concepts.

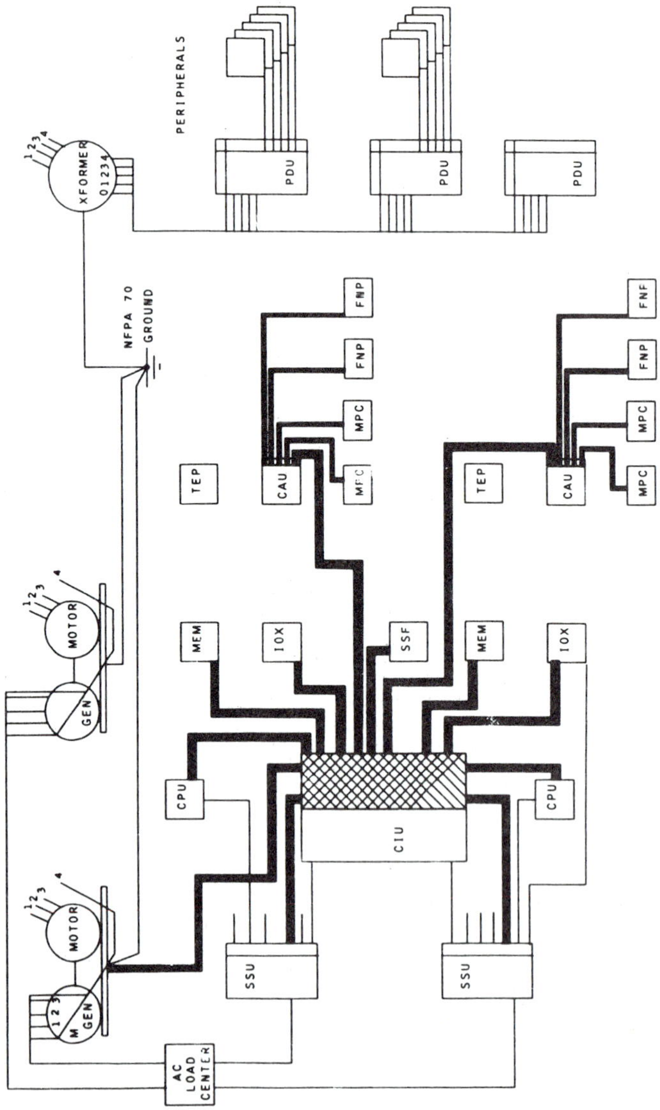

FIGURE 4-8. SYSTEM GROUND DIAGRAM

SITE POWER REQUIREMENTS 4-17 DN02-01

Figure 15-9 Graphics in a manual for specialists can be quite technical. Compare this diagram to the one in Figure 15-1.

illustrate the level of technical detail required in both the text and the graphic aids of an operation manual addressed to specialists.

THE SERVICE MANUAL

The service manual differs in purpose from the operation manual. The operation manual provides clear instructions for a product's use and care, while the service manual details repair and maintenance procedures for the product. Almost all service manuals are written for trained specialists or *advanced* do-it-yourselfers.

When developing a service manual, you can assume that your audience is technically sophisticated and very familiar with the product. It is also safe to assume that they already know how to operate the equipment properly. What they need from a service manual is the means to detect and solve problems. Therefore, a service manual does not include an introduction to the product; however it does include

- product specifications, such as capacities for lubricants and cooling agents.
- technical background on the function and operation of the product and its subsystems.
- routine maintenance procedures and recommendations for servicing.
- trouble-shooting charts.
- repair procedures.
- model change information.
- parts lists.
- safety considerations.
- an index.

Detailed specifications should be placed in the technical background sections, in procedure descriptions, and in tables and other graphic aids. Make sure that the numbers used in specifications do not get lost in sections of text and do not interrupt the flow of your description. You may want to highlight these numbers by using boldface type or separating them from surrounding text with extra space, or you may want to set specification numbers apart from the text in separate boxes, tables, or charts.

Because the service technician must always keep current, the service manual should be loose-leaf bound to facilitate updating. Readers will need to know about any new models that come out or any discoveries made about the product after its release. You may also wish to update service

TROUBLESHOOTING CHART

Symptom	Cause	Remedy
Indicator Light not "On"	Computer not "On"	Make sure power switch is in "On" position
	Power cable not plugged in	Check power socket for loose or disconnected power cable.
	Power supply not plugged in	Check connection with wall outlet
	Bad fuse in computer	Take system to authorized dealer for replacement of fuse
	TV on wrong channel	Check other channel for picture (3 or 4)
	Incorrect hookup	Computer hooks up to VHF antenna terminals
	Video cable not plugged in	Check TV output cable connection
	Computer set for wrong channel	Set computer for same channel as TV (3 or 4)

Figure 15-10 The trouble-shooting chart is organized by symptom.

personnel on likely modifications that customers make to the product during its field use. This kind of information is usually fed back to the manufacturer through sales representatives and local dealerships.

The service manual, like the operation manual, should be arranged from general to specific. However, the pace should be quicker, and *procedures* rather than descriptions should be emphasized. Use the imperative voice predominantly. Graphic aids can be detailed, but still must be clear, appropriate, and well integrated into the text.

Help your readers locate information by organizing it within product systems and subsystems (such as brakes and wheels, engine and clutch, and body and interior for a car service manual) rather than by categories of information (start-up, maintenance, storage). To further help your readers detect and repair problems, include a trouble-shooting list organized by *symptoms* rather than system parts (see Figure 15-10). Also, use comprehensive cross-referencing to help readers diagnose problems more easily.

A parts list names and describes every component, no matter how small. It helps readers identify and order the parts that need replacing. The list should include the manufacturer's number, a brief description, and the quantity of that part used in the specific mechanism being serviced. If possible you should refer readers to a graphic aid within the manual that shows the part. Also, organize the parts list by type (e.g., condensers, resistors, capacitors).

The index of a service manual should reflect the procedure orientation of this document (Figure 15-11). For instance, the citations under the entry *brake caliper* include *disassembling and assembling, removing and installing,* and *repairing*.

CHAPTER SUMMARY

Manufacturers are responsible for providing operation and service manuals to their customers. Thus, manual writing is a key technical writing skill. Most manuals follow a general-to-specific arrangement of information as they explain the features, functions, and procedures related to the safe operation or servicing of equipment. Manuals rely heavily on both mechanism and procedure description techniques. Graphic aids are also important to help illustrate the physical and functional qualities of the mechanism.

Manuals written for the general public cannot be technical. They must be easy to use and easy to understand. Manuals prepared for specialists can be far more technical and, depending on the complexity of the product or operating procedures, can be published in several volumes.

EXERCISES

1. Go to a local computer equipment store that caters to personal computer users. Examine five user guides developed by five different manufacturers for similar products. Rank the manuals according to the degree to which they would influence

INDEX 1

INDEX

Subjects are indexed by section number in **bold**, followed by the page number(s) within the section in ordinary type, followed by the subsection(s) in **bold** in parentheses. Thus, **5**:6 (**3.2**) refers to page 6 in section **5. Engine and Clutch**, in subsection **3.2**.

Figure 15-11 The index of a service manual should be another tool to help the specialist perform maintenance and diagnose problems.

your purchasing decision. Defend your ranking system by discussing the clarity of language, the arrangement of information, the use of graphic aids, the accessibility of information, and the usefulness of the back matter.

2. Think of three pieces of equipment that you purchased recently (video cassette recorder, telephone, compact disk player, vacuum cleaner, microwave oven, television set, automobile, motorcycle, bicycle). Consider how much time you spent with each manual and the conditions under which the manual was used. How helpful was each manual to you in setting up the equipment, operating the product, maintaining it in proper working order, and storing it? Compare the manuals for these three pieces of equipment according to these criteria plus any others you feel are appropriate.

3. If one or more of the manuals you analyzed for Exercise 2 proved inadequate to your needs, what improvements can you suggest? Write these suggestions down and send them to the manufacturer.

WRITING ASSIGNMENT

Select a piece of equipment related to your major field (microscope, hypodermic needle, Bunsen burner) that could be used by either a general consumer or a specialist. Prepare a portion of an operation manual addressed to the general consumer that includes the following sections:

- preface
- introduction
- overview mechanism description
- theory of operation (procedure description)
- two specific operating procedures

Next, prepare an operation manual covering these identical sections but addressed to the specialist. After completing this, prepare the outline (only) for a service manual for this piece of equipment. For each of these writing exercises, prepare a complete written situational analysis.

NOTE: Apply the editing criteria at the end of Part III to your assignment before you submit it to your instructor.

C H A P T E R

16

Promotions

CHAPTER PLAN

Large organizations generally employ an advertising agency to help them announce products and generate sales; smaller companies frequently rely on their own personnel to do this job. In either case, technical staff members are important members of the marketing team because they know the product's strengths and weaknesses, methods of use, and user traits. Whether you actually write the promotional material or review material someone else has produced, it is important to be familiar with some of the basics of product promotion. This chapter will

- survey the purpose of advertising and sales promotion in a technical environment.
- discuss the basics of product promotion.
- introduce promotion techniques.

THE ROLE OF PRODUCT PROMOTION

It is unlikely that technical staff members will be heavily involved in consumer-oriented advertising; therefore, we will focus on business-to-business promotion, wherein a manufacturer or service attempts to sell products and services to organizations, agencies, and institutions.

Business-to-business advertising enables advertisers to make initial contacts with prospective customers, introduce products and make them desirable, compete in the marketplace, establish and reinforce company image, and even ask for orders. Although most of these goals can also be accomplished by sales representatives, the cost of one-on-one selling is very high (currently estimated at $200 per sales call). Advertising can reach more people more often than a single sales representative or even an entire sales force. Similarly, advertising can make its way into the offices of key decision makers who might refuse to meet with sales representatives. And, perhaps best of all, advertising can search out prospects unknown to sales representatives.

Business-to-business advertising appears primarily in trade journals (such as *Computerworld, Plant Engineering, Material Handling Engineering*). Readers subscribe to these magazines and newspapers in order to stay current in their industries. Although advertisers in trade journals deal with a more attentive audience than do advertisers in consumer magazines, bear in mind that these journals carry many ads about similar products by different manufacturers. Making your company's product stand out in this environment can be difficult. Later in this chapter, we will look at specific tactics you can employ to distinguish your company's products from their competitors.

Another promotional format is direct mail, which, as its name suggests, is a message sent directly to a prospective customer via the mail. Unlike the consumer junk mail each of us receives almost every day, direct mail in the business environment is usually an expensive proposition that is customized to individual markets and customers. Often, direct mail can actually help generate an order or significantly shorten the sales cycle; other times its objective is to keep the company name in front of customers; still other times, the manufacturer simply wishes to keep customers informed of important events or product announcements.

Brochures, a third type of promotional format, can provide more detailed product and service information than can be carried effectively in advertising. Therefore, they help to substantiate advertising claims. They are used as follow-up material to inquiries generated by advertising or as leave-behind materials in person-to-person sales calls. These documents must reinforce the message and content of the company's advertisements as well as move prospective customers closer to the buying decision. Because the cost of a typical brochure is relatively high, most manufacturers prefer to give them only to qualified leads rather than to the broader audience reached by advertising.

In all promotions a manufacturer's claims must be honest and representative. "Truth in advertising" laws in the United States levy stiff fines against those who undertake deceptive promotions; our court system is clogged with suits arising from the misrepresentation of the capabilities, cost, quality, or features of a product. If you are ever responsible for supplying an advertising agency with information about a product, you will have to ensure the veracity of that information; you will also have to make sure that your information has been correctly presented in the ad itself. And if you ever write promotional material, you will have that much more control over its accuracy and truthfulness.

UNDERSTANDING THE BASICS OF PRODUCT PROMOTION

Effective product promotion depends on three preparatory activities: conducting a situational analysis, positioning the product and the company in the marketplace, and coordinating all materials into a consistent campaign.

The situational analysis is frequently supported by extensive market and product research. This research serves as the foundation for establishing a unique or strong place—that is, a *position*—for the product in its competitive environment. Once the situation has been analyzed and the product has been positioned, a promotional campaign to present the product to its target audience can be planned.

CONDUCTING A SITUATIONAL ANALYSIS

By now, you should be comfortable with developing a situational analysis before beginning any technical writing assignment. In promotional writing, you need to know

- your role
- the objectives
- the audience
- the marketplace
- the product
- the company
- the promotional format

Your Role

If you are the writer of an advertisement, direct mail package, or brochure, you are the voice of your company. Speak directly to your readers in a tone that demonstrates your company's understanding of their problems. It is also your role to convince readers that the product being discussed is an appropriate solution to these problems. Whether you are the writer or the reviewer of a promotional document, it is your responsibility to ensure the accuracy and sufficiency of the information.

As part of this portion of the situational analysis, identify all the key resource people with whom you must work. For instance, who selects the products to be promoted? Who analyzes the marketplace? Who sets sales objectives? Who identifies the journals in which advertisements appear or the exhibits in which the company participates? Who selects the audiences to be reached and the promotion schedule? Who has final approval? With these questions answered, you may have a clearer definition of your own responsibilities and the ways in which you must interact with other marketing team members.

The Objectives

The objectives of a promotional campaign and the objectives of the individual materials within that campaign should be defined in measurable terms at this stage of preparation. The objectives (similar to the purpose statements we have been using thus far) should state as precisely as possible what is to be achieved, the degree of achievement expected, and the time frame.

A typical campaign can have as its objective one or more of the following:

- to create awareness of a product
- to establish or change an attitude about a product
- to generate interest in a product
- to sell or to shorten the sales cycle of a product
- to gather data about potential buyers

To Create Awareness Materials with this objective disseminate information: they tell readers something they did not know before. Change of attitude or action do not necessarily result from reading these materials; however, the audience's awareness of the company or its product is dramatically heightened. To test the effectiveness of this kind of advertising, it is necessary first to establish an awareness quotient that measures the degree to which a particular kind of audience is aware of the product or the company. After the campaign is introduced, another awareness survey is conducted to determine the degree of change. Objectives in these cases are stated in terms of change percentages.

For instance, a major computer manufacturer conducted an awareness survey among its customers that asked key executives to name the first computer company they thought of when they thought about computers (*top-of-mind awareness*). Even though these respondents were customers of the company conducting the research, the results showed that only 30% thought first of that company. After a year of combined print advertising (in both trade and management publications), direct mail, and regional seminars, the company tested its customers again and found that it had gained more than 20 points in the top-of-mind awareness category.

To Establish or Change an Attitude Promotional materials with this intent try to get the audience to view the product, service, or company in a certain light. Also known as image advertising, this kind of campaign must be integrated very carefully with existing company materials so that any transition comes about smoothly and consistently.

One environmental controls company that had been thought of as conservative and low-risk wished to change its image so that people would think of it as an aggressive, high-technology corporation (which in fact it had always been). It began a series of attention-grabbing advertisements in *Scientific American, Science Digest, Science 85,* and *High Technology* that touted some of the company's more exciting advances (high-density microchips, top-level data security rankings) and contributions to medical technology, avionics, defense, and robotics. Within a year, attitudes began to adjust. People were at first surprised by the avant-garde role this company had been quietly playing; then they came to expect the company to participate consistently in high-tech projects.

Information on attitude is determined by before and after attitude surveys. Objectives are stated in terms of the degree of change anticipated and are often expressed in numerical values.

To Generate Interest Here a manufacturer wishes to have readers inquire about a product. Readers may be given a local sales representative's phone number, a toll-free number, or a business address (often on a business reply card). The objectives in such a case are usually stated in number of inquiries anticipated. An advertiser might test the type of device that works best in generating inquiries by running several versions of the same advertisement, each asking for a different response.

To Sell or to Shorten a Sales Cycle Some manufacturers do not have a sales force and do all of their selling through mail order or through manufacturer's representatives in chain stores. Other companies must use their sales force to push more expensive products that require one-on-one meetings and, therefore, have to let some of their less expensive products be supported by direct mail and advertising. Still other companies know that product promotion can cut four or five months out of a selling situation that might otherwise run as many as eighteen months. Materials must be developed that lead readers to buy the product. Usually, the advertisement or promotional package will have an order form or a number to phone for placing orders. Objectives in these cases are stated in the number of sales expected. Success is tested against that number.

To Gather Data Some promotional materials are meant to elicit information from readers rather than provide it to them. These materials usually take the form of questionnaires that can be distributed at trade shows, conferences, exhibits, or through direct mailings. They help manufacturers determine customer preferences, product and service requirements, and satisfaction levels. By asking customers their opinions, the manufacturer also promotes the company by demonstrating concern and responsiveness. Objectives in these cases are stated in terms of the number of responses received.

The Audience

The next part of your situational analysis should focus on your audience so that you can speak with authority about its problems and needs. To discover audience characteristics and requirements, you might speak directly with current customers and potential customers. Many companies hire research companies to perform this aspect of their situational analysis before launching a campaign for a product. Other companies do not contract for specific research but purchase instead the results of studies performed by demographic and financial service organizations such as Dun & Bradstreet, Dow Jones, or McGraw-Hill.

Another means of learning about the audience is to examine the media in which advertising could be placed. You might know from your sales representatives, for instance, that information about climate-control devices for large buildings should reach major real estate investors, builders, architects, planners, and top-level business management. It would be wise, then, to study journals that claim such readers among their subscribers (such as *Industrial Design* and *Architectural Digest*). These magazines can supply you with circulation figures, readership totals (those who actually read the journal—usually a larger number than subscribers), and a basic demographic profile. You can also study these magazines to determine the kinds of issues editors have deemed important to their readers and the kinds of ads they run. What other products are being advertised? What are the problems these advertisers address? What is the vocabulary level and the degree of detail?

Frequently, a manufacturer discovers that several promotional pieces are required to reach different readers. For instance, in a brochure touting a major computing breakthrough, new circuitry is described to top-level executives (nontechnical audience) in the following manner:

> This new circuitry is a major efficiency booster. Requiring 40% less power than conventional types, the circuitry is five times faster and requires considerably less space. It can improve processing speed to better serve the rapidly changing demands of the large organization. It also helps minimize heat build-up within the cabinetry for longer component life and greater reliability.*

The same circuitry is described this way in a product brief addressing data processing professionals (technical audience):

> First the system implements its logic functions by means of CML. This type of circuitry utilizes a steady stream of current to achieve higher switching performance than TTL, which turns current on and off. Because CML does not require time lags for voltage recharging, it is typically five times faster than conventional TTL. With CML, processing speed improves significantly. . . .**

The first example stresses the benefits of the new technology and stays away from technical terminology. The second example describes the way in which the circuitry works. The writer uses acronyms for circuitry types and assumes the audience will understand these technologies.

The Marketplace

In addition to knowing the audiences your promotional materials must address, you should know the marketplace in which your product competes. Identify all other manufacturers of similar products. Consider the costs, features, acceptance, benefits, and advantages of these competitive offerings. What are the economic factors at play in the marketplace? For instance, will there soon be a glut of your product, or is your company introducing a needed but new kind of product? This information and considerably more ought to be available to you from the marketing management of your company because it is more than likely they studied the marketplace prior to developing the product or readying it for announcement. This marketplace information will be especially valuable to you when you are called on to devise a unique or strong position for the product in its competitive environment. Without directly criticizing your competition,

*DPS 88 brochure (Waltham, MA: Honeywell Information Systems, Inc., 1984).

**DPS 88 product brief (Waltham, MA: Honeywell Information Systems, Inc., 1984).

you can offer solutions to certain competitors' problems that your product avoids or corrects.

The Product

A thorough understanding of the product to be promoted is essential to the situational analysis. It is a good idea to be familiar with the following product characteristics before you write or review promotional materials:

- physical characteristics (size, shape, weight, density, texture, power requirements, color, construction, parts, sound, packaging, aroma, flavor, speed)
- functional characteristics (what the product does and how it does it)
- qualitative characteristics (attractiveness, efficiency, convenience, uniqueness, durability, comfort, competitiveness, reliability, serviceability availability)
- quantitative characteristics (price, warranty, service programs, distribution, special offers)

Then translate these features and functions into *benefits to the customer.* By knowing the product well, you will be better able to emphasize product strengths and minimize product weaknesses.

The Company

Knowing the company is also important to the situational analysis because everything that is advertised or otherwise promoted is a reflection of the company. It is not wise, for instance, to use a cartoon approach to advertising if the company wishes to have a serious, business-oriented image. The awareness quotient measured at the beginning of most promotional campaigns will help you understand how the company is perceived and what your audience's expectations of it might be. Unless a dramatic change in image is appropriate, all advertising should be in keeping with the image already established for the company.

The Promotional Format

How the promotional materials should be packaged is another decision that should be made in the situational analysis. Should an advertisement, brochure, sample kit, brief, specification sheet, or direct mail package be developed? Should a combination of these be used, and if so, how and when in the overall sales cycle? After we look at advertising, direct mail, and other promotional materials separately, you will be in a better position to make such decisions.

A situational analysis can consider many other important factors, including budget, schedule (when an ad or promotion appears), reach (numbers of readers), and frequency (how often an ad might appear)—all elements beyond the scope of this book. A bibliography appears at the end of this chapter, should you be interested in learning more about advertising and sales promotion.

POSITIONING THE PRODUCT

Trade press advertising, direct mail packages, and brochures must work together to support the image of the product and the company producing it. This image—the particular place a product holds in its competitive environment—is known in advertising spheres as the product's *position*. Determining and achieving a product's appropriate position in the marketplace requires a great deal of research. You need to know current and past user behavior, customer requirements, levels of satisfaction with existing products, and the correspondence between customer needs and the product's benefits. Research should reflect the input of the sales force as well as key customers. It should reflect competitive conditions as well as *hinge factors* (those events or conditions that can change the market and therefore the product's position).

Although as a technical staff member you will probably not be called upon to develop the market analysis, you must work from the facts uncovered. Promotional materials developed in a vacuum—that is, without benefit of this information—are unlikely to succeed.

Finding the right position for a product is not always easy. For instance, a well-known drain opener was first positioned as a chemical that could eliminate the germs that lurk in stopped-up pipes. Positioned this way the product did not fare well. After additional research, the manufacturer learned that customers did not really care about lurking germs; rather, they cared about speedy results. Because the product also works very rapidly, the company was able to alter its position from a drain opener that disinfects to a drain opener that acts rapidly. Within weeks, sales climbed. The company stayed with this advertising stance and continues to enjoy the largest share of the market.

The following questions will help you determine an appropriate position for products you may be promoting someday. All answers should be based on well-documented, objective research.

1. *What are the unique characteristics of the product?* For instance, Honeywell is the only major computer vendor that offers tandem (i.e., duplicate) configurations on all of its large-scale computers. IBM, in contrast, offers this important dimension on only its top-of-the-line models.

2. *Are the unique characteristics of importance to the consumer or the industry?* Many of Honeywell's customers cannot afford any down time (time during which the computer system is not running). They do not consider tandem configuration a luxury; they consider it a necessity.

3. *If the product is not necessarily unique, what are some of the things it does best?* For instance, the Volkswagen Rabbit GTI can accelerate quickly, hug the road on sharp curves, and maneuver in and out of tight spaces. The car is also fuel efficient, reliable, and comfortable.

4. *What are the most important requirements placed on the product by its customers?* Volkswagen has been known as a producer of sensible cars for a long time. It was losing out to flashier cars in the marketplace because consumers were demonstrating an interest in sports cars. Not wanting to abandon its staid reputation entirely, VW positioned its GTI as the "sensible sports car."

5. *What are the real and proven benefits the audience can derive from the use of the product or service?* It is one thing for a new car, for instance, to have fancy styling. That feature may appeal to a consumer's ego but really does little to demonstrate or prove a car's worth. It is preferable to turn the feature into a benefit, as in this example:

> Sleek, aerodynamic lines attract attention. But more importantly, they allow the Sedan to cut through the wind with a drag coefficient of just .30. That means more miles per gallon and less noise per mile.

PLANNING A CAMPAIGN

A promotional campaign is a program comprised of several different materials and is conducted over a specified period. A campaign for a technical product should accomplish at least the following objectives:

- introduce or provide essential information about the product
- position the product in the marketplace
- generate interest in the product
- support the overall company image

Because customers of technical products can often make purchases without a visit from a sales representative, another objective of a promotional campaign is to generate sales.

Although few companies can afford a media blitz, promotional materials always should be coordinated with one another. And although budget constraints can often hinder a campaign's effectiveness, the ideal marketing program includes a logical mix of advertising, direct mail, and promotional brochures. In the promotion of technical products, there is usually little need to do television advertising (although electronics and automotive industries have been doing so for years). Consult a media planner to help you establish an effective mix of promotional vehicles, the number of pieces, and the frequency of the messages that will best reach and stimulate the target audience.

ELEMENTS OF TRADE PRESS ADVERTISING

Trade press advertising is a major element in any media campaign. Literally hundreds of newspapers and magazines that cater to specific interests exist, by means of which advertisers can capture the attention of key decision makers and users of their products. In the computer industry alone, for instance, journals and newspapers include *Computerworld, Datamation, IEEE Spectrum, InfoSystems, Electronic News, MIS Week, Computer Decisions, Micro Communications,* and dozens more. By analyzing the circulation audit (a report that details the composition of a journal's audience), advertisers can select the journals with the highest proportion of potential customers in which to place their ads.

In smaller companies, you may actually write and place advertisements for products you help develop. In other cases, you may be held accountable for supplying instructions to an advertising agency and approving its creative efforts. Although there is no foolproof formula for developing memorable and effective advertising, the key elements to consider are the headline, graphics, body copy, and layout. All of these elements must work together in a way that leaves a memorable, positive impression.

HEADLINES

Headlines are the most important element of any print advertisement. According to extensive research conducted by such firms as the Starch Readership Service and Gallup and Robinson, five times as many people read the headline as read the body copy of an advertisement. If you do not catch your reader's attention with your headline, you might not have a second chance. Therefore, your take-away message (what you want your readers to remember) should be readily apparent in the headline. Here are some tips for writing effective headlines:

1. *Promise a benefit to readers.* Let them know what they can expect if they use the product. Naturally, be certain that your claim is true before you promote it.

Data Processing Managers—Decrease Your Applications Development Backlog with Productivity Tools from Computech

2. *Reveal exciting news to readers.* Use key words such as *announcing, introducing, new, revolutionary,* or *innovative,* which identify the newsworthiness of your announcement. These kinds of headlines usually have high recall (many readers remember them over other advertisements in the same journal or newspaper).

Introducing a New Way to Secure Insulation in a Dry Wall

3. *Tell readers how to do something of interest to them or how something has been done.* This headline acts as the lead-in to an advertisement with a tutorial tone. By the end of the ad, readers should understand that your product is the foundation of the recommended process.

How to Fly a Big Airplane with the Tips of Your Fingers

4. *Offer advice to your readers.* Most readers are always on the lookout for helpful tips. In an advertisement of this nature, however, be certain that you offer valuable information that readers are not likely to gain elsewhere.

Guidelines for the Energy-Conscious Builder

5. *Use an imperative verb to motivate readers to stay with the advertisement and ultimately to buy the product or service.* Although this kind of advertisement connotes confidence on the advertiser's part, it usually is a hard sell—a stance that can frighten some readers away.

Take a Chance on Our Software

6. *Feature a special offer, event, or price.* The offer must not be misleading, although important details and qualifications can be reserved for the body copy.

Limited Offer: Buy Four Micronet PCs for Your Office and We'll Give You One for Your Home

GRAPHICS

The graphics of an advertisement—line art, photographs, or drawings—must work together with the copy, headline, and layout to create the desired effect. According to David Ogilvy, founder of a leading advertising agency, most copywriters think in terms of words and devote little time to planning their illustrations. Yet often the illustration occupies more space than the copy. It should work just as hard to sell the product as the text and should telegraph the same promise made in the headline. To create effective graphics you should:

- illustrate in color whenever possible.
- keep illustrations simple (e.g., avoid crowd scenes in photographs or cluttered line art).
- use captions underneath or at the side of most photos or art (your reader's attention will be automatically drawn to this text).
- represent the product accurately (e.g., don't pose a very tall model next to high computer cabinets to make them seem smaller).
- let the photography or art tell a story whenever possible.*

Figure 16-1 offers a good example of the effective combination of headline and graphic. The headline makes a unique offer to readers, one that proved irresistible. (Macintosh sales soared during the run of this offer.) The graphic is the primary thrust of the ad, yet it is simple and intriguing. The gloved hand ties in directly with the headline; the "mouse" control device, however, creates a visual twist. The mouse shows the product, as do the monitor and keyboard inset at the lower right.

BODY COPY

Believe it or not, only 5% of a journal's readership reads the body copy of an advertisement. This might sound like an insignificant number of people, but 5% of the readership of a journal like *Telecommunication Products and Technology* (total circulation 56,000) is 2,800 people. Those who do take the time to read the copy are far more likely to be real prospects than those who do not. Therefore, these readers should be rewarded for their interest with informative, stimulating material.

Although the jury is still out on whether advertisement copy should be long or short, in consumer-oriented magazines and newspapers, short ads seem to work best. In technical journals, long copy seems to work best. Readers like to know an advertiser has something to say, and longer copy implies you have a significant statement to make.

Here are some tips to help you develop interesting body copy.

1. *Tell readers what the product will do for them.*

Quarus Lets You Be Everywhere You Need to Be Without Going Anywhere

As travel budgets dwindle and competition grows more fierce, you've got to cover a lot of territory at a minimal cost. QUARUS, an interactive

*David Ogilvy, *Confessions of an Advertising Man* (New York: Atheneum, 1984).

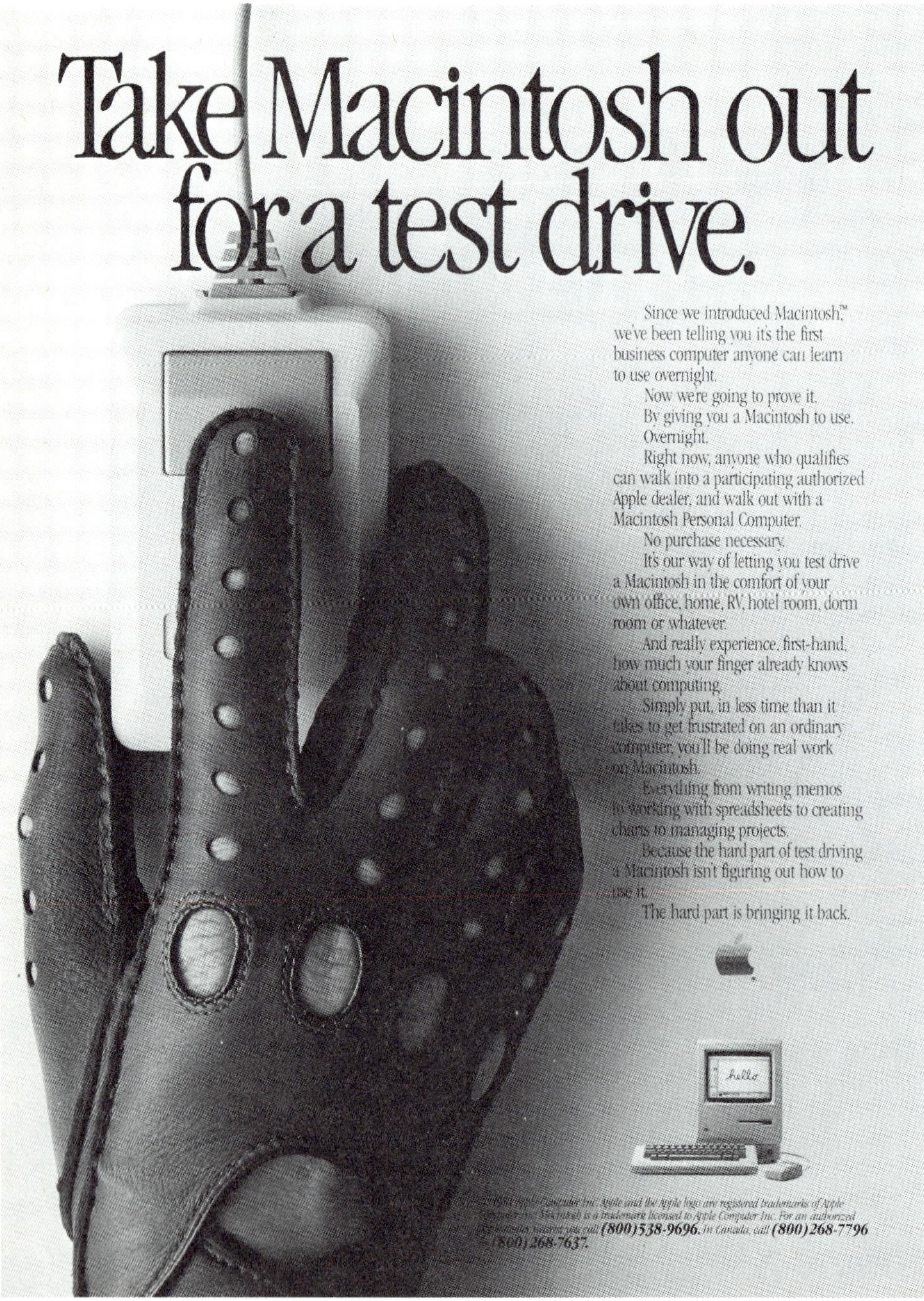

Figure 16-1. Headline and graphic work together to create an effective ad.

meeting system, lets you conduct meetings all over the world without catching any planes, trains, or buses.

Through the sophisticated technology of QUARUS, you can call a meeting anytime, secure a list of participants, keep minutes automatically, and more. To find out how QUARUS will save you travel time and travel dollars, call your local QUARUS representative.

2. *Tell a story.*

When Sam Johnson Needed Bypass Surgery, a New Technique Helped Him Bypass the Need for Surgery

Last year Sam Johnson's coronary arteries were so blocked that his doctor told him he needed immediate bypass surgery. Sam's doctor called a heart specialist at Mercy Medical to schedule an emergency consultation. The specialist told Sam's doctor about cardiovascular angioplasty, a procedure that opens blocked arteries without surgery. Sam underwent the procedure and went home the next day, much healthier and happier—and with hospital bills that he could afford.

Come talk to us at Mercy Medical. We know that solutions to health problems aren't always surgical.

3. *Use a testimonial.* Testimonial ads are often written from the first-person point of view, with the speaker endorsing the product.

I May Be a Fool to Jump out of Planes, but I Know Enough to Trust the Best—Excellor

I used to jump out of planes because I was told to. Now it's just because I want to. And since I'm calling the shots, I'm calling the chute as well. Excellor parachutes. I wouldn't trust my life to anything else.

—Bud Peters, Air Force Colonel (Ret.)

4. *Answer a question posed in the headline.*

What Are You Doing to Reduce Your Backlog?

If you are a typical data processing manager, you are faced with a two- to three-year backlog of programming requests. What are you doing about it, besides wringing your hands?

XYZ Systems can help. Our programming support products allow you to develop complete applications in house. Imagine—system-generated, error-free code and automatic, detailed documentation.

Give XYZ Systems a call today at (800) 555-4765.

5. *Discuss special offers.* These include markdowns, limited conditions or pricing, coupons, sweepstakes, or rebate programs. Many copywriters think that special offers are boring, but customers do not share this view. If you have a promotion of this kind to make, the body copy should focus on it and motivate the reader to take advantage of it.

For What Your Company Is Currently Paying in Maintenance Charges, It Can Have a New System with New Potential

For a limited time only, take advantage of our budget-matching offer. For the same fee your company is currently paying in monthly maintenance charges for its computer system, it can enjoy the use of advanced hardware and software. New productivity tools for end users are yours at no additional charge. Free migration support is available to help you make the move painlessly. For more information, speak with your Computech representative today or call 800-555-1111 toll free.

LAYOUT

The layout, the way in which headline, graphics, and body copy are arranged, can enhance an ad's readability and effectiveness. Poor layout, in contrast, can make an ad fail. Usually, layout is the domain of an art director, but you should know some of the basic principles of layout so that you can evaluate the ads presented to you. Work with the art director in writing and planning the advertisement so that all the parts of the ad "play" well together.

The amount of space an ad will occupy is an important consideration. An ad that looks good in a two-page spread will not necessarily work in a quarter-page corner. It is also important to consider the context within which the ad will appear. Will it be opposite the editorial content of the journal or newspaper? If so, white space will be very important to provide readers with visual relief from the dark, text-heavy opposite page.

The placement of headlines, copy blocks, graphics, subheads, photo captions, and logos must be balanced—distributed in an eye-pleasing, inviting way. Other important principles of good layout are unity and proportion. Unity is the degree to which the entire ad holds together as a single, integral piece. In general, the simpler the layout, the more unified the look of the ad. Proportion is the relationship of elements to each other. In an ad containing a headline, a photo, and copy blocks, one or more of these elements will be dominant. Consider the Macintosh ad in Figure 16-1: both the headline and photograph are very large, whereas the body copy block is small and off to the right. Nonetheless, the entire advertising space is used in a balanced way so that there are no real empty areas.

DIRECT MAIL ADVERTISING

Direct mail advertising, as its name suggests, is a form of advertising sent directly through the mail to targeted individuals. It is usually designed to motivate recipients to respond in a specific way (e.g., request additional information, have a sales representative call on them, or make a purchase). Its effectiveness depends on its use in appropriate situations and on its ability to complement other forms of promotion in a media campaign.

Direct mail has several different uses. It can help convey product information to prospects before a salesperson visits. In the case of established products or customers, fieldwork by sales personnel may be eliminated or reduced by having the manufacturer solicit purchases through the mail. As follow-up to a sale, direct mail can help maintain customer satisfaction and continue to keep the manufacturer's name in the reader's mind. Its three most important advantages are:

1. *Elimination of wasted appeals.* Direct mail helps an advertiser communicate only with those people most interested in a product or service, although the degree to which direct mail allows this control depends on the accuracy of the mailing list. The cost of communicating with prospects via direct mail is usually lower than that for national print advertising.
2. *Establishment of a personal tone.* Effective direct mail personalizes each letter or package with the properly spelled name and title of the recipient. Frequently, the sales representative's signature appears at the bottom of the letter, and a business card is enclosed. Recipients feel like they are getting personalized attention and are thereby more inclined to make purchases from the company giving them this attention.
3. *Elimination of competitive distractions.* When it is read, direct mail is not in competition with the advertising or editorial content found in other print media, such as magazines.

The most noteworthy disadvantage of direct mail is that it arrives unrequested; thus, the recipient may have no interest in the message. Chances of ignoring average kinds of direct mail packages are high, so catchy techniques and careful identification of recipients are required. Getting past the executive secretary's mail-screening procedures to the actual desk of the high-level manager requires crafty, creative packaging.

Not too long ago, Lear Jet launched a direct mail campaign for its private executive jet product line by sending very large posters in mailing tubes to the CEOs of *Fortune* 500 companies (the likely customers of such high-priced, sophisticated products). The posters were life-size reproductions of the cockpits of the planes as seen from the pilot's position, including a breathtaking glimpse of a landing field lit up at night. Because the poster was so spectacular, the mailer was taken into the executives' office. The executives, in turn, so enjoyed the poster that they took the time to read the accompanying letter. Executives soon called the sales representatives requesting additional copies of the posters for their friends and family members. The sales representatives used the occasion of personally delivering additional posters to discuss the product with the executives. The representatives initiated calls, of course, if they did not hear from the executives. Sales figures soared as a result of this effective direct mail campaign.

How can you as a technical staff member help create an equally effective direct mail campaign? To see how, let's examine the components of a direct mail package.

THE SALES LETTER

The sales letter is used as a lead-in to the other materials included in the direct mail envelope or package. These letters can be used to make direct sales, to announce products and special offers, to generate good will, to describe changes in the marketplace or organization, or to solicit information.

To satisfy the purpose of the mailing, the sales letter must attract and keep the reader's attention. Thus, sales letters are usually held to a single page. Because you are writing to people who have not asked you to do so, you cannot take up too much of their time while making your points.

Start with a headline that grabs the reader. The headline is not set in large, flashy letters as it would be in a print advertisement; nonetheless, it should be made eye-catching through the use of capital letters, underlining, color, italics, or centering. The tips for writing effective headlines given earlier in this chapter apply to headlines of sales letters as well. A few additional suggestions will help you start your letters off with a punch:

1. *Present a startling fact.*

Seventy-five percent of all children who died in car accidents last year were not wearing seat belts.

2. *Refer to an upcoming event.*

Our Decision Support seminar in Phoenix will help you see the whole picture.

3. *Negate a commonly held belief.*

Dieting is not the best way to get in shape.

The first paragraph of the sales letter should tie the headline to the rest of the letter. It should be no more than two or three short sentences.

The rest of the letter gets down to business and stays there. For instance, the second paragraph should create a desire for the product in accordance with customer needs. Generally, audiences are attracted by such appeals as:

adventure	efficiency	quality
appearance	excitement	refinement
challenge	flexibility	responsibility
comfort	intelligence	status
convenience	investment	thrift
confidence	power	uniqueness
durability	pride	versatility
ease of use	problem solving	

Above all, when you are advertising technical products to business or industrial audiences, the appeal should be as informative and factual as possible. For example, let readers know how economical a computer is (thrift), how long it can run continuously without failure (durability), how stable and well engineered it is (confidence), and how fast it is (power).

The final paragraph of your sales letter should motivate readers to complete a specific action. They should be asked to call a toll-free number, to clip a coupon, to fill in an order blank, to return a postpaid card, or to take a trial offer. Letters are more effective if the action is spelled out; do not leave it to the reader's initiative to find a postpaid card in the package and send it in.

Close the letter with the signature, typed name, and title of an individual. Use a postscript to reinforce a particular message if possible, because the eye is drawn to such a final note.

The sales letter in Figure 16-2 illustrates these structural elements.

CATALOGUES

Direct mail packages frequently include product catalogues that display a full range of offerings. Whether a catalogue is comprehensive or is devoted only to specific kinds of items, all catalogues share certain characteristics:

- They are full of photos of the advertised products. Especially popular products are placed on the cover.
- Each product is described briefly.
- A catalogue or order number is cited for each product.
- The price of each product is listed.

AMCHEM

Kick Summer Pests Out of Your House!

Dear Homeowner:

You can rid your house of summer pests more effectively than professional spraying services can. AND you can do it for less than half the cost!

SPIKE-100, a solid pesticide stick, works better than any spray available against roaches, spiders, termites, ticks, silverfish, fleas, and ants. The revolutionary solid stick makes SPIKE-100 so safe to handle and use that you can feel confident doing the job yourself. No masks, aprons, or even gloves are necessary. Just hammer a dry SPIKE-100 stick into the ground at the base of your house foundation. (The number of sticks you will require depends on the size of your home.) Then simply water the area according to the package directions. In minutes, the dormant chemicals are activated and released into the house foundation.

SPIKE-100 lasts weeks longer than sprays because it is absorbed directly into the house. Sprays are diffused into the air; their ability to penetrate insect nests and trails is limited. SPIKE-100 is a premeasured application of pesticide, so there is never any doubt about the right amount. Government tests have proven

Many catalogues also include an order form to facilitate readers' purchases.

Catalogues are fairly expensive to produce because of the writing time, photography, and printing costs involved. However, few other documents can better portray the range and depth of a manufacturer's offerings. Catalogues also have a certain irresistible quality. Rarely is one simply tossed out without the recipient at least peering inside. Frequently, catalogues enjoy a long shelf life: people keep them around for use in deciding on special purchases.

OTHER INCLUSIONS

Manufacturers of products that can be sampled should include a free sample whenever possible. When a product is better than its competition, there is no more convincing argument than the recipient's own judgment of the that SPIKE-100 is considerably safer to use than conventional sprays--and that's important when children and pets are in the home. And, perhaps best of all, SPIKE-100 costs less per application than most leading sprays.

So, say good bye to bugs and to your pesticide control service as well. SPIKE-100 is available from your local AmChem representative:

Dewey Butler
Cardinal Hardware
1400 Mission Road
Portsmouth, VA 23703
(804) 555-3409

Call today for a free demonstration and additional information.

Sincerely,

Arnold Stault

Arnold Stault
AmChem Regional Manager

Figure 16-2 A good sales letter attracts and keeps the reader's attention.

product's performance. A sample should never go out without an explanatory sales letter or a circular enclosed in the package. This document should direct the reader's attention to the sample, highlight the product's features, explain any special offers, and call the reader to action.

Free gifts are also powerful attention getters. Sending a personalized pen and pencil set, a coffee mug, a key chain, a paperweight, or another relatively low-cost item usually ensures that a good portion of your message will be read. The higher the value of the premium item, the more likely it will be retained by recipients. Therefore, the item should carry the manufacturer's logo to serve as a constant reminder.

Depending on the action an advertiser wants recipients to take, a direct mail package may also carry a postpaid response card. This device, sometimes a perforated portion of a brochure, can be used for orders, subscriptions, enrollments, or requests for sales calls, demonstrations, and further information. Another possible inclusion in the direct mail package is a coupon, an especially effective means of stimulating sales. Even expensive items can use this device by offering discounts or special options on a product.

Believe it or not, the envelope for a direct mail package is one of its most important components. Unless the envelope is sufficiently inviting, readers will toss the mail out unopened. When dealing with technical audiences, it is rare to include anything but a typewritten personalized name, title, and address, although it is possible to add a brief message at the lower corner inviting readers to open the envelope. You may also use an attractively colored envelope or a window that reveals part of the envelope's contents.

BROCHURES

The primary sales promotion vehicle for most manufacturers is the brochure. This document is distributed to specific individuals who have demonstrated interest in a product or who are likely prospects. Brochures are sometimes also included in direct mail packages. Unlike print advertising, which targets broad audiences, sales promotion brochures are given out on a piecemeal basis, often at trade shows and exhibits or during sales calls. Sometimes they are sent out in response to inquiries for further information generated by print ads. Brochures can provide more detailed information than sales letters or trade press advertising. A letter may be 250 words long; an ad may be even shorter—but a brochure can be as long as 16 pages.

Brochures are designed to motivate customers and prospects to make a purchase. This objective is accomplished through their content, language, and format.

CONTENT

Because readers usually approach a brochure wishing to learn about a product, you, as the writer, must provide a clear, concise, accurate description of that product. Using the needs of the audience as a guideline, organize your description to emphasize the most important features of the product. These six steps will help you develop a successful brochure:

1. *Address the audience's needs, interests, and goals explicitly*. For instance, the brochure might begin with a scenario of a problem facing the reader.
2. *Describe the product or service so that its ability to serve a need or solve a problem is emphasized.*
3. *Clearly identify the key benefits or solutions provided by this product or service.*
4. *Link each benefit or solution with a specific product feature*. In the case of a fuel-efficient car, for example, the brochure might link this benefit to the

engine design, aerodynamic curvature of the hood, lighter-weight chassis, manual transmission, and front-wheel drive. The skills you acquired in developing process and mechanism descriptions (Chapters 8 and 9) will help you write effectively about the way in which the product is constructed and used.

5. *Stipulate limitations that apply to the product or its use.* These might include availability, warranty duration, exception conditions, or prerequisites.

6. *Close the brochure with a call for action.* Suggest, for instance, that the reader call a sales representative or dealer to find out more about the product. A list of local distributors might be included if the brochure is meant to bring the reader to the point of purchase.

LANGUAGE

Ideally, the language of a brochure should involve readers. Emphasize direct address and avoid passive constructions. Make your language rhythmic by alternating long and short sentences. Sentences should be concise, with important ideas placed in the independent clauses. Rely on jargon only when all of your readers will be certain to understand it. Finally, avoid unpleasantries: "never pitch from the negative" is a basic tenet of sales promotion. Choose positive, believable words and phrases.

The tone should be casual yet respectful, as if you are speaking comfortably to your reader. The personal voice should demonstrate that the reader is important to the manufacturer.

DESIGN

The way in which a brochure will be used affects its design. For instance, if it is to be part of a direct mail campaign, it should be small enough to fit into a standard business envelope. If it is to be distributed personally during sales calls or at trade shows, it can have the dimensions of a typical magazine.

Graphic aids are a must in all brochures because they help engage the interest of the reader. Work closely with the art director to establish a careful balance between visuals and the text. Graphic aids should directly corroborate information presented in the text.

Wherever appropriate, captions should be used for photographs and drawings to explain what is being demonstrated. Researchers have determined that 90% of a brochure's readers will read the captions but only 30% will read the entire document. Therefore, be certain that captions are informative and that by reading *only* the captions readers will still have a good idea of the product. Use the captions, as well as the graphic aids, to reinforce the key points being made in the text.

Headlines, heads, and subheads are also important elements. They move readers along in a logical sequence as key points are made. Many readers look only at the headings in a document, reading sections of interest but not others. Therefore, headings should not only be eye catching, they should also impart valuable information.

CHAPTER SUMMARY

As a technical staff member or technical writer, you may very well be asked to promote products or services made by your organization or to help in the development or review of promotions. When approaching technical audiences, you generally are limited to print advertising, direct mail, and brochures. Each of these formats has a different purpose and is used at a different point in the sales cycle.

Print advertising is used to announce products to a wide audience and to generate interest. Direct mail can be used to stimulate sales and to personalize sales pitches to prospective customers as well as to shorten the overall sales cycle. Brochures are used primarily as follow-up information, in response to print advertising, as giveaway materials at trade shows, or as documents to leave behind after a sales call.

ADDITIONAL READING

If you are interested in learning more about promotion techniques, the following books are good sources of information. You should also look at magazines such as *Advertising Age* and *Business Marketing,* which are available in your university library.

Boone, Louis E., and David L. Kurtz. *Contemporary Marketing*. 4th ed. Hinsdale, IL: Dryden Press, 1983.

Bovee, Courtland, and William F. Arens. *Contemporary Advertising*. 2nd ed. Homewood, IL: Richard D. Irwin, Inc., 1986.

Caples, John. *Tested Advertising Methods*. 4th ed. Englewood Cliffs, NJ: Prentice-Hall, 1974.

Dunn, S. Watson, and Arnold M. Barban. *Advertising: Its Role in Modern Marketing*. 5th ed. Hinsdale, IL: Dryden Press, 1982.

Moore, Lois K., and Daniel L. Plung, eds. *Marketing Technical Ideas and Products Successfully*. New York: Institute of Electrical and Electronics Engineers, 1985.

Nickels, William G. *Marketing Communication and Promotion: Text and Cases*. 3rd ed. Columbus, OH: Grid Publishing, 1984.

Ogilvy, David. *Confessions of an Advertising Man.* New York: Atheneum Publishers, 1984.

Pride, William M., and O. C. Ferrell. *Marketing: Basic Concepts and Decisions.* 5th ed. Boston: Houghton Mifflin Company, 1987.

Robinson, William A. *Best Sales Promotions.* 3rd ed. Chicago: Crain Books, 1979.

Roman, Kenneth, and Jane Maas. *How to Advertise.* New York: St. Martin's Press, 1977.

Sawyer, Howard G. *Business-to-Business Advertising.* Chicago: Crain Books, 1978.

Stone, Bob. *Successful Direct Marketing Methods.* 3rd ed. Chicago: Crain Books, 1986.

Ulanoff, Stanley M. *Advertising in America.* New York: Hastings House, 1977.

EXERCISES

1. With your classmates and instructor, devise a set of editing criteria that covers each of the promotional formats you are asked to develop in the writing assignment below. These criteria will be used to judge the success of your assignment.
2. Select a trade journal related to your field of study. Examine its ads. Using the guidelines for effective print advertising discussed in this chapter, critique the ads. Decide on the best and worst ads. Be prepared to discuss your findings.
3. Examine a trade journal in your field. Looking *only* at the ads, create a list of audience characteristics. Consider age, interests, level of responsibility, financial status, occupation, and gender. How closely does this list of characteristics match your understanding of the readers of this magazine?
4. Using the methods of situational analysis and product description you learned in Chapters 2 and 9, construct five different headlines suitable for a print ad in one of the trade publications you analyzed in Exercises 1 and 2.

WRITING ASSIGNMENT

This is a team project for groups of five. Using a product description assignment from Chapter 9 and the editing criteria you developed in Exercise 1, develop a media campaign that includes at least two different print ads, a direct mail package, and a brochure. For your situational analysis, assume that your readers have a technical background, are decision makers for their organizations, and are cost-conscious. Assume also that the marketplace is fiercely competitive but your product offers distinct advantages.

C H A P T E R

17

Oral Presentations

CHAPTER PLAN

Technical personnel in business, industry, government, and research are frequently called on to share their expertise through oral presentations. Many people dislike speaking before groups, but public speaking is a skill that can be learned like any other. If you prepare by knowing your subject and your audience, and if you master some basic preparation and delivery techniques, you will be able to control the nervousness these situations create and channel it into useful energy and enthusiasm. To help you sharpen your public speaking skills, this chapter will

- consider the advantages and disadvantages of oral presentations.
- analyze the requirements of different speaking situations.
- survey various preparation and delivery techniques.

ADVANTAGES AND DISADVANTAGES OF ORAL PRESENTATIONS

It's a fact of life: technical personnel make speeches, brief people at meetings, run seminars, and present conference papers, whether they like speaking in public or not. These oral presentations have many of the same objectives that written ones do, but oral communication offers at least three advantages over written communication.

First, the *immediacy* of oral presentations allows you to tailor information to the precise moment and audience. If you see that listeners seem lost or confused, you can stop and offer further clarification. If you pay close attention to the temperament of each audience, you can establish, change, or reinforce attitudes. Even in very formal presentations, like those made at a symposium or conference, you can alter your presentation on the spur of the moment to accommodate the mood in the room (boredom, hostility), a common experience (the traffic jam just outside the hotel), or a current event (the resignation of a key member of the organization that morning). A question and answer period at the end of a presentation provides another opportunity to impart very current information.

Consistency is the second advantage of the oral presentation. Many people hear the same message at the same time and under the same conditions. People are not isolated from others receiving the same information (as they would be, for instance, if they were reading memos in individual offices). The odds are better that the people receiving your message will interpret it in a similar way since they can observe each other's reactions and hear each other's questions.

The third advantage of an oral presentation is the degree of *involvement* it affords a speaker and an audience. Naturally, involvement varies according to the presentation environment, but for the most part it is safe to say that people are more likely to participate in a process when they receive information verbally than when they see it in writing. In a group setting, one idea often leads to another; in a written context, information may remain static.

To be truly effective as a speaker, however, you must also compensate for some of the disadvantages of oral presentations. These include speech anxiety, the limits of audience retention, and distractions.

In everyday conversations, most of us speak comfortably and naturally to another individual or to a small group. Yet when we are faced with the possibility of speaking in front of a large group of people (or even a small group of our supervisors), some of us fall apart.

If you have ever had to speak in front of a group, you probably know the feelings of insecurity and nervousness that can suddenly seize you. You probably know the sweaty palms, the lightheadedness, the queasy stomach, the stammering, the quavering voice, the dry mouth, the ringing ears, and the shaking hands. When these things happen to you, remember that most beginning speakers experience the same symptoms. Remember, too, that your audience most likely will be totally unaware of any anxiety you might be feeling. These qualifications aside, though, the best remedy for nervousness is *preparation.* The more you research your topic, the more carefully you develop your ideas, and the more you practice your delivery, the less nervous you will be.

Another disadvantage of oral presentations is the average listener's inability to retain a lot of information. According to experts, only 11% of what is heard is remembered for more than an hour. Therefore, you must build into your speech devices that enhance retention—repetition, vivid examples, concrete language, a logical organization, a maximum of three or four major points, and the flexibility for stopping to answer questions if you need to.

To complicate matters further, audiences are easily distracted during an oral presentation. Capturing attention is difficult; maintaining it is an art. Environmental noise—whirring air conditioners, ringing telephones, traffic sounds—can quickly divert an audience's attention from what you are saying. In addition, each member of your audience will be carrying on an interior monologue—wondering what to have for dinner, reliving an argument, thinking about an assignment. Your speech will have to compete with both kinds of distractions.

PREPARING FOR AN ORAL PRESENTATION

You can use the same techniques to analyze your audience and determine the scope of your topic that you use for written communications. But because the *purpose* of an oral presentation is different from that of a written document, we will spend some time examining this factor. Two other factors that we have not encountered previously need to be analyzed as well: the physical facilities and the occasion, or circumstance, of the talk. When you have considered all of these factors, you will be able to prepare a clear and interesting presentation.

DETERMINING THE PURPOSE OF YOUR SPEECH

The purpose of an oral presentation should closely match the concerns of the audience: give them exactly what they want. Attention will dwindle rapidly once it becomes clear you will not be moving in the direction your audience anticipates. Conversely, listeners will be on the edge of their seats if you state up front that you will be addressing their needs.

Although speeches are made for lots of different reasons, there are five basic reasons for making a presentation about a technical subject: to inform, to instruct, to persuade, to motivate, or to solve a problem. As in writing, recognizing your purpose will help you focus the information you must present. A clear purpose will also help you select examples to support your points and to clarify issues.

To Inform

When the primary purpose of a presentation is to inform the audience, ask yourself: "What does the audience already know about the subject?" Don't bore listeners with unnecessary background information, but at the same time, don't assume more knowledge of the subject than they might have. Also, ask yourself how timely the information they have is. Has it been months since they've had an update, explanation, or review? Has it been only a few weeks or perhaps even days? Additionally, ask why the audience wants the information. What will listeners do with the information you provide? Answers to these questions will help you decide what your audience must know.

A supervisor of a research project called a meeting to discuss the direction she was planning to take in the next year. All the people in the room were well acquainted with the history of the project and its recent activity, yet the speaker rambled on and on about where the research had led them, rather than where it was headed. The atmosphere in the room became so leaden that people began whispering to one another to keep awake.

What was required in this case was a simple summary statement such as, "We've all watched the Sonora project grow in complexity this year. Now I would like to tell you about some of the exciting plans my group has for focusing our studies and publishing our findings." Armed with this kind of brief, direct introduction, the speaker would have established rapport with the audience through an acknowledgment of their shared experience. She would have gained their appreciation by getting right to the point, and she would have let them know what she intended to share with them during her presentation.

To Instruct

When the primary purpose of a presentation is to instruct an audience, the level of knowledge and expertise your audience possesses is again important. However, in this case your intention is to help listeners develop specific skills or achieve a certain degree of subject literacy. Therefore, mere oral presentation is insufficient. Your talk must rely heavily on accurate, interesting examples that *demonstrate* the points being discussed, and you will probably want to encourage participation through the use of graphic aids, exercises, handouts, and even reading material, if necessary.

The best example of this kind of oral presentation is a class. In a class, the instructor begins with a review of the key points made in the last class, then previews the class discussion for that day. Very likely, there are handouts to examine, which demonstrate the principles being stressed in that lesson. There may even be homework assignments to draw on during class discussion. Portions of the instructor's oral presentation are in a lecture format, while other portions are simply open-ended, with the students actively contributing ideas and questions. A good instructor will end the class with a recapitulation of the conclusions made during the class and a strong emphasis on any activity (e.g., homework) expected of students.

It is easy to see the role that graphic aids and group participation can play in the classroom environment. Consider, for example, how ineffective a CPR class might be if the instructor did not bring a working human model along. Similarly, computer service technicians would be hard-pressed to fix computer equipment if their classes did not include practice equipment. Even though oral instruction occurs in circumstances less formal than the classroom, the visual and participative elements remain important factors. For instance, supervisors teach new operating procedures or methods to their workers by actually demonstrating the point of instruction and then answering questions. Engineers describe the benefits of a new design by holding up a schematic diagram for a group of visiting customers to see. Vice presidents depend on charts and graphs to explain the results of new research to a group of press agents.

To Persuade

When persuasion is the purpose of your presentation, you are acknowledging a difficult situation: your listeners do not share your view of the subject. Situations that call for persuasion can range from seeking further funding for a project to selling merchandise to lobbying for a design change to bidding for a construction contract. Your audience may be open-minded, indifferent, wary, cynical, or even hostile. Therefore, the first question you should ask is "How persuasive must the talk be in order to accommodate the audience's attitude toward the subject?" The answer to this question will guide you in selecting evidence and emphasis.

Situations that involve persuasion present a challenge and excitement unmatched by other speaking situations. Inexperienced speakers, however, can be haunted by a voice that whispers, "What if they don't agree with me by the time I finish?" For those of you who do not welcome such a challenge, the following suggestions are offered.

First, acknowledge the audience's attitude as you understand it. Ask for corroboration or further explanation if you want the audience to participate. Adjust your perspective on the basis of new information, if necessary. Such a beginning will demonstrate either your appreciation for the listeners' point of view or your willingness to understand it better. Then, let them know that you do not share their thinking.

However, do *not* tell them that you intend to persuade them that your opinion is better, your methods more practical, your products more beneficial, or your solution more effective. Your arguments themselves will have to prove those conclusions to your audience. In fact, if you begin by telling an audience that you are out to change their minds, they may become obstinate and decide not to listen to a word you say. Each point you present will be met with resistance rather than curiosity.

Instead, explain how you developed your own perceptions and why they seem appropriate. Introduce points in simple-to-complex order, so that your listeners agree with smaller issues before they are faced with a major one. Try to determine the level of audience acceptance before you move to another point. String ideas together so that the transitions between them are gradual and smooth.

If you follow this advice, you should be able to see alert looks on the faces of your listeners. Whether they are agreeing or disagreeing, at least they are *listening*. There should be an increasing number of heads nodding agreement as the talk progresses. And, if you're really lucky, some converts may come forth during a question and answer period to corroborate the points raised in the talk. In this case, audience testimony will help reinforce your own point of view.

To Motivate

When motivation is the goal of a presentation, you want your listeners to execute certain actions or behave in a specific way. Examples of motivational talks include the manager of an arid land management project asking listeners to conserve water in the summer months by watering their lawns just before sunrise or at sunset, a highway patrol officer convincing listeners to use seat belts whenever they drive so that fatalities and injuries can be reduced, or a nutritionist suggesting that listeners minimize their consumption of red meat and egg yolks to reduce their serum cholesterol levels.

Sometimes all you have to do is tell your audience what has to be done, why they should do it, and when it has to be done. Such a task is direct, clear, and uncomplicated, and your audience may respond out of respect for the chain of command, job commitment, fear, or trust. Few circumstances are so cut and dried, however. It is usually more effective to appeal to the values of your audience: why would they want to do what you're asking of them?

For instance, the vice president of a company that had received a lot of customer complaints met with employees. He did not simply say, "From now on, each one of you should make improving product quality your number one priority." Too many people in the audience would have said to themselves, "I am a white-collar worker; I don't make products. He's not really talking to me." Instead, the vice president began by explaining how many complaints the company had received. He followed with an ominous prediction that they would be out of the business before long if things didn't get better. By then he had everyone's attention. He had made the problem concrete and very real to each person in the room. They all wanted to know what could be done to prevent such a prediction from coming true.

The vice president had done something that is basic to the motivational speech: he created a *need*—the need for economic survival, in this case. And he explained the benefit his listeners would receive if they were willing to go along with him—continued job security. The best motivational presentations, then, create a need and provide a reward for executing the actions.

To Solve a Problem

When the primary purpose of a presentation is to solve a problem (rather than present a preformulated solution to a problem), the forum takes on a more participative tone than the other situations we have explored thus far. For instance, a design engineer may be faced with an inefficient circuitry diagram for the power requirements placed on a new computer system; he or she may explain the problem and solicit suggestions for change. The vice president of a company receiving an abundance of customer complaints may hold a series of follow-up meetings with specific departments to determine the individual roles they can play in ameliorating the situation.

Sometimes when you find yourself in such a situation, you have certain recommendations you believe could solve the problem. However, it is more likely that you are looking for help from the audience in finding the best solution. The classic pattern for handling this kind of situation is:

1. Define the problem as you see it. Call for corroboration or additional input at this point. Any pertinent background information might also be presented at the outset. The members of the audience must agree on the nature of the problem before any legitimate solutions can be proposed or explored.
2. Examine the known or possible causes of the problem. If causes can be isolated, it may be possible to eliminate them. If the causes are eliminated or minimized, chances are the problems will be, too. Consider the analogy of a physician treating a patient with back pain. It would be possible to treat the symptom with medication, perhaps muscle relaxants, but such a treatment would not guarantee that the problem would not recur. If the physician

treats the cause, in this case a slipped disk, the symptom will be eliminated and recurrence is unlikely. Again, ask for input from the audience.

3. Begin a compilation of possible solutions. Present these to get the ball rolling, but solicit other solutions from the audience as well. You may wish to have a blank flip chart or a chalkboard to keep track of suggestions.

4. Try to determine the best solutions. Act as a conductor rather than a commander, letting the audience argue, test, and agree on the best solution.

5. Formulate an implementation strategy. Having agreed on a solution to the problem, you and your audience should cooperatively implement the plan. It may be necessary to hold a separate gathering to discuss different implementation strategies. In any case, those in attendance at the implementation discussions should agree on the best strategy before action is taken.

MAKING THE MOST OF AVAILABLE FACILITIES

You may not always be able to choose the location of a presentation, but there is no excuse for not finding out about the facilities before you speak. The conference chairperson, your host, the meeting planner, the hotel concierge, or previous speakers are all sources of information on the quality and variety of services, accommodations, and equipment available.

You should know beforehand if there will be 200 straight-backed, wooden folding chairs or 200 side-armed desks or 200 well-cushioned armchairs. In the first instance, listeners might become uncomfortable a few minutes into your talk; you'll want to be relatively brief. In the second case, listeners will be able to take notes easily, lean forward on the desk tops to ease discomfort, or stretch their legs out to the side. In this situation you can afford to take a little more time than you can in the first. And if you are to be addressing people sitting comfortably in armchairs, you will want to be especially lively in your presentation to avoid any tendency on the part of the listeners to daydream. You should know how those chairs will be arranged as well, so that you can plan your gestures, your graphic aids, and the amount of voice projection required. In smaller groups, seats might be arranged around a conference table or in workshop clusters. In larger groups they might be arranged in amphitheater, classroom, or auditorium style. Will you be speaking from a dais, an isolated lectern, a stationary stand-up microphone, a stage, a platform, or will you be able to pace naturally in front of the room?

What's the lighting like in the room? If you are going to be showing slides, for instance, can you arrange for a spotlight to be on you so that you don't become a disembodied voice in the darkness? Are the right kind of slide projector and screen available, or do you have to supply your own? Will you be needing an operator to advance the carousel for you?

What kind of decor is in the room? If the room is ornate, as hotel meeting facilities frequently are, both you and your graphic aids will have to compete with brocade wallpaper, gilded windows, fancy mirrors, and chandeliers. Perhaps the opposite will be true, and you will be presenting in a small, wood-paneled, windowless conference room. It is easy for a presentation to lose its verve in a stifling situation like that.

Will people be able to hear you well? You might find yourself in a conference center with half walls or portable walls, which allow noise to filter into your area from other meetings or assemblies. Will you need a microphone? If so, will you be wearing a small one attached to your shirt or blouse or using a stand-up microphone?

Knowing the answers to *all* these questions before you develop your talk will allow you to factor them into your presentation. And this kind of preparation will leave little room for the unnerving surprises that can upset your sense of confidence.

ACKNOWLEDGING A SPECIAL SITUATION

Sometimes, in addition to simply making a presentation, you must acknowledge the special event or set of circumstances that brings your group together, such as the centennial anniversary of your organization, the death of a senior officer, the introduction of a new product, an annual sales conference, a year-end review, or the reorganization of a division. In each case, you are wise to factor this event into your remarks to help establish a common ground for both you and your listeners. Your introduction should allude to the event immediately:

"Thank you for this opportunity to speak with you on this momentous occasion."

"It's a pleasure to share the excitement of this conference with all of you."

"Now, as we face the coming year, let's look back for a moment to assess our progress in the last year."

Depending on the importance of the event, it may be woven into the body of the speech as well:

"This point has particular significance for us as we approach our one hundredth year in the business."

"The reorganization of our division has been engineered to streamline our managerial staff and to strengthen our stand in the marketplace."

The event should also be a legitimate part of your conclusion. By alluding to the occasion a final time, you continue to acknowledge the common bond between you and your audience. Tie in a reference to the occasion with a reinforcement of your primary message. Here is an example:

"And so, as we celebrate the three hundred and fifty-fifth birthday of Anton van Leeuwenhoek, we realize how very far we have come through the benefits of his invention, the microscope. Now, with the aid of super-microscopes, our research laboratory has recently begun to unravel the mystery of our bodies' immune system. The discovery of lymphokines, hormones that direct the fight against disease, is both a fitting tribute to Leeuwenhoek and the beginning of a revolution in medicine. We should all be proud of the role we have played in this discovery, but we should also recognize that we merely have unlocked a different door and much awaits us beyond this new threshold."

DEVELOPING THE PRESENTATION

In the case of formal oral presentations, such as conferences, symposiums, keynote speeches, and lectures, it is usually best to prepare a complete written copy of your speech. This written document differs from a report, essay, article, or proposal in several important ways:

1. It reflects the cadence and word choice of spoken English. You avoid multisyllabic words, highly complex sentence structures, and tongue-twisting phrases.
2. Its structure is more repetitive because your audience is listening rather than reading. For points to be remembered, they must be presented emphatically and then reiterated.
3. It contains less technical detail than its report counterpart. The speech format does not allow your audience to retain complicated facts, figures, or concepts as well as the report.
4. It allows for audience input.
5. Its examples are generally more vivid, both to maintain interest and to stress key points.

Often, the organization sponsoring the conference or lecture prints a set of proceedings that includes the text of presentations. Even if your speech will not be published, however, a complete written statement allows you to develop a clear, focused presentation. Edit and revise your speech until you are confident it is the most effective it can be. As we will see in a moment, the written speech is never memorized and is rarely read verbatim, but it is an important part of an extensive preparation process.

More than any of the written work we have examined thus far, a speech requires simplicity and clarity of structure and language. Because listeners must be able to follow and *remember* the development of ideas, use a simple sentence structure such as the easy-to-grasp subject-verb-object arrangement of information. The following sections will give you detailed advice on organizing and writing a speech.

ORGANIZING THE SPEECH

As in the case of all good writing, an oral presentation begins with a solid outline. The speaker uses an outline to get his or her ideas down on paper, to focus the topic, and to determine relationships between bits of information.

The speech outline should be developed according to the standard three-part structure of introduction, body, and conclusion. In a memo to young reporters, Edward R. Murrow once paraphrased Aristotle, giving advice that applies to each of these three sections: "First you tell them what you are going to tell them. Then you tell them. Then you tell them what you've told them."

In other words, the introduction should present the topic to be discussed and outline a plan of development so that listeners can anticipate the order in which information will be discussed. It explains the purpose of the speech as well, if that is not obvious to the audience.

In the body of the oral presentation, provide all the information you promised the audience in your introduction. The way in which the body of the speech is organized will affect such factors as the audience's interest, understanding, retention, reaction, and motivation. The body of the speech should contain only three or four main points that explain or expand the topic introduced at the outset of the talk.

The conclusion reinforces the purpose of the speech by presenting a summary, reiterating key points, making recommendations or a call to action, or posing questions for further consideration. The conclusion should always be relevant and brief.

As you develop your presentation outline, take into account this three-part guideline. In addition, decide on the best possible organizational strategy during the outline stage. In Chapter 4 we examined a variety of organizational strategies that you might want to review; they apply to speeches as well as to written communication.

GENERATING INTEREST

Having made the many preparatory decisions about your speech, it is now time to begin writing. Your first task is to get the audience's attention. You want people to listen intently, to relate to you, and to care about what you say to them.

Some of the suggestions for writing effective headlines and sales letters we reviewed in Chapter 16 are also suitable for generating interest in oral presentations, such as

- promising a benefit
- revealing exciting news
- telling listeners how to do something of interest to them
- offering valuable advice
- presenting a startling fact
- referring to an upcoming event
- negating a commonly held belief

Add to this list these other suggestions for effective beginnings:

- identifying the problem you intend to solve
- referring to the occasion of the speech
- referring to a common experience
- introducing your subject with a *relevant* anecdote or joke
- presenting observations you may have made about the audience as you awaited your chance to speak to them. For instance, you might say: "While I was waiting to speak to you, I couldn't help notice the camaraderie this group exhibits."
- quoting a relevant and recognizable source
- referring to a current event of great importance
- alluding to the speaker who spoke before you or to the host who just introduced you
- identifying with your audience on a level pertinent to your speech. A computer expert might say, for example, "Like most of you, I am a concerned parent—one who has noticed the rapid growth in popularity of video games. And I'm worried."

No matter what kind of beginning you choose, it should be appropriate to the audience and to the purpose of the discussion. It should also establish a respectful and sincere tone that is carried through the entire presentation.

No matter how terrified or insecure you might feel during the first few minutes of your presentation, do not admit your nervousness (it will make your audience uneasy) and do not apologize (your audience may not have found any fault with you, so don't focus negative attention on yourself).

Make full use of the introduction. The first three minutes of any presentation are crucial to capturing your audience's attention.

MAINTAINING INTEREST

The level of audience interest will drop off after the first three to five minutes, no matter how exciting the speaker is. However, no speaker can afford to let the level of interest drop too low, or none of the points to be made will be understood, appreciated, or retained.

Interest levels can be buoyed up by certain techniques. It is possible to amplify interest by arranging your information in a certain way. If a scientific audience is being addressed, for instance, the speaker would be wise to use a deductive approach, presenting a thesis and proving it point by point. Other audiences might remain more interested if they participate in the discovery process; in these cases an inductive approach might work best.

Other ways of stimulating interest have been discussed throughout this book in reference to writing. They apply equally well to speaking. In most cases, the effect of these techniques is to make the speech concrete, to make it easier for listeners to understand or relate to what is being said. Here are a few reminders.

1. *Examples* are a good way of getting and maintaining listeners' attention. Consider the effectiveness of a speaker's use of the space shuttle *Challenger* disaster when discussing the importance of establishing verifiable quality standards in high-technology products. Listeners will likely recall their own personal reaction to this tragedy.
2. *Personal experience* is often used effectively. In a speech on driving safety addressed to a high school audience, a speaker shared her experience of an automobile accident. The fact that she lived to tell about it made her a convincing advocate of safe driving tactics.
3. *Graphic aids* help capture an audience's attention by adding a visual dimension to your speech; they enable the audience to visualize the points being discussed; and they help avoid misunderstandings.
4. Used sparingly, *rhetorical questions* can awaken a daydreaming audience. Asking "How would you feel if that happened to you?" is a great way to get people to pay attention.
5. In many cases, a *demonstration* of the idea you are presenting is possible. This may be done by calling for volunteers from the audience, by bringing a product or working model with you, or by showing a film that proves the point. Such a tactic not only helps maintain interest, it builds credibility.

6. The appropriate use of *colorful language*—metaphors, similes, analogies, allusions, and personification—helps to keep an audience's interest focused on you.
7. The use of *transitions*—clear statements that introduce and summarize ideas—helps listeners keep up with you as you move from point to point. You can also use them to indicate the significance of ideas, relationships, or logical progressions.

Also important to maintaining interest are such factors as eye contact, body language, voice control, enthusiasm, spontaneity, appearance, and presence. These areas will be discussed more fully later in the chapter.

None of these methods used alone will be enough to keep an audience interested in what you have to say. A mix of elements is essential to controlling pacing and momentum.

As you move into the concluding portion of your speech, let the audience know by using such markers as "in conclusion," "to summarize," "finally." The audience's attention will soar once again on hearing these magic words. Take advantage of this renewed interest by reiterating the points most important for them to retain. Use this time to call for any required action. Close with a thank you, and if appropriate, open the floor for questions or discussion.

USING GRAPHIC AIDS

As you develop your presentation, decide whether your talk could benefit from graphic aids. The decision should be based on the needs of your audience, the complexity of the information being discussed, your own comfort in using visuals, the room in which you will be presenting, the availability of any required equipment, production resources and time constraints, and the size of the group.

If you use graphic aids, be sure that each is as simple as possible, is easy to see, and is clear. Graphic aids should be attractive—if time and money permit, it may even be wise to have them prepared professionally. Be sure also to give yourself cues in your notes so that you know when to introduce or change them as you speak.

Formats for oral presentations can include overhead transparencies, slides, flip charts, working models or products, films, or videotapes. Chalkboards and blank flip charts are also often used during presentations to summarize key points or ideas from the audience. Films, videotapes, and slides offer opportunities to expand the scope of your talk, but they can also divert attention away from you. The darkness they require may, in addition, prevent listeners from seeing you clearly or taking notes. As you work on your presentation, decide whether these disadvantages outweigh the advantages.

SAMPLE ORAL PRESENTATION

The following excerpts from an oral presentation demonstrate the principles discussed in this chapter. The speaker is the division general manager for a corporation that manufactures undersea submersible equipment. She has been charged with giving a group of major stockholders a technology update. To best understand this talk, you need to know that this line of the business has been very successful over the past two decades. It has grown from a $20-million-per-year venture to a well-established $114-million-per-year business. Unfortunately, the major customers for such products, oil companies, are having to cut back on their speculative undersea exploration activities because of significantly lower oil revenues in recent months. The speaker wishes to assure investors that the company is stable. However, new technologies are necessary to make products more attractive to cost-conscious customers. Additionally, these technological advances will help the company venture into other marketplaces, such as scientific research and military applications.

[1] Good morning. It's a pleasure to meet with you today to brief you on our current and future work in small unmanned submarines, popularly known as submersibles. First, for those of you who are new to our annual stockholders' meeting, I will give you a little background on this relatively new technology. Next, we'll look at the individual product lines with which our company has had considerable success this past year, touching on some of the large contracts that have been negotiated since we last got together. Then, it will be important for us to focus on some of the trends our analysts see developing in our primary marketplace, the oil industry. Finally, I would like to share with you some of the exciting new technologies we are researching and beginning to develop. Please feel free to stop me with questions at any time.

1. Introductory remarks: outline of what will be discussed

Unlike human divers, who can routinely dive only to about 1000 feet, unmanned submarines, or submersibles, can operate at depths ranging from 3000 to 15,000 feet—without risking lives.[2] [3] Also unlike human divers, submersibles do not have to surface after five or six hours. Rather,

2. First part begins: background information
3. Comparison

[4] they can stay submerged for days or even weeks. You might be most familiar with the work of submersibles from the publicity about the recent discovery of the *Titanic* at the bottom of 13,000 feet of icy waters. No humans could have survived such explorations. Also, submersibles were used to recover the black box from the cockpit of a downed Air India jetliner off the coast of Ireland not too long ago.

4. Examples

[Place overhead transparency of drawing of submersible on the projector.][5]

5. Cue to visual aid

[6] As you can see from this line drawing of a typical device, submersibles are controlled from the safety and comfort of a surface support ship by means of a two-way cable. This cable transmits both electric power and commands from the surface ship to the remotely operated vehicle (ROV). It is also able to transmit video signals and sensor data back to the ship for processing.

6. Mechanism description

[7] Submersibles are used primarily as underwater eyes and as service technicians to inspect and maintain offshore oil-drilling rigs, undersea pipelines, and telephone or power cables. Additionally, navies and marine scientists are beginning to make use of this technology to study the geology and biology of the deep ocean. Advances in robotics and sensory technology also enable some types of submersibles to perform delicate repair and assembly tasks, such as the recovery of lost gear, collection of samples, and repair of oil rig components. . . .

7. Purpose of submersibles

Today, submersibles represent an $800 million market for undersea services and offshore support equipment.[8] Depending on the sophistication of individual devices, the price of each unit ranges from $200,000 to over $1 million. We foresee this marketplace growing to $1 billion within the next decade. To do so, however, newer technologies will be necessary. I will touch on these in a few moments, but first let's look at the performance of our specific product lines.

8. Second part begins: individual product lines

[9] This past year we discontinued production of our oldest product line, the ROV 78. This older equipment provides visual surveillance capabilities only and has been superseded by products that offer both visual and manipulative functions at comparable cost. We still offer the owners of these devices complete service and maintenance contracts, which constitute a $600,000 revenue opportunity for us.

9. First product

Additionally, we have begun a buy-back program to induce the owners of this equipment to upgrade to more efficient machines.

[10] Our ROV 82 model is doing very well, with two multimillion dollar contracts having been negotiated in the past year with ABC Oil and XYZ Oil. This model is designed for offshore drilling support to a depth of 5000 feet. The unit is equipped with searchlights, a video camera, sonar, and a manipulator/gripper arm for repair and maintenance operations. . . .

> 10. Second product

Now let's take a look at a trend our analysts see developing in the oil industry and its subsequent effect on our business.[11] As you probably know, the price per barrel of oil has dropped significantly since 1982, so that, for the time being, it is more cost-effective to import oil than it is to develop additional offshore drilling sites. This trend has caused a cutback in the number of orders for exploratory equipment. Basically, we face a much tougher market than we did a few years ago. Our analysts believe this trend will last for at least the next two years. . . .

> 11. Third part begins: industry trends

To help stimulate sales in a somewhat sluggish marketplace, we have begun development of a technologically advanced submersible that can do far more than just look around.[12] This new machine will be able to perform sophisticated maintenance procedures, such as component replacement; it will be able to clean debilitating marine growth from undersea drilling platforms; and it will be much easier to operate than our current models.

> 12. Fourth part begins: new technology

[Place overhead transparency of new submersible design on projector.][13]

> 13. Cue to visual aid

Additionally, as you can see from this drawing, the new submersible will eliminate the requirement for a lengthy attachment cable. Using sonar and fiber optics, these remotely operated vehicles will no longer run the risk of being snagged on ocean floor debris. We also hope that the fine-motor movements this machine is capable of performing will help us capture the attention of the scientific research community and the military. By expanding our marketplace, we will offset the cutbacks in the oil industry I mentioned earlier.

Our market projections for this product. . . .

[14] In closing, I would like to emphasize the hearty growth our division has enjoyed this past year. We are taking seri-

> 14. Conclusion

ous steps toward reversing a possible reduction of customer requirements by enhancing the capabilities of our product line and by venturing into scientific and military markets. I am very optimistic that next year, when we get together again, we will be able to report significant progress in the development of these products and headway in marketing them to new customers. Thank you.

DELIVERING YOUR SPEECH

At this point, you have developed your written statement and are ready to practice, preferably in front of other people. Practice is necessary for several reasons: it helps you become a more confident speaker; it makes you a more natural speaker; it enables you to become comfortable with the graphic aids you intend to use; and it helps you pace yourself.

The best kind of delivery style is an extemporaneous one. You structure your talk according to your written document but deliver your information in an enthusiastic, spontaneous way. The more you practice your presentation, the more comfortable you'll be with this style of delivery. Never *read* your written document unless you are presenting a paper at a technical symposium. This kind of presentation can be extremely boring. Never try to memorize your presentation either; you will seem stiff and unnatural. An extemporaneous delivery allows you the flexibility of modifying your talk to suit your audience. It also encourages greater rapport with your audience.

Here is a brief list of tips to help you present yourself well when you are making a speech.

1. Look tidy, professional, and conservative. Try to radiate confidence. Face the audience, keep shoulders and hips level, and maintain a pleasant expression (not a grin) on your face.
2. Be positive, objective, and forceful. Never use coarse or abusive language and avoid an overly casual approach.
3. Prepare notes, highlight the copy of your speech, or prepare an outline once you have practiced your speech several times. Have quotations, facts and figures, and examples at your fingertips in a separate notebook or piece of paper.
4. Stand erect with both feet on the floor. Don't slouch or shuffle your feet. Hands should be at your sides when not used for gesturing; they should not be held behind your back or in your pockets.

5. Use hands and facial expressions for emphasis. Keep arm and hand movements natural; don't exaggerate.

6. Do not wander aimlessly about the platform or make nervous movements (like scratching yourself or jingling keys or jewelry). If you want to stand in a different place, move when you make a transition from one part of your speech to another.

7. Look at the audience, not at the wall, the floor, or the lectern. Make eye contact with the front row, then with people sitting to the sides, middle, and back of the room.

8. Avoid speech tics ("uh," "you know"); simply pause if you must gather your thoughts. Pace your words so that they are easy to hear and understand. Project your voice well; a low-pitched voice is usually best understood. Speak in solid statements rather than trailing questions.

9. Inspect the speech location. Notice, for instance, how the furniture is arranged, where the electrical equipment is controlled, how the projection screen is operated. Doing so will enable you to move freely from your seat to the lectern and to arrange your notes and other materials comfortably.

10. Turn your nervousness into positive energy. Take a deep breath before you begin and breathe slowly. Pause and look around you if you feel a bit shaky. *Act* confident even if you're not; acting confident usually helps you *feel* confident.

SURVIVING THE QUESTION AND ANSWER PERIOD

Frequently, an oral presentation is followed by a question and answer period. Prepare for questions by challenging each of the ideas you present as you prepare and practice your speech. Be your own devil's advocate. This tactic will also help you fill in any gaps your speech might have.

Sometimes during a question and answer period, no one in the audience will be willing to pose a question. In this case, pose one yourself to get the ball rolling. You might mention a question asked in a previous talk (if you have made this presentation before). Or you might present a question that had occurred to you when you first heard about your topic. This tactic usually breaks the ice.

Other times, you may be overwhelmed with questions. Act delighted by the response. Call on as many individuals as time will allow (acknowledge questioners by describing their clothes rather than by pointing to them). Repeat or paraphrase a question before you begin answering. If necessary, invite some people to stay after and talk if you do not have a chance to get to their questions.

An uncomfortable moment may come when you don't know the answer to a question. It is usually best to admit that you don't know the answer. Offer to look into the question. The worst possible situation (but one you can survive) is when you are faced with a hostile question. You will be most effective if you answer it calmly and tactfully.

CHAPTER SUMMARY

Technical personnel often have to speak in public. Each public speaking situation requires careful preparation, which takes into consideration such familiar factors as audience, purpose, and scope, but which also requires attention to facilities, equipment, and any special events. The best antidote to nervousness is adequate preparation. Many of the development principles you learned for written communication apply to oral presentations as well. A well-developed written statement and frequent practice will save all presenters—novice and experienced alike—from speech anxiety.

An extemporaneous delivery is most effective because it seems spontaneous and enthusiastic (even though it is actually rehearsed extensively). When you are speaking, your words, appearance, body language, gestures, eye contact, and voice control all affect how your ideas are received.

ADDITIONAL READING

For more information and advice on oral presentations, consult one of these texts:

Jeffries, James R., and Jefferson D. Bates. *The Executive's Guide to Meetings, Conferences, and Audiovisual Presentations.* New York: McGraw-Hill, 1982.

Mambert, A., and D. Ehninger. *Principles of Speech Communication.* 8th ed. Glenview, IL: Scott, Foresman, 1980.

Morrisey, G. *Effective Business and Technical Presentations.* 2nd ed. Reading, MA: Addison-Wesley, 1975.

Schmid, Calvin F., and Stanton E. Schmid. *Handbook of Graphic Presentation.* 2nd ed. New York: John Wiley & Sons, 1979.

Simmons, Sylvia H. *New Speaker's Handbook.* New York: Dial Press, 1972.

Snyder, Elayne, and Jane Field. *Speak for Yourself—With Confidence.* New York: New American Library, 1983.

Wilcox, R. *Oral Reporting in Business and Industry*. Englewood Cliffs, NJ: Prentice-Hall, 1967.

Zelko, H., and F. Dance. *Business and Professional Speech Communication*. 2nd ed. New York: Holt, 1978.

EXERCISES

1. Contact someone currently working in your field to find out the kinds of oral presentations he or she must make.
2. Attend a public meeting on a subject of interest to you. Evaluate whether the needs of the audience were met by the speaker. If an element of persuasion was involved, was the speaker able to persuade the audience to accept his or her perspective? What criticisms of the speaker's performance would you make?
3. In a class within your major field of study, consider the way in which your professor presents information. How well prepared is he or she? How much class participation is encouraged? Are the objectives of the class session clear? Are the objectives met? Are the objectives suited to the needs of the audience? How many major points are brought up in a single session? Are a sufficient number of examples used to clarify information?
4. Watch a television broadcast of a political event. How do the speakers appear? How do they gesture? Is their voice control good? Are the speakers flexible in their reaction to spontaneous events? Do they seem well rehearsed? Do they seem well informed?

SPEAKING ASSIGNMENT

Select a topic about which you are well informed and prepare a ten-minute oral presentation to deliver to your technical writing class. You might find that your formal report or your mechanism or process description assignments will lend themselves to this speaking assignment. Practice your presentation so that you can deliver it extemporaneously (with the help of an outline, a highlighted copy, or note cards). Use graphic aids as appropriate. Prepare for a brief question and answer period following your talk.

NOTE: Apply the editing criteria at the end of Part III to your assignment before you submit it to your instructor.

CHAPTER

18

Getting a Job

CHAPTER PLAN

One of the most exciting, yet intimidating, times in your life is when you go job hunting. This is true whether you are looking for your first full-time job, changing jobs, or changing careers. Unless you are a particularly lucky or exceptionally talented individual, *getting a job is a job unto itself.* It takes planning, insight, determination, confidence—and, above all, persistence. And your technical writing skills can play a key role.

To help you learn effective job-seeking skills, this chapter will

- show you how to analyze your strengths and weaknesses in light of particular job goals.
- list ways of finding job openings.
- discuss how to prepare résumés, application letters, and follow-up correspondence.
- acquaint you with effective interviewing strategies.

PREPARING FOR THE HUNT

Getting a good job requires help: help from books like this one and others listed at the end of this chapter, help from job placement offices at your school or in your community, and help from those who know you well enough to speak about your qualifications. Although much of the advice presented here addresses the special needs of first-time job seekers, mature candidates can benefit from it as well.

Without exception, these sources expect job applicants to demonstrate a clear knowledge of themselves, their career area and the marketplace, the requirements of the job they are seeking, and the firm to which they are applying. In addition, employers also expect an impressive résumé, an appealing application letter, a properly filled in application form, a solid performance during an interview, and a courteous follow-up letter.

By examining each of these expectations systematically, you will be far better prepared for the job of getting a job. Even if your career start is still several years away, practice with the methods and materials presented in the remainder of this chapter can help you with summer employment, part-time work during the school year, or even independent study projects.

KNOWING YOURSELF

You live with yourself day in and day out, but how much time do you take to understand who you are, how you function, what you like and dislike, what you need, and what you need to avoid? If you are like most people, you spend very little time thinking about these things. Before you can get a job, however, you really need to know yourself inside and out. You need to feel comfortable with your many strengths, and you must be aware of your weaknesses or limitations.

To accomplish this analysis, try some brainstorming. Take lots of paper and a pen to a quiet place where you cannot be disturbed for at least an hour. Then, in no special order, write down everything you can think about yourself regarding the following characteristics:

job experience	education
course work	special projects
recognition	interests
skills	volunteer work
accomplishments	goals
written work	hobbies
availability	salary requirements
expectations	values

After you have completed this brainstorming exercise, consult your school transcripts, family albums, and other records to fill in any gaps. The longer the list, the better; no bit of information is too insignificant to include at this point. Then, using this long list of characteristics, focus the items by performing the following steps:

1. Place each attribute in one of the categories from the preceding list.
2. Within each category, determine which characteristic is your strongest. That is, which accomplishment or activity speaks most favorably of your abilities. If the category is "volunteer work," for instance, you may decide that your work in helping to register voters in the last election is more impressive than the walkathon you participated in six years ago.
3. Eliminate abstract words or phrases by developing personal, concrete examples of each quality or characteristic. For instance, if one of your values is being *responsible,* translate that abstraction into measurable traits such as: "supervising 15 children (ages 10–12) during summer camp."

KNOWING YOUR CAREER AREA AND THE MARKETPLACE

As a college student majoring in a specific subject, you may already have a good idea of your intended career area. As someone in the work force thinking of changing jobs, or even careers, you may feel the same way. Even if you are comfortable with the issues, technology, and developments in your career area, consider strengthening your grasp of them by performing the following tasks:

1. Survey the leading trade journals in your intended field. What are the controversial topics and what are the arguments on *both* sides? Most importantly, where do *you* stand in relation to these issues?
2. Look at the classified advertisements in these journals and ask yourself the following questions:

- What jobs are being advertised most frequently?
- How many entry-level positions are advertised?
- What are the salary ranges advertised?
- In what cities are the most jobs being listed? Am I interested in relocating to any of these places? If so, what is attractive about these locations?
- Which companies are running the most ads or the biggest ads? Would I like to work for any of these organizations? Why or why not?

3. Consult the Bureau of Labor Statistics *Occupational Outlook Handbook* (Bulletin 2200) to determine trends in the marketplace. Is your field growing, remaining static, or shrinking? What are the factors that affect it?

4. Speak with people you know who already hold positions in your desired area. What do they see as the positive aspects of the profession? What complaints do they have? Do they feel that the field has promise for you? Why or why not?

5. Speak with job counselors at your school's placement office or with government employment agency personnel about job availability and marketplace trends. How many qualified candidates do they believe apply for individual positions? Show them your list of characteristics and ask what they believe your strengths and limitations to be.

Now re-examine the list of your own characteristics. How do your qualifications stack up against the available positions? Do your strengths match marketplace needs? With answers to these questions in hand, you are now ready to consider seeking specific jobs or kinds of jobs.

KNOWING THE REQUIREMENTS OF A SPECIFIC JOB

The first word of advice in seeking a specific job is that you be realistic. Someone fresh out of college is most likely not going to land a management or executive position. Management jobs are secured after experience with a firm and through contact and cooperation with others. Set your sights on a job that will allow you to demonstrate your proficiency in your field and will lead you along your desired career path. The old adage "You have to learn to crawl before you can walk" has bearing here. Many great scientists began as lab assistants. Presidents of banks began as tellers. Directors of communications began as technical writers. Don't overshoot your mark; you will not only be disappointed (which has a debilitating effect on your motivation), you will be wasting valuable time.

Consider your strengths and weaknesses and determine the kind of job for which you are best qualified. Don't set your sights too low, either, or you will find yourself being rejected by prospective employers who believe you are overqualified. With a general category of job in mind, find out the requirements of such positions. One of the best ways to do this is to consult the *Occupational Outlook Handbook,* which describes almost every kind of job available in the United States today. Here is the listing for the position of technical writer:

> Develops, writes, and edits material for reports, manuals, briefs, proposals, instruction books, catalogs, and related technical and administrative publications concerned with work methods and procedures, installation,

operation, and maintenance of machinery and other equipment. Receives assignment from supervisor. Observes production, developmental, and experimental activities to determine operating procedure and detail. Interviews production and engineering personnel and reads journals, reports, and other material to become familiar with product technologies and production methods. Reviews manufacturers' and trade catalogs, drawings and other data relative to operation, maintenance, and service of equipment. Studies blueprints, sketches, drawings, parts lists, specifications, mockups, and product samples to integrate and delineate technology, operating procedure, and production sequence and detail. Organizes material and completes writing assignment according to set standards regarding order, clarity, conciseness, style, and terminology. Reviews published materials and recommends revisions or changes in scope, format, content, and methods of reproduction and binding. May maintain records and files of work and revisions. May select photographs, drawings, sketches, diagrams, and charts to illustrate material. May assist in layout of materials for publication. May arrange for typing, duplication, and distribution of material. May write speeches, articles, and public or employee relations releases. May edit, standardize, or make changes to material prepared by other writers or plant personnel.*

With the very specific information provided in descriptions of this nature, review your list of characteristics one more time to determine the essential qualities and experience that you have. At this point, focus the list to include only your most important and best characteristics—the ones you feel qualify you for the kind of job you are seeking.

UNCOVERING JOB OPPORTUNITIES

With all of this preparatory work under your belt, you are now ready to uncover jobs that are suitable to you and for which you are well suited. Even with the increasing competition for jobs, good jobs still exist, and *you can get one of them*. Here is where being systematic and persistent can be especially important. Above all, do not become discouraged if nothing appeals to you the first time you look in the local paper. Many other job sources exist besides this one, and you should leave no stone unturned. Here are just a few of the other sources for job listings or information:

- the classified section of the *Wall Street Journal* (available at university and public libraries). Concentrate on Tuesday, Wednesday, and Thursday papers.

*Bureau of Labor Statistics, *Occupational Outlook Handbook* (Washington, D.C.: U.S. Government Printing Office, 1984).

- the Sunday classified section of the *New York Times* (also available at university and public libraries).
- your college's job placement office.
- city, county, state, and federal job agencies.
- notices in newsletters of student and professional organizations in your field.
- the bulletin board in the administrative office of your major field of study.
- community organizations (such as the Chamber of Commerce, which can tell you of new businesses in town or those about to locate in town).
- professors who have kept in touch with former students now in the work force.
- friends and colleagues who already are in the work force.
- personnel departments of companies or organizations for which you would like to work.

As possibilities surface, keep exact records. For each job in which you are interested, note

- its precise title
- the source that carried the notice (or person who told you about it)
- the date on which the ad appeared
- the closing date for filing applications
- the date on which you sent your application
- the name, address, and contact information pertinent to the job
- the status of your application

Place this information on a simple chart like the one in Figure 18-1. That will enable you to track many job applications at once. Remember: many tries are usually required to secure a job that offers you the opportunity to grow.

For each job you are interested in, re-evaluate your list of personal characteristics. Does the list indicate a sufficient number of traits and experiences to make you a viable candidate for such a job? Do you have the skills, course work, goals, values, and abilities to do the job well? Do you have the level of competence required for this job? Be severe with yourself—personnel officers will have an abundance of applicants. Are you merely going to be another applicant, or are you truly able to be a *candidate*? The more realistic you are at this point, the less disappointment you will face. Although you certainly do not want to sell yourself short, do not fool yourself either. Focus on jobs you stand a strong chance of securing.

JOB TITLE	SOURCE OF AD	DATE OF AD	CLOSING DATE	APPL. DATE	CONTACT	STATUS

Figure 18-1 A tracking sheet organizes vital job-search information.

Before you can send out an application letter or résumé, there is one more step you should follow to enhance your chances of getting the job you want.

KNOWING THE ORGANIZATION TO WHICH YOU ARE APPLYING

After making your preliminary selections, research each company so that you can determine which you would most like to work for. Personnel officers expect candidates to have done their homework regarding their companies. Job seekers should know a great deal about an organization *before* they apply. Frequently updated sources include annual reports and fact books, *Standard and Poor's Index* (in the library), the *Fortune* 500 listing, the *Business Week* annual report issue (March), press coverage (consult the indices of the *New York Times* and the *Wall Street Journal*), and college placement office profiles.

For each company or organization you have under consideration, research the following information:

history
size
number of employees
executive officers
locations
policies
financial standing
job distribution
product line
research and development activity
subsidiaries
special projects

If you have specific organizations in mind for which you would like to work, request copies of their annual reports (see letters of inquiry, Chapter 12). Other places you might check are the *Standard Directory of Advertisers* (companies are sorted by product; company officers are named); Moody's manuals, Dun and Bradstreet's directories, and Thomas's *Register of American Manufacturers*.

MATERIALS TO HELP YOU GET A JOB

Whether you are seeking a job by interviewing with recruiters on your college campus or are applying by mail for positions, it is important to present a strong impression through a well-written résumé and application letter. All of the preparation you have just completed will now help you to create attractive, informative documents to set you apart from your competition.

THE RÉSUMÉ

A résumé is a categorized list of personal experiences, education, and characteristics that highlights the key attributes of the applicant. The brainstorming sessions you have recently completed will provide you with the content for your résumé, and the examples in this text will provide you with some ideas for categories.

College students are often concerned that their lack of work experience will handicap them in their job search. It may be comforting to know that corporate personnel officers understand this concern. They recognize that recent college graduates have yet to develop their job skills, and that most often students have not previously held jobs similar to the one for which they are applying. What is important for a student to demonstrate, then, is his or her background and character. Emphasis should be placed on education, activities, part-time jobs, volunteer work, recognition, memberships, and special projects. For instance, if you have taken courses of value to the job being sought, the résumé can indicate a degree of familiarity with the career area. If you have had part-time jobs throughout college, the résumé presents the image of a responsible and industrious person. Club offices held demonstrate leadership and organizing abilities. Membership in professional organizations shows a commitment to a field as well as an ability to participate and cooperate with others. Special projects imply seriousness, creativity, and self-motivation. The point is that job experience is not the primary criterion upon which hiring decisions are made for entry level positions. Therefore, it is important for you to learn how to represent yourself to your best advantage.

Although there are a variety of acceptable formats for résumés, they share similar constraints. Above all, résumés should be attractive, easy to read, informative, and succinct. In an application package, the résumé will be the most important document. It may be scanned before the application letter is read; it may be distributed on its own to reviewers other than the personnel director who receives it initially. Therefore, a résumé should be self-sufficient, and it should be impressive in its own right. Here are some ways of achieving this effect:

1. Type your résumé on standard 8½″ × 11″ bond, nonerasable white paper. The typeface should be sharp and attractive so that the final results will be appealing. Spacing should facilitate readability; hence, select margins of at least one inch on all sides and provide white space between a category heading, its amplifying information, and the next category.
2. Keep your résumé to one page if possible. For entry level positions, the résumé should *never* be longer than two pages. The samples shown are about the length the average graduating senior can expect his or her résumé to be. Absolutely no inaccurate information or typographic errors should appear on the résumé. Mistakes make you look sloppy and careless.
3. Organize your résumé to best highlight your own characteristics in light of the job requirements. The two basic organizing techniques are reverse chronological and functional.

Although other methods exist (such as analytical, creative, mixed), most prospective employers are looking for a conservative, clear presentation of information. They do not wish to be distracted by the mode of presentation; they just want the facts.

The reverse chronological classification, or historical arrangement, presents information in major categories such as "work history" and "education," with work history usually appearing first. The information within these categories is presented in backward chronological order, with the current year appearing first and the earliest year appearing last.

The functional arrangement also uses major categories, but information within categories is prioritized so as to best display your talents. For instance, instead of listing previous jobs according to the year, they are listed according to the job type ("construction crew supervisor" rather than "Summer 1985"). If you held a more interesting job two summers ago than the one you held last summer, you can emphasize it with a more noticeable position in the résumé. It is possible to use a chronological listing by function if you wish, if it is not necessary to emphasize a particular job or degree.

Category names should be as informative and specific as possible. For instance, rather than using a title like "Employment," or "Work History," you might consider something more specific, such as "Job-Related Experience" (which includes volunteer work) and "Other Work Experience." The two categories help guide the reader to important information and also emphasize your work experience. Rather than using a category title like "Hobbies," you might choose "Special Interests." Type category headings in capital letters or underline them to help them stand out on the page. You also may wish to center each heading to make it even more emphatic.

You can be creative when selecting appropriate category titles, but be certain that you provide the information important to the reader (work experience, education, interests, and affiliations). Likewise, select titles that are serious and appealing.

To see this advice in practice, let us look at two résumés, one illustrating reverse chronological arrangement (Figure 18-2), and the other functional arrangement (Figure 18-3).

Notice in these two versions of the same résumé that the applicant's name is in capital letters followed by address and phone number. It is set apart from the rest of the résumé for clarity and emphasis (two possible placements are illustrated in our samples). The category "Position Objective" is centered at the top of the list so that the person receiving the résumé instantly knows the position being sought. The date of availability appears below the objective. This piece of information helps the personnel director sort the applications. If the job opening must be filled immediately and the applicant is available immediately, the director will pay greater attention to this résumé than one in which the applicant is not available for three or four months. Such information, then, is both a service to the recipient and an asset for the applicant.

Generally, the following information is cited in the rest of the résumé:

- education: date of matriculation, name of the university, degree received or anticipated, job-related course work, and class standing
- work experience: date of employment, job title, location of job, and responsibilities
- special projects: date, description of project
- organizations: date of affiliation, name of organization, any posts held
- special interests: description of area of interest
- references: full name of person (get the person's permission first), mailing address, and phone. Cite only enough references to demonstrate the pertinent qualities and experiences you have had, usually three or four. Less than three references makes it look as though you can't get anyone to vouch for you. If you cannot come up with at least three references immediately, change the last line of the résumé to read: "References available upon request."

Finally, it is a good idea to work with the college placement office to develop a dossier, or file, of all of your references, course transcripts, and even writing samples. The dossier can be sent out to the prospective employer from the placement office and will have greater credibility than if sent to the employer by the applicant. If you do develop a dossier, be sure to cite it on the last line of the résumé.

THE APPLICATION LETTER

Most job seekers will do at least some of their applying through the mail. Therefore, it is essential to be able to write a strong letter that makes a positive first impression. These days hardly anyone gets a decent job simply by writing. Instead, you must rely on the letter and résumé to separate you from the large number of other applicants for a single position. That letter must motivate the reader to want to interview you.

Writing an impressive letter—one that will make the reader want to meet you in person—is a difficult task. Although this letter is termed an *application* letter, it would be more appropriate to call it an *interview request* letter, because its primary purpose is to secure an interview. The letter, therefore, must be lively and informative. It must be the most effective "sales" letter you have ever written.

Here are some pointers for developing an effective application letter.

1. It is best to apply for a job that has been announced, although there are occasions when someone has told you of a job not yet posted, or someone recommends that you speak with someone about the *possibility* of a position. Address the letter to the person who will be deciding on the candidate. (Find out who this is by contacting the company and asking them directly whom to send the letter to.) For each application, type an *original* letter; *never* send a copy of your letter. Use the same typeface and paper as used on the résumé.

2. In the first paragraph

- cite the source in which you read or heard about the job (do not cite rumors).
- briefly indicate your current status.
- provide an overview of your basic experience and qualities that qualify you for the position.

3. In the second paragraph

- establish your own unique personality by referring to an interesting item that is listed in your résumé. This item should be important to both your prospective employer and to your own sense of accomplishment.

CHARLES A. REGINO
1478 Fountain Boulevard
Montana Springs, Arkansas 33987
(501) 555-2095

POSITION OBJECTIVE
Junior Technical Writer

Available: June 8, 1987

EDUCATION

September 1983 to present: Trenton State University
Trenton, New Jersey 07098

B.S. in Computer Science (will be awarded April 1987). Minor: Technical Communication.
Relevant course work: Programming Fundamentals of BASIC, APL, FORTRAN, and COBOL-85; Introduction to Technical Writing; Technical Report Writing.
Participated in intramural soccer and field hockey.
Class standing: 42 of 712.

JOB-RELATED EXPERIENCE

September 1985 to present: Work/study clerk
Department of Computer Science
College of Engineering
Trenton State University

Assisted in revision and maintenance of a computerized general ledger system for the university's purchasing department.

Summer 1986: Technical writer - special project
Dean of Students Office
Trenton State University

Developed a student budgeting procedures booklet to be used by Dean of Students Office for student orientation to campus life.

January 1985 - April 1985: Technical writing intern
Carleton Turbine Engine Company
42 Smith Street
Trenton, New Jersey 07098

Researched and developed a user manual on a computer-assisted design system.

Regino, 2

OTHER WORK EXPERIENCE

Summer 1985: Store clerk
Registry Arms Gift Shop
789 Westwood Avenue
Montana Springs, Arkansas 33987

Ran cash register, maintained inventory, served customers.

Summer 1984: Camp counselor
Ahwatukee Boys' Camp
Montana Springs, Arkansas 33987

Supervised groups of 12 boys (ages 8 - 10) for four consecutive two-week sessions.

AFFILIATIONS AND SPECIAL INTERESTS

October 1984 to present: Brother in Chi Delta Mu fraternity
Trenton State University

Elected treasurer for 1986/87 school year.

September 1983 to present: Member of Computer Science Club
Trenton State University

Elected secretary for Spring 1986 semester.

December 1982 to present: Amateur photographer

Own and operate Canon T60 35mm camera. Prefer nature photography.

REFERENCES

Professor N. M. Marcus; Department of Computer Science; College of Business; Trenton State University; Trenton, New Jersey 07098; (201) 555-8366.

Professor Susan Stacy; Dean of Students Office; Trenton State University; Trenton, New Jersey 07098; (201) 555-8821.

Martin Carlotti; Carleton Turbine Engine Company; 42 Smith Street; Trenton, New Jersey 07098; (201) 555-7354.

COMPLETE DOSSIER AVAILABLE UPON REQUEST

Figure 18-2 Reverse chronological arrangement of information.

CHARLES A. REGINO
1478 Fountain Boulevard
Montana Springs, Arkansas 33987
(501) 555-2095

POSITION OBJECTIVE
Junior Technical Writer

Available: June 8, 1987

WORK HISTORY

Technical Writing Intern

Researched and developed a user manual on a computer-assisted design system.

Carleton Turbine Engine Company. 42 Smith Street. Trenton, New Jersey 07098. January 1985 - April 1985.

Technical Writer - Special Project

Developed a procedures booklet on student budgeting for use by the Dean of Students Office in the student orientation program.

Dean of Students Office. Trenton State University. Summer 1986.

Work/Study Clerk

Assisted in revision and maintenance of a computerized general ledger system for the university's purchasing department.

Department of Computer Science. College of Engineering. Trenton State University. September 1985 to present.

Store Clerk

Ran the cash register, maintained inventory, and served customers.

Registry Arms Gift Shop. 789 Westwood Avenue. Montana Springs, Arkansas 33987. Summer 1985.

Camp Counselor

Supervised groups of 12 boys (ages 8 - 10) for four consecutive two-week sessions.

Ahwatukee Boys' Camp. Montana Springs, Arkansas 33987. Summer 1984.

Regino, 2

EDUCATION

B.S. in Computer Science (will be awarded April 1987)

Minor in Technical Communication. Trenton State University. Trenton, New Jersey 07098.

Relevant course work: Programming Fundamentals of BASIC, APL, FORTRAN, COBOL-85; Introduction to Technical Writing; Technical Report Writing.

Participated in intramural soccer and field hockey.

Class standing: 44 of 712. September 1983 to present.

AFFILIATIONS AND SPECIAL INTERESTS

Fraternity Brother

Chi Delta Mu fraternity. Trenton State University. Elected Treasurer for 1986/87 school year. October 1984 to present.

Club Member

Computer Science Club. Trenton State University. Elected secretary for Spring 1986 semester. September 1983 to present.

Amateur Photographer

Own and operate a Canon T60 35mm camera. Prefer nature photography. December 1982 to present.

REFERENCES

Professor N. M. Marcus; Department of Computer Science; College of Business; Trenton State University; Trenton, New Jersey 07098; (201) 555-8366.

Professor Susan Stacy; Dean of Students Office; Trenton State University; Trenton, New Jersey 07098; (201) 555-8821.

Martin Carlotti; Carleton Turbine Engine Company; 42 Smith Street; Trenton, New Jersey 07098; (201) 555-7354.

COMPLETE DOSSIER AVAILABLE UPON REQUEST

Figure 18-3 Functional arrangement of information.

1478 Fountain Boulevard
Montana Springs, Arkansas 33987

April 7, 1987

Mr. Albert Poindexter
Poindexter Computing Consultants
57 East Washington Street
Little Rock, Arkansas 33984

Dear Mr. Poindexter:

I am writing to you regarding the Junior Technical Writer position advertised in the April 5, 1987, issue of the Little Rock Herald. I will graduate shortly with a B.S. degree in computer science from Trenton State University. I feel my education and my work experience make me a strong candidate for the job.

As listed on my résumé, I have taken a variety of computer science courses. I have also become familiar with such equipment as the IBM 3380 used by Trenton State. Note that I've combined this

4. In the third paragraph

- discuss job experiences directly related to this position.
- discuss course work, special projects, personal reading, or interests related to the position.

5. In the final paragraph

- state your availability for an interview but give reasonable boundaries for contacting and meeting with you.
- thank the reader for his or her time, consideration, and cooperation.

Discuss your information in an order that will complement the job qualifications. Use words or phrases that make the reader glance at your résumé education with both course work and work experience in technical communication.

I have worked as a technical writing intern for the Carleton Turbine Engine Company in Trenton. I have also been a clerk in the computer science department, where I helped revise and maintain a computerized general ledger system for the university's purchasing department. This system is a significant improvement over the previous ledger and is still in use today. My special budget project for the Dean of Students Office has also proved valuable.

I would very much like to apply my skills to a position with your firm. I will be returning to Montana Springs within two weeks and will be available for a personal interview anytime thereafter. I can be contacted at the above address or at (501) 555-2095. I look forward to hearing from you and thank you for your consideration.

Sincerely,

Charles A. Regino

Charles A. Regino

Encl.: Résumé

Figure 18-4 The application letter helps set you apart from your competition.

a second or third time. Use a simple sentence on occasion to break up the discussion and to emphasize the items you want to impress on the reader. Make your letter suggest competence, confidence, and a genuine interest in the position for which you are applying. Employers look for potential employees who will be an asset to their organization and who are at the same time good to work with and good to know. They shy away from applicants who seem to promise problems or an overinflated ego.

Consider the letter written to accompany Charles Regino's résumé (Figure 18-4). He is applying for a job within his level of competence and with a company that can offer him substantial career growth. His attitude is reserved yet confident. He points to areas that will surely be of interest to the reader and are directly related to the job he is seeking.

Note that the first paragraph of the letter alludes to a specific source that had advertised the position, the date of the ad, and the title of the position described. Such information helps ensure that the applicant's letter is placed in the right category of incoming inquiries, since it is likely that the employer has several ads running simultaneously. In addition to saving a prospective employer time, such a reference helps an organization ascertain the effectiveness of its advertising. For instance, if the majority of applications are in response to an ad placed in the local newspaper, while very few are spawned by the ad appearing in a key trade journal, it may be wise to eliminate the trade press ad in future recruiting efforts.

Directly after this introductory information, the student informs the reader of his current status ("I will graduate . . ."). The first paragraph ends with a confident yet not egotistic statement that the applicant believes he is a strong candidate for the job. At this point, the prospective employer decides whether to read further, so it is a good idea to indicate right away that you have relevant experiences or qualities. Such an overview statement serves as the launching pad for the rest of the letter.

The second paragraph begins with a reference to the résumé. At this point, the reader might very well look at the résumé to spot the listed item on the page. In doing so, he may glance at the other categories as well. The writer has given the résumé greater exposure simply by including this short transitional phrase.

The second and third paragraphs stress previous training and practical experience the applicant believes are relevant to the job. He does not merely list this information; where appropriate, he discusses the special significance of his work (the general ledger system that is an improvement over the previous one).

The closing paragraph stresses his desire to work for the firm. It suggests having a personal interview and provides contact information. Finally, it thanks the reader for his consideration. The enclosure line reminds the reader that a résumé is included.

Remember, the application letter is your only means of distinguishing yourself from others. Résumés, although extremely important, are impersonal and factual. The information may vary from list to list, but basic schooling and experience may indeed be very similar. In reading résumé after résumé, the personnel officer finds they all look pretty much the same. Therefore, take advantage of the letter to demonstrate your *unique* strengths, convictions, thoughts, and values. In doing so, you will be exhibiting your personality rather than letting the facts speak for themselves.

THE INTERVIEW

Let us assume that your application letter and résumé *do* help you secure an interview, or that you attend a job fair on your college campus and line some interviews up in that way. The most important thing to remember about an interview is that it provides you with as much of an opportunity to learn about the company as it provides the company a chance to learn about you. You should be prepared to ask questions, to demonstrate interest in the company, and to display a certain degree of assertiveness. Let us consider some ways of preparing for an interview as well as conducting yourself during it.

First, try to anticipate what is expected of you. Having researched the job and the company, prepare a list of ten logical questions an interviewer ought to ask you. Although you should not have pat answers, you should consider the way in which you would answer each of these questions. Here is a list of questions most interviewers will ask:

- To what degree are you experienced in this field?
- What are your career goals?
- What kinds of jobs have you held? What did you like or dislike about them? What did you find most rewarding?
- Why are you interested in working for our firm?
- Are you a veteran? Did the military provide you with job-related training?
- What is your educational background? How do you believe it prepares you for this position? Which courses do you believe will prove the most valuable to you in this position and why?
- What do you believe are your strong (and/or weak) points?
- Can we contact a former employer, a professor, or your references?
- Can you travel extensively?
- Are you willing to relocate?

Do not prepare short answers to these questions. There is nothing less appealing to an interviewer than a stiff, abrupt reply. Consider your own experiences and course work and include them as part of your answers. Likewise, demonstrate your knowledge of the company and job by tailoring your responses to meet the needs of the interviewer. End your answers with a related question about the company, if possible. Practice answering questions in this manner with fellow students, friends, or family members.

Just as there are questions you ought to be prepared to answer, there are questions that you should *not* necessarily answer. These questions violate your civil rights and can prejudice the interviewer. Technically, your getting a job should be based on your ability to perform it rather than on *any* aspects of your personal life. Here is a sample of questions that should not be asked:

- Are you married? What does your wife (husband) do? How long have you been married?

- How old are you? When were you born?
- Do you have children? How old are your children? Will you have trouble finding someone to watch your children if you come to work for us?
- What church do you attend? Would your going to church prevent you from working on Sundays if necessary?
- Do you rent or own your home? Do you live with your parents or friends?
- Have you ever been arrested? (Note: an employer *can* ask if you have been convicted of a felony.)
- Do you drink or take drugs?
- How much do you weigh? How tall are you?
- Have your wages ever been attached or garnished? Have you ever declared bankruptcy? Are you currently in debt?
- Is English your natural tongue? Where were you born? Is English spoken at home?

In all fairness, you may feel perfectly comfortable answering any or all of these questions—you must use your own judgment. However, if you feel the question is unnecessary, you should parry it with a statement such as, "I would prefer to concentrate on my qualifications for the job rather than on my personal life," or with a nonbelligerent question such as, "Could you explain how this question is relevant to my being able to do a good job?" If, in fact, the question turns out to be somewhat related to your abilities, you may wish to answer. For instance, an interviewer may ask your age because it is necessary that you be twenty-one years old in order to maintain inventories of alcoholic beverages or combine controlled substances in a laboratory. You may be asked about your physical condition in a job where you will be required to handle heavy equipment. In any case, *you* must be the one to decide whether you answer or not.

Much of an interview focuses on how you conduct yourself rather than on your answers to questions. Interviewers look at your body language, your appearance, your demeanor. They measure how nervous you are, whether you have a sense of humor, whether you would be able to get along with the other people already on the job. Therefore, it is important for you to acquaint yourself with the interview experience *before* you go out for an important interview. Practicing with friends, although valuable, is not the same as the actual interview experience. Purposely accept a few interviews for jobs that do not appeal to you greatly to gain the benefit of the experience. Learn to do the following things:

1. Appear well-groomed. Conservative, clean, tailored clothing is important, but do not overdress (a three-piece suit for a technician's job interview is inappropriate). Hair should be neatly combed; if you have a beard, trim it

carefully. Women should have attractively manicured hands (but should not wear a gaudy shade of nail polish). Wear a minimal amount of jewelry. If you look like you care about your appearance, you imply that you will be conscientious about your work.

2. Sit comfortably. Do not sit at the edge of the chair; nor should you slouch or lean. Although crossing your legs is appropriate, do not lay one leg at a right angle across the other. If your legs are not crossed, keep them together with your feet flat on the floor. Do not fidget or squirm. Do not shake your foot or drum your fingers. Sit still but not rigidly. If you can project an air of calmness, your interviewer will feel that you are at ease with yourself and feel confident about being there.

3. Be aware of your body language. Shake the person's hand firmly as you introduce yourself. Do not wait for the interviewer to speak first. If you are called into the office where the interviewer is already present, stride assuredly toward the person with a sincere smile. Use confident gestures. Avoid arms crossed in front of your chest or an uplifted fist placed defensively behind your head. Avoid an overly casual gesture such as both arms (elbows out) placed behind your head. Keep your hands or fingers away from your mouth and face. Remember to smile. Establish good eye contact, but do not stare. Such confident, assertive body language demonstrates your strength and seriousness.

4. Bring a sample of your work with you to the interview. It may be a writing sample from a course you took, a set of laboratory results, blueprints, or a model. Unless you are asked to display this material earlier, wait until the end of the interview. Explain briefly what it is and what it demonstrates. Then leave it behind with the interviewer so he or she can look it over at leisure. This sample will further reinforce your candidacy. If the sample is something that must be returned, arrange for an appropriate time to retrieve it personally (thereby giving yourself an automatic second interview opportunity).

The more practice you receive in such situations, the more relaxed you will be in crucial interviews. Preparation is the most important aspect of interviewing. If you develop a sufficiently impressive application letter and résumé and follow them up with a strong interview performance, your chances of obtaining the job are increased significantly. You have demonstrated your ability to handle yourself in a variety of difficult situations.

FOLLOW-UP ACTIVITY

The day after an interview, send a short note to the person who interviewed you. This note should thank the reader for the interview and should provide

1478 Fountain Boulevard
Montana Springs, Arkansas 33987

April 21, 1987

Mr. Albert Poindexter
Poindexter Computing Consultants
57 East Washington Street
Little Rock, Arkansas 33984

Dear Mr. Poindexter:

It was a pleasure meeting you yesterday. I appreciated the opportunity to interview with you and to see your facilities. As I mentioned during our talk, the job you described sounds exactly like the kind of position I would like to begin my career. I look forward to hearing from you in the very near future.

Sincerely,

Charles A. Regino

Charles A. Regino

Figure 18-5 The follow-up letter demonstrates your continuing interest.

any additional information you may have neglected to raise during the interview. Last, it should let the reader know you are looking forward to hearing from him or her in the near future. Figure 18-5 shows the note that Charles Regino sent to Mr. Poindexter after completing the interview.

After the interview, you may decide you are no longer interested in being a candidate. If such is the case, you owe it to the prospective employer to notify the company and withdraw your name as a candidate as soon as possible. Such a letter might look like one in Figure 18-6.

Of course, that wonderful day will come when a company with which

1478 Fountain Boulevard
Montana Springs, Arkansas 33987

April 21, 1987

Mr. Albert Poindexter
Poindexter Computing Consultants
57 East Washington Street
Little Rock, Arkansas 33984

Dear Mr. Poindexter:

I very much enjoyed speaking with you yesterday, and I thank you for giving me the opportunity to interview for the Junior Technical Writer position. However, upon hearing more about the job, I find that it would require less of my computer science background than I had anticipated.

Please withdraw my name as an active candidate. However, please keep me in mind if you have a job in the future requiring more detailed computer knowledge. Thank you again for your consideration.

Sincerely,

Charles A. Regino

Charles A. Regino

Figure 18-6 A polite notice of your withdrawal as a job candidate will keep your options open in the future.

you have interviewed offers you a job. If you wish to accept the position, you can call the company and tell them immediately. Discuss any questions you might have and then confirm the conversation with a written acceptance letter. If you do not wish to accept the job, politely refuse it in writing. Offer the employer a clear, understandable reason. Figure 18-7

1478 Fountain Boulevard
Montana Springs, Arkansas 33987

May 6, 1987

Mr. Albert Poindexter
Poindexter Computing Consultants
57 East Washington Street
Little Rock, Arkansas 33984

Dear Mr. Poindexter:

I am writing to accept the Junior Technical Writer position you offered me in your letter of May 4, 1987. As I said in my phone conversation with you earlier today, I will be able to start work on June 8, 1987.

Thank you so much for this opportunity. I look forward to a long and satisfying experience with your firm.

Sincerely,

Charles A. Regino

Charles A. Regino

Figure 18-7 Accepting the long-awaited position.

illustrates an acceptance letter; Figure 18-8 is a rejection letter that specifically leaves the door open for a future opportunity.

CHAPTER SUMMARY

Job hunting calls into play all the technical writing and oral presentation skills you have learned in this book. Application letters, résumés, and inter-

1478 Fountain Boulevard
Montana Springs, Arkansas 33987

May 6, 1987

Mr. Albert Poindexter
Poindexter Computing Consultants
57 East Washington Street
Little Rock, Arkansas 33984

Dear Mr. Poindexter:

I am unable to accept the position of Junior Technical Writer you offered me in your letter of May 4, 1987. Just last week, I was offered a position at Stanford University as a technical writer in their computer office. The distinct advantage that this position offers is the chance to go to graduate school on a reduced tuition plan.

I greatly appreciate the opportunity you offered me. I hope that when I have completed graduate school you might consider me for a more senior position.

Sincerely,

Charles A. Regino

Charles A. Regino

Figure 18-8 Politely rejecting an offered position.

views can work together to help you get a job in your chosen career area. To be successful, however, you must first scrutinize your job aspirations, experience, values, strengths, and needs. Likewise, careful research into appropriate jobs and organizations is required before you can make an informed decision about your career goals.

ADDITIONAL READING

For more information on getting a job, consult one of these texts:

Bolles, Richard N. *What Color Is Your Parachute: A Practical Manual for Job-Hunters and Career-Changers.* 2nd ed. Berkeley, CA: Ten Speed Press, 1984.

German, Donald R., and Joan W. German. *How to Find a Job When Jobs Are Hard to Find.* New York: AMACOM, 1983.

Lewis, Adele. *How to Write Better Résumés.* 2nd ed. New York: Barron's Educational Series, Inc., 1983.

Price, Jonathan. *How to Find Work.* New York: Signet, 1983.

Salmon, Richard. *The Job Hunter's Guide to Eight Great American Cities.* Cambridge, MA: Brattle Publications, 1978.

Straub, Joseph. *The Job Hunt: How to Compete and Win.* Englewood Cliffs, NJ: Prentice-Hall, 1981.

EXERCISE

Prepare for the job-seeking experience by doing the following:

- Select a specific job or job type for which you are realistically qualified (given the experience and course work you will have *by the time you will actually be looking for permanent work*).
- Create a list of primary and secondary job requirements.
- Create a detailed list of your own qualifications and experience.
- Match your own qualifications with the job requirements.
- Locate a specific announcement of an actual job.
- Research the organization announcing the position.

WRITING ASSIGNMENT

Write a strong application letter and résumé for the job you researched in the preceding exercise. Attach a copy of the ad and your organization research to this assignment so that your instructor can evaluate the degree to which you have satisfied the needs of the prospective employer. After these assignments have been graded, your instructor may select several strong "candidates" to interview. (If you are not selected for an interview, it may mean that your application package was not sufficiently impressive to warrant one.) If you are selected to interview, prepare responses to questions likely to be asked, conduct yourself properly during the interview, and remember to bring a sample of your work.

NOTE: Apply the editing criteria at the end of Part III to your assignment before you submit it to your instructor.

EDITING CRITERIA

The following editing criteria, in conjunction with those concerned with the mechanics of writing (see the end of Part I), should be applied to each of your writing assignments *before* you submit it for workshop analysis or for grading. Your fellow students and your instructor will use these same criteria in their evaluations of your work. Criteria specific to each topic we have discussed in Part III appear after the set of general concerns.

GENERAL CONCERNS

1. Is the situational analysis appropriate to the assignment? If not, why not? What changes should be made? How will these changes affect the document?
2. Does the writer maintain a consistent role? If not, identify areas of inconsistency and make corrections.
3. Is the level of detail appropriate to the needs of the audience? If not, what changes are required?
4. Is word choice appropriate to the writing situation? If not, identify and correct the problems.
5. Will the audience understand this document? If not, why not? Identify and correct places where an explanation should be expanded or where additional information is required.
6. Does this document adhere to all format specifications? If not, identify and fix all incorrect areas.
7. Is the topic being discussed clearly identified at the outset? If not, make necessary changes.
8. Is the arrangement of information appropriate to the writing situation? Why or why not? If not, make any necessary changes.

LETTERS AND MEMOS

1. Does the writer supply the complete return address and date in the heading?
2. Is a complete inside address used?
3. Is the salutation appropriate?
4. Does the introductory paragraph identify the topic to be discussed and make reference to any previous discussion or context?
5. Do the body paragraphs identify and expand on the key points the writer ought to be making?
6. Is the order of points in the body of the letter or memo logical?

7. Does the closing paragraph recapitulate the main idea of the correspondence?

8. Does the writer clearly state any required action on the reader's part in the closing paragraph?

9. Is the complimentary close appropriate?

10. If enclosures are included, is reference to them made correctly?

11. Does the letter or memo get and maintain the reader's interest?

12. Is the reader supplied with sufficient information to understand and act on the content of the correspondence?

PROPOSALS

1. Does the title page of the proposal include the writer's name, title, department, proposal topic, and date?

2. Does the proposal include a table of contents?

3. Does the statement of the problem indicate that the writer understands the nature of the problem? Does it reflect the interests of the audience?

4. Does the background information include only enough material to acquaint the reader with the context of the problem? Does it demonstrate the writer's ability to absorb and present data related to the proposal topic?

5. Does the statement of objectives indicate what will be accomplished if the proposal is accepted? Does it supply a means of measuring the methods, schedules, costs, and conclusions presented later?

6. Do the possible solutions present realistic alternatives to the proposed plan of action? Does this section focus clearly on the point the writer intends to pursue in the plan of action section?

7. Does the plan of action detail the steps to be taken in implementing a solution to the problem? Does it follow naturally from the previous section of the proposal?

8. Does the qualifications section portray the proposal writer's credentials effectively?

9. Do the facilities and resources cited seem adequate for the proposal?

10. Is the proposed schedule realistic?

11. Do the costs seem in line with the objectives of the plan of action?

12. Do the conclusions and recommendations follow logically from the information presented in the proposal?

13. Does the preliminary bibliography provide the proposal writer with sufficient exposure to the topic in order to begin development of the formal report?

14. If there is an appendix, has its existence been mentioned in the proposal itself?

REPORTS

1. Does the letter of transmittal include the full report title, the date on which specifications were agreed, the purpose of the report, its major conclusions and recommendations, and any special acknowledgments?

2. Is the cover made of heavy stock? Is the label centered? Does it include the full title of the report, the author's name and affiliation, the date of submission, and the report status?

3. Is the title page on the same stock as the rest of the report? Is it in the same typeface? Does it include the full title of the report, the author's name and affiliation, the date of submission, the report status, and a statement of authorization? Is all such information centered both across and down the page?

4. Is the abstract informative or indicative? If it is informative, does it include the purpose, method, problem, scope, key discussion points, primary results, key conclusions and recommendations, and any unique report features? Is the third-person voice maintained throughout? Is the abstract self-sufficient? Are abbreviations used sparingly? Is telegraphic style avoided? Is the abstract an accurate reflection of the report's content?

5. Do the acknowledgments cite the person or persons making the contribution, the nature of the contribution, and the effect of the contribution?

6. Does the table of contents list the major headings and key subheadings of the report? Does the order of these headings reflect the order followed in the report? Are the page references correct? If necessary, are periods or lines used between the headings and the page numbers to ensure clarity?

7. If figures or tables appear in the report, is a list of tables or figures included? Do these lists include all of the figures or tables in the report? Are the title and number of each figure or table correct? Does the order of the figures or tables reflect the order followed in the report? If necessary, are periods or lines used between the titles and the page numbers to ensure clarity?

8. Is the problem clearly stated in the introduction? Does the introduction accurately reflect the situation being considered? Is the purpose statement clearly presented? Is it an accurate assessment of what is undertaken? If this is a secondary research report, does the writer explain the method of obtaining information? If this is a primary research report, does the writer define the methods and materials used in the experiment or in the compilation of data? Does the method statement foster credibility? Is the scope clearly stated? Does it accurately reflect any and all constraints placed on this report? If a background statement is appropriate, is it clearly stated? Does it adequately prepare readers for discussion?

9. Is the order of subsections in the discussion section effective? Is each section relatively self-sufficient? If not, is there clear reference to other sections that complement dependent sections? Does the writer provide am-

ple information with which to draw results (if appropriate), conclusions, and recommendations?

10. Is the results section a comprehensive statement of possible results? Do subsections follow the order established in the discussion? Are all results logical? If a result is unexplainable, does the writer indicate this? Do the results corroborate any conclusions or recommendations made?

11. Are key facts integrated and easy to follow in the summary? Are logical combinations of facts made? Do the facts cited here lead the reader naturally to the conclusions and recommendations made?

12. Are conclusions ordered in terms of significance to the audience? Are all possible conclusions drawn? Are all conclusions based solely on information provided in the discussion section of the report? Are all conclusions logical?

13. Are all recommendations feasible? Are they logically drawn from the discussion and conclusions sections of the report? Are they ordered to suit audience needs?

14. Does the bibliography adhere to the format prescribed in Chapter 14 or to the format established in your field of study? Do footnotes follow the specifications given?

15. Are bulky materials included in the appendix? Are they appropriately labeled, paginated, and cited in the table of contents? Are supplemental materials included? Are they truly of interest or importance to readers? Are they appropriately labeled, paginated, and cited in the table of contents?

16. Are graphic aids used wherever they are required? Are explanatory captions or titles used? Do the graphics enhance the written message? Are figures and tables listed separately, with page numbers, after the table of contents?

MANUALS

1. Is the cover durable? Is the title of the manual descriptive? If the manual is part of a series within a related product line, is this made clear?

2. Does the title page carry the full title? If a special set of instructions is required, does it appear on this page? Are any mandatory notices included? Does the manufacturer's logo appear?

3. Does the preface assist readers in using the manual? Are related documents mentioned? Do the legal disclaimer, copyright date, and order number appear on this page?

4. Are any warnings applicable to the entire manual provided on a page dedicated to this purpose?

5. Does the table of contents list the major section and subsection headings? Are titles and page numbers accurate? Are separate listings provided for figures and tables?

6. Does the introduction explain the purpose and scope of the document? Does it provide any required background information, definitions, and prerequisites?

7. Does the body of the manual follow a general-to-specific arrangement of information?

8. Does the reader receive sufficient information about each procedure to execute it without further assistance?

9. Is information cross-referenced throughout the document?

10. Is there a glossary, index, parts lists, and/or response device in the back matter?

11. Has the writer used headings, subheadings, paragraph sets, numbering, boldface type, boxes, and graphic aids appropriately to increase the clarity of the manual?

PROMOTIONS

Together with your classmates and instructor, devise a set of editing criteria applicable to the promotional materials you write.

ORAL PRESENTATIONS

1. Has the speaker selected a topic of interest to the audience?
2. Does the introduction capture the listeners' attention?
3. Is the purpose of the talk clear? Does it satisfy the needs of the audience?
4. Does the speaker maintain a satisfactory level of audience interest throughout the presentation?
5. Does the speaker use graphic aids as appropriate? If so, are these simple, easy to see, readable, clear?
6. Does the speaker use suitable examples to demonstrate or to clarify points?
7. Is the language concrete?
8. Are clear transitions provided to guide the listener through the talk?
9. Does the body of the talk follow a logical sequence of ideas?
10. Is sentence structure relatively simple?
11. Does the speaker seem confident?
12. Does the speaker establish and maintain eye contact?
13. Is the delivery tone sincere and respectful?
14. Is the speaker's appearance suitable to the speaking environment?
15. Are the speaker's gestures natural?
16. Does the speaker make effective use of the conclusion to reiterate key points, call for action, or offer suggestions?

GETTING A JOB

Résumé

1. Is the résumé typed neatly on white bond paper?
2. Are the applicant's full name, address, and phone number given at the head of the résumé?
3. Is the position objective stated?
4. Is the date of availability stated?
5. Are category labels precise and informative?
6. Does the résumé's arrangement complement the applicant's strengths?
7. Is only relevant information supplied?
8. Is sufficient information supplied?
9. Is the content of the résumé impressive enough to make the applicant a strong candidate?

Application Letter

1. Does the writer supply the complete address and date in the heading?
2. Is a complete inside address used?
3. Is the salutation appropriate?
4. Does the introductory paragraph contain

- the name of the specific job for which the applicant is applying?
- the source of the job announcement?
- a statement of the applicant's status?
- a summary of the applicant's qualifications?

5. Do the following paragraphs

- discuss the important information listed on the résumé?
- discuss previous job experiences and/or course work related to the position?
- establish the writer's personality?
- distinguish this applicant from others?
- explain the applicant's interest in the field, position, company?

6. In the closing paragraph, does the writer attempt to secure an interview in a polite yet assertive manner?
7. Is standard letter format used?
8. Does the letter make you interested in meeting the writer?

APPENDIX

I

Formal Report

Tolerance of *Plantago ovata* to Three Herbicides

P. J. Specht

May 1986

Final Report

TOLERANCE OF *PLANTAGO OVATA* TO THREE HERBICIDES

P. J. Specht
Department of Plant Sciences
University of Arizona

Authorized by K. C. Hamilton

May 1986

Final Report

ABSTRACT

Greenhouse tests were conducted to determine the tolerance of two varieties of Plantago ovata Forsk. to postemergence applications at various rates of three herbicides.

Paraquat (1,1′-dimethyl-4,4′-bipyridinium ion), diquat (6,7-dihydrodipyrido[1,2-a:2′,1′-c]pyrazinediium ion), and diuron [3-(3,4-dichlorophenyl)-1,1-dimethylurea] were applied over-the-top on young plantago seedlings. Tests were run with and repeated without including X-77 surfactant in the spray mixture.

Plantago seedlings usually tolerated rates of paraquat up to and including 0.5% v/v; rates of diquat up to and including 0.25% v/v; and rates of diuron up to and including 0.56 kg/ha. In tests in which diuron and paraquat were applied as a tank mix, seedlings tolerated the highest rates applied in combination (0/5% paraquat plus 0.42 kg/ha diuron).

No apparent difference was noticed in comparing the responses of the two varieties to herbicide treatments. Inclusion or exclusion of surfactant from the spray mixture had no consistent effect on results.

ACKNOWLEDGMENTS

I would like to acknowledge Dr. K. C. Hamilton for his support on this project. Listening to him (when I would) and working with him over the last several years have provided me with an invaluable opportunity to learn. His friendship, encouragement, criticism, and advice are greatly appreciated. His assistance during the development of this report has enabled me to interpret experimentation results so that they prove most useful to my readers.

TABLE OF CONTENTS

LIST OF TABLES

1. INTRODUCTION

1.1. Statement of the Problem

India has a virtual monopoly on the production and sale of plantago seed husk. Political, economic, or natural disorders could cut off the world supply of this crop at any time. Hence, it is of importance to American manufacturers (who purchase most of the plantago exported from India) to consider means of producing plantago commercially in the United States.

Development of an effective chemical weed control program will be a major consideration if plantago is to be grown economically in the United States. The cost of hand-weeding (the method used in India) would make commercial development here impossible. Mechanical cultivation is costly in terms of time and fuel.

1.2. Statement of Purpose

The primary objectives of this document are to define the nature and importance of plantago, explain the effectiveness of three herbicides used in testing, and to report the results of 23 tests on two varieties of Plantago ovata ("AZ 316" and "AZ 329") performed between January 26, 1986, and May 12, 1986, at the University of Arizona's Campbell Avenue Farm. The objective of the testing was to determine this plant's tolerance of three herbicides: diuron, diquat, and paraquat, germane to an initial study (conducted by the University of Arizona's Department of Plant Sciences) of this plant's potential commercial production in the United States.

1.3. Statement of Scope

This report will outline the history and importance of plantago as well as cite all testing and results pertaining to the above-mentioned study of plant tolerance. It is written to all members of the Plant Sciences Department for their use in determining the overall suitability of this plant to cultivation in the western United States.

1.4. Method of Acquiring Data

Current and recent secondary sources were used to conduct the literature review portions of this report. These were found at the Science and Engineering library of the University of Arizona and in the reference collection of the Department of Plant Sciences in the College of Agriculture.

Primary sources of information were gained during extensive greenhouse testing conducted at the University of Arizona. Detailed description of the methods and materials applied during experimentation will be presented in the Discussion section of the report.

1.5. Background

Write it out in report

[This subsection covers the properties and use of Plantago ovata Forsk.]

2. DISCUSSION

2.1. Nomenclature and Taxonomy

[This subsection lists the several common and scientific names by which Plantago ovata is known.]

2.2. Characteristics of Plantago

Plantago ovata Forsk. (Plantaginaceae) is an annual herb. In nature, the height of plantago varies from 2.5 cm to 10 cm. Under cultivation, 30 to 45 cm is the normal height of plantago. The plants are acaulescent and covered with a thick pubescence.[5] The leaves are narrowly linear, 7.5 to 20 cm long and 0.6 cm wide. Many flowering shoots arise from the base of a single plantago plant. The inflorescence is a closely bracketed spike of small, greenish flowers, cylindrical to ovoid or globular in shape, and 1.2 to 4 cm long. The flowers are protogynous and they mature from the base of the spike upwards. The seeds are small (2 to 3 mm long, 1 to 1.5 mm wide) and boat-shaped. Seed color varies from brown to pinkish white, and is usually pinkish-gray. The concave side (hilum side) of a plantago seed is covered with a thin white membrane.

2.3. Economic Importance to India

[This subsection discusses the economic importance of this crop to In-

- 2 -

dia, which produces about 75% of the Plantago ovata grown in the world.]

2.4. Agricultural Production of Plantago

Plantago grows on a variety of soil types and does best commercially on light, well-drained soils. A sandy loam soil is ideal.[9]

In India, plantago is planted between October and December when the average daytime temperature reaches 27 to 40° C. The average length of the growing season is 120 days. Research has been done to determine the feasibility of growing more than one crop per year in India. Joshi, et al., reported success with two growing seasons.[10] The first began at the beginning of November and ended in mid-April. The second was planted at the end of July and harvested at the end of December. A third attempted season was unsuccessful due to disease promoted by the wet monsoon season in September at the time of seed maturation. In other research, Iyengar, et al., found that moving the planting date past December 8th reduced plantago yields.[11]

Various seeding rates for plantago have been reported in India; the average rate is about 9 kg/ha. The seed is usually broadcast on the flat by hand, then raked lightly to cover the seed. A light irrigation is applied after planting, with subsequent irrigations every 2 to 3 weeks until all the flowering spikes are out. Irrigation is then discontinued as the seed matures. Plantago has a low nutrient requirement. Fertilizer rates of 50 kg/ha of nitrate and 25 kg/ha of phosphate are adequate. The average yield of plantago seed is 1000 kg/ha.[12]

[At this point, the report discusses weather conditions conducive to growing Plantago ovata. It also mentions that decades of hand-weeding in India have eliminated most encroachment of weed species there.]

2.5. Herbicides Selected for Testing

Three herbicides, noted for their effectiveness with crops planted in the southwestern United States, have been selected for testing with plantago under greenhouse conditions. Discussion will focus on the behavior of diuron, diquat, and paraquat as weed control agents.

2.5.1. Diuron

The herbicide diuron is available under the trade names Karmex and Karmex DL. Formulations of diuron include wettable powder, granules, and dilute formulations of diuron alone or in combination with other herbicides. This chemical has a proven low toxicity to fish and mammals. It is a soil-active herbicide that is greatly enhanced by the addition of a surfactant. Diuron is usually sprayed on the soil before the emergence of annual crops or applied as a postemergence spray directed to the base of the crop plant. The usual carrier for diuron is water. Good mechanical agitation during application is necessary to keep diuron in suspension. Diuron can also be sprayed in an oil suspension or, in the granular formulation, broadcast directly on the soil.[14]

The use of diuron in herbicide programs has been very effective in cotton crops. Split applications of diuron, or using diuron in combination with other herbicide treatments, have increased weed control, increased cotton yields, and reduced crop injury, as compared to the effects of a single, preplant application of the chemical.

Diuron has not been demonstrated to accumulate in the soil when applied, even repeatedly, at selective rates. Residual activity of diuron does not usually affect tolerant crops, but cases of diuron residues injuring cotton seedlings have been reported.[15] Frequent irrigations and flood irrigation instead of furrow may decrease the length of persistence of diuron in desert soils. Addition of organic matter to the soil may also increase the rate of diuron degradation. Broadleaf plantain (Plantago major), which is more tolerant to diuron than cotton, degraded the chemical more quickly than cotton.[16]

[Subsection 2.5.2 discusses, in parallel order to subsection 2.5.1, the trade names, forms, toxicity levels, application mode, carrier, effectiveness, accumulation rate, and degradation of two other herbicides, diquat and paraquat.]

2.6. Testing Methods and Materials

Twenty-three tests were conducted to determine survival and growth responses of two varieties of Plantago ovata ("AZ 316" and "AZ 329") to three herbicides: diuron, diquat, and paraquat. The tests were completed in a greenhouse at the University of Arizona's Campbell Avenue

Farm between January 26, 1986, and May 12, 1986. No supplemental lighting was used in the greenhouse.

2.6.1. Common Materials and Procedures

Many of the materials used and procedures followed are common to all of the tests included in this report. Following a general summary of these common materials and procedures there will be a brief description of each test.

Plants were grown in 400 cc plastic pots. Two holes were made in the bottom of each pot to allow drainage and uptake of water. Small cotton plugs were placed in the holes to prevent loss of soil. A small amount of fertilizer (18-24-6) was placed in each pot. These pots were filled with air-dried soil (60% sand, 25% silt, 15% clay, 1% organic matter) that had been passed through a 6.5 mm mesh screen. The pots were surface-irrigated and allowed to drain. Twenty-five plantago seeds were scattered on the moist soil surface and covered with 0.3 cm dry soil. The pots were placed on an aluminum baking tray and covered loosely with a transparent plastic sheet until the seeds germinated. Emerging seedlings were thinned to three or four per pot.

Approximately 3.5 weeks after planting, the pots were sorted for uniform seedling height, stage of development, and vigor. Various rates of herbicides were applied from a height of 30 cm over-the-top. Herbicide applications were made with a knapsack sprayer through a single-nozzle wand fitted with an 8003 nozzle tip. Herbicides were applied in 380 liters per hectare of water at 2.8 kg/cm^2 pressure. After treatment, the pots were replaced on the baking trays and sub-irrigated as needed.

Tests of single herbicides consisted of five treatments (including an untreated check). Tests of combinations of herbicides consisted of nine treatments (including an untreated check). Each treatment was replicated four times. A single pot containing three or four plants was a replication. There were 12 to 16 plants in each treatment. The pots were arranged on the trays in a randomized complete block design.

Tests were completed when herbicide symptoms were obvious, usually 1 week after treatment. Seedling survival data were determined by counting live plants. A plant was considered alive if it had at least one green, growing leaf. Seedling height was determined by measuring the longest leaf or uninjured portion of a leaf. Measurement was made by extending the leaf and measuring up from the soil surface.

Root development was assessed by turning the soil out of a pot and making a visual evaluation of the roots at the bottom of the pot. A numerical value of from 0 to 10 was assigned to each pot. A rating of 10 corresponded to normal root development. A rating of 0 was given when no roots were visible. Data for height and root development were analyzed according to Duncan's Multiple Range Test at the 5% level of confidence. Survival data were not analyzed because of the limited number of data.

2.6.2. Description of Tests

Paraquat with X-77 on Plantago "AZ 316" (Trial I)

Plantago "AZ 316" seeds were planted on March 2, 1986. Emerging seedlings were thinned to four per pot. Plants were treated with paraquat (0, 0.125, 0.25, 0.5, or 0.75% v/v) plus 0.5% X-77 surfactant (alkylarylpolyoxyethylene glycols, free fatty acids, and isopropanol) 25 days after planting. At treatment the average plant was 6 cm tall and had six true leaves. Data were recorded 7 days after treatment.

Paraquat with X-77 on Plantago "AZ 316" (Trial II)

Plantago "AZ 316" seeds were planted on April 3, 1986. Emerging seedlings were thinned to four per pot. Plants were treated with paraquat (0, 0.125, 0.25, 0.5, or 0.75% v/v) plus X-77 surfactant (0.5% v/v) 24 days after planting. At treatment the average plant was 11 cm tall and had four true leaves. Data were recorded 7 days after treatment.

[The report continues with descriptions of the following tests:
Paraquat with X-77 on plantago "AZ 329" (Trial I)
Paraquat with X-77 on plantago "AZ 329" (Trial II)
Diquat with X-77 on plantago "AZ 316" (Trial I)
Diquat with X-77 on plantago "AZ 316" (Trial II)
Diquat with X-77 on plantago "AZ 329" (Trial I)
Diquat with X-77 on plantago "AZ 329" (Trial II)
Diuron with X-77 on plantago "AZ 316" (Trial I)
Diuron with X-77 on plantago "AZ 316" (Trial II)
Diuron with X-77 on plantago "AZ 329" (Trial I)
Diuron with X-77 on plantago "AZ 329" (Trial II)
Paraquat and Diuron with X-77 on plantago "AZ 316"

Paraquat and Diuron without X-77 on plantago "AZ 316"
Paraquat and Diuron with X-77 on plantago "AZ 329"
Paraquat and Diuron without X-77 on plantago "AZ 329"
In each description the writer provides information about the strain tested, the date of the test, the chemical with which plantago was treated, the status at the start of the application, the time of application, and the data recorded.]

3. RESULTS

Seedling survival, height, and root development were the criteria used to evaluate and compare plantago response to various herbicide treatments. Following a discussion of the use of surfactants with paraquat, diquat, and diuron, the results of tests of these three herbicides on plantago will be presented.

3.1. Use of Surfactants

The inclusion of a surfactant in the spray mixture can increase the phytotoxicity of paraquat, diquat, and diuron to some plant species. Surfactants can increase the phytotoxicity of these herbicides by increasing the water solubility of the herbicide and by providing better contact between the herbicide and the leaf surface. Increased solubility and leaf contact can increase the phytotoxicity by increasing the amount of an herbicide actually absorbed by the leaf.

Each test in this report was repeated at least once without including X-77 surfactant in the spray mixture. Excluding X-77 from the treatments usually made no difference in the response of plantago to paraquat, diquat, and diuron. Unless otherwise indicated, the following results are limited to tests including X-77 in the spray mixture.

3.2. Tests with a Single Herbicide

3.2.1. Effects on Seedling Survival

Paraquat usually reduced survival of plantago "AZ 316" (Table 1) and "AZ 329" (Table 2) seedlings at rates higher than 0.25%.

The results with diquat were less consistent. In some tests, rates of diquat higher than 0.125% caused reduction in the survival of seed-

Table 1. Response of plantago "AZ 316" seedlings to paraquat.

Treatment	Response		
Paraquat (% v/v)	Survival (% of check)	Height* (cm)	Root rating* 10 = normal** 0 = none
Trial I:			
untreated check	100	11 a	8 a
0.125	100	11 a	7 a
0.25	88	7 b	4 a
0.5	44	3 c	3 a
0.75	38	3 c	4 a
Trial II:			
untreated check	100	15 a	7 a
0.125	100	14 a	7 a
0.25	100	15 a	7 a
0.5	100	11 b	5 a
0.75	75	8 c	5 a

*Values in a column within a trial followed by the same letter do not differ significantly.

**Visual evaluation at bottom of pot.

Table 2. Response of plantago "AZ 329" seedlings to paraquat.

Treatment	Response		
Paraquat (% v/v)	Survival (% of check)	Height* (cm)	Root rating* 10 = normal** 0 = none
Trial I:			
untreated check	100	8 a	8 a
0.125	100	7 a	5 b
0.25	100	8 a	4 bc
0.5	81	4 b	3 bc
0.75	88	4 b	2 c
Trial II:			
untreated check	100	12 a	7 a
0.125	100	12 a	7 a
0.25	81	8 b	5 a
0.5	69	3 c	7 a
0.75	38	2 c	3 b

*Values in a column within a trial followed by the same letter do not differ significantly.

**Visual evaluation at bottom of pot.

[Tables 3 through 6, omitted here, depict the responses of "AZ 316" and "AZ 329" seedlings to diquat and diuron.]

lings of both plantago varieties. In other tests, rates as high as 0.25% diquat did not reduce seedling survival (Tables 3 and 4).

Diuron did not greatly reduce the survival of seedlings of either variety at any of the rates tested (Tables 5 and 6).

3.2.2. Effects on Seedling Height

In some tests, treatment with paraquat at rates of 0.25% or higher significantly reduced the height of "AZ 316" and "AZ 329" plantago seedlings. In all tests, paraquat at rates of 0.5% or higher reduced seedling height (Tables 1 and 2).

Diquat reduced the height of both varieties of plantago seedlings in most tests at rates higher than 0.125% (Tables 3 and 4). In the first trial of diquat on "AZ 316," all treatments significantly reduced seedling height (Table 3).

Diuron reduced the height of plantago "AZ 316" seedlings in one trial. This reduction occurred at the highest rate of diuron tested (0.56 kg/ha) (Table 5). Diuron did not significantly reduce the height of "AZ 329" seedlings at any of the rates tested (Table 6).

Some burning of foliage was observed at all rates of paraquat, diquat, and diuron. Increases in degree of leaf burn with increasing rates of these herbicides contributed to the significant differences in seedling height observed in some tests.

3.2.3. Effects on Root Development

Paraquat did not reduce root development of plantago "AZ 316" seedlings at any of the rates tested (Table 1). Paraquat reduced "AZ 329" root development at all rates in the first trial and at the highest rate only (0.75%) in the second trial (Table 2).

Plantago "AZ 316" root development was significantly reduced at two of the four rates of diquat tested (0.25% and 0.5%) in the first trial. Root development was not reduced at the highest rate of diquat (0.75%) in this trial. When this test was repeated (Trial II), no significant reduction of "AZ 316" root development was observed with any of the diquat treatments (Table 3). All diquat treatments reduced root development of plantago "AZ 329" seedlings in the first trial. In the second trial no significant reduction of "AZ 329" root development was observed at any of the rates of diquat tested (Table 4).

Treatment with diuron did not reduce root development of either plantago variety at any rate tested (Tables 5 and 6).

3.3. Tests with Two Herbicides

The following discussion is of tests combining paraquat and diuron. These two herbicides were selected because of the apparent greater tolerance of P. ovata to paraquat rather than diquat (the other surface-applied chemical). Separate tests were run for each of the plantago varieties ("AZ 316" and "AZ 329"). Separate tests were run with and without X-77 surfactant in the spray mixture.

[In subsections 3.3.1 and 3.3.2 the report discusses the results of testing with two herbicides, paraquat and diuron, on both strains of Plantago. Each subsection covers survival, height, and root development (paralleling the order of the discussions of single-herbicide tests). Tables 7 through 10, omitted here, depict the results.]

4. CONCLUSIONS

4.1. Comparison of Effects on Varieties

The two varieties of plantago examined in this experimentation project, "AZ 316" and "AZ 329," were evaluated in separate tests. The data resulting from these tests have been combined in a series of tables within this section so that appropriate conclusions can be drawn and discussed. Values in these tables are averages of two trials. The combined data have not as yet been statistically analyzed.

With each herbicide tested, it appears that any noticeable reductions in percent survival, height, or root development begin at the same rate for both varieties. In general, symptoms appear in both varieties after treatment with 0.5% paraquat (reduction in root development at 0.25% paraquat) (Table 11) or 0.25% diquat (Table 12). Diuron did not appear to affect survival and height of either variety at any rate. Root development responses to diuron treatment were not consistent (Table 13).

4.2. Comparison of Effects of Paraquat and Diquat

Paraquat is more phytotoxic to most plants than diquat. When the tests presented in this report were begun, it was expected that plantago seedlings would be more tolerant to treatment with diquat than treat-

Table 11. Comparison of response of plantago varieties "AZ 316" and "AZ 329" to paraquat.

Treatment	Response*					
	Survival (% of check)		Height (cm)		Root rating 10=normal** 0=none	
Paraquat (% v/v)	"AZ 316"	"AZ 329"	"AZ 316"	"AZ 329"	"AZ 316"	"AZ 329"
Untreated check	100	100	13	10	8	8
0.125	100	100	12	10	7	6
0.25	94	91	11	8	5	4
0.5	72	75	7	3	4	5
0.75	57	63	6	3	4	2

*Values included in this table are averages of two trials.

**Visual evaluation at bottom of pot.

[Tables 12 through 15, omitted here, provide more comparison data.]

ment with paraquat. However, in all tests included here, diquat appeared to injure seedlings more and at lower rates than paraquat. For example, diquat began to reduce "AZ 316" survival at the 0.125% rate; paraquat began to reduce "AZ 316" survival at the 0.25% rate. At the 0.5% rate, survival of plants treated with diquat was 41%, whereas survival of plants treated with paraquat was 72% (Table 14).

Data from several tests have been combined in Tables 14 and 15 to illustrate this comparison. As in the previous comparison of varieties, this discussion is of tendencies rather than statistical differences. The results of tests with paraquat or diquat suggest that plantago is more tolerant of treatment with paraquat than diquat. On this basis, paraquat was selected for further tests in combination with diuron.

One consistent trend is observed in comparing the results of the test of combinations of paraquat and diuron with X-77 surfactant to results of a similar test without surfactant. Apparently, root development of "AZ 329" seedlings may not be reduced by these herbicide treatments if X-77 is not included in the treatments. A closer examination of the effects of including surfactants in treatments on plantago could be an interesting subject for future research.

[In subsection 4.3, three possible reasons for an antagonism of paraquat and diuron are cited, including inhibition of the rate with which paraquat and diquat form free radicals, reaction of the plant to other chemicals in the soil, and the effect of clay colloids present in the wettable powder form of the herbicides. The author concludes that additional research may be necessary to determine a cause for this potential antagonism.]

5. RECOMMENDATIONS

5.1. Use of Paraquat and Diuron in Herbicide Programs for Plantago in Southwestern United States

The results of these greenhouse tests suggest that herbicide programs including the use of paraquat and diuron may be effective for weed control in plantago in the southwestern United States. It appears that plantago may tolerate rates of paraquat up to 0.5% and diuron up to 0.56 kg/ha when either of these herbicides is applied alone.

Diquat is more injurious to plantago than paraquat, but could possibly be used (at lower rates than paraquat) to replace paraquat depending on its ability to control the weeds in a given location.

On the basis of survival, plantago seedlings satisfactorily tolerated all greenhouse combination tests with paraquat and diuron up to the highest rates tested. Survival would be the main criterion for deciding the value of testing these herbicides in a field situation.

Field plot research with paraquat and/or diuron has been conducted in Arizona over the last three years. Results indicate that plantago is very tolerant of paraquat at rates up to and including 0.5%. In some locations, and certain years, plantago stands were reduced by use of diuron at 0.56 kg/ha due to light soils and/or abnormally high rainfall.

5.2. Suggestions for Further Research

Future research in determining an effective weed control program for plantago in Arizona may include screening of herbicides:

- To find a soil-active broadleaf herbicide which is safe for use in plantago regardless of soil type and rainfall.
- To find a contact herbicide which would be "safer" in the minds of the public than paraquat.

FOOTNOTES

[1] R. L. Whistler, Jr., and C. R. Smith, Industrial Gums: Polysaccharides and their Derivatives, 1973.
[2] B. B. Hyde, "Mucilage-producing cells in the seed coat of Plantago ovata: Developmental fine-structure," American Journal of Botany, 1970.
[3] The United States Dispensatory, 21st ed., 1926.
[4] C. Chandler, "Improvement of plantago for mucilage production and growth in the United States," Boyce Thompson Inst., 1954.
[5] J. M. Modi, et al., "Isabgol: A dollar earner of North Guarat," Indian Farm, 1974.
[6] I. J. Basset and B. R. Baum, "Conspecificity of Plantago fastigiata of North America with P. ovata of the Old World," Canadian Journal of Botany, 1969.
[7] Ibid.
[8] R. N. Chopra, et al., Indigenous Drugs of India, 1958.
[9] Whistler, 1973.
[10] P. Joshi, et al., "A short note on cultivation of Plantago ovata in Maharashtra," Indian Journal of Agronomy, 1970.
[11] M. A. Iyengar, et al., "Some factors affecting yields of Plantago ovata in Maharashtra," Indian Journal of Agronomy, 1965.
[12] Whistler, 1973.
[13] T. E. Russell, "Plantago wilt," Phytopathology, 1975.
[14] Weed Science Society of America, Herbicide Handbook of the Weed Science Society of America, 3rd ed., 1974.
[15] K. C. Hamilton, "Preplanting applications of diuron in cotton," Weeds, 1966.

[16] Ibid.
[17] R. H. Springett, "The bipyridylium herbicides: Their properties and use," Outlines Agricultural, 1965.
[18] Weed Science Society, 1974.
[19] G. N. Giannopolitis and S. K. Ries, "In vitro production of superoxide radical from paraquat and its interactions with monuron and diuron," Weed Science, 1977.
[20] Ibid.
[21] R. P. Upchurch, et al., "Soil sterilization properties of monuron, diuron, simazine, and isocil," Weed Science, 1968.

APPENDIX

II

Handbook and Style Guide

At the end of each part in this textbook, you have been asked to apply a set of editing criteria to your written assignments. For instance, you asked yourself such questions as:

- Are grammar, spelling, punctuation, and mechanics correct?
- Are sentence structures effective?
- Is the style appropriate to the writing situation?

Naturally, these questions, and others like them, can be answered only when you have a working knowledge of what is *correct, effective,* and *appropriate.*

After years of exposure to the rules and standards of writing, most of us have an innate feel for what is right and what is not, even if we are not able to cite the specific regulation. Despite these years of training, we do occasionally have trouble knowing the correct way to proceed. This handbook reviews some common trouble spots and provides guidelines for improving your writing.

The first section treats proper punctuation. Writers of technical documents frequently find themselves dealing with sophisticated, complex ideas and sentence structures. Something as seemingly insignificant as an omitted comma can make the difference between clarity and confusion.

The second section covers mechanics such as abbreviations, capitalization, numbers, and spelling. When these elements appear incorrectly, readers might conclude that the writer was not concerned with precision. As a result, the writer's credibility might be diminished.

The third section cites common sentence errors such as sentence fragments, comma splices, lack of subject-verb agreement, and misplaced modification. You will find the guidelines presented here especially helpful as you edit your rough drafts. Since rough drafts are usually written quickly, they often contain a lot of sentence errors. If you read your written work aloud, you should be able to spot these kinds of errors readily, because, for the most part, they sound strange or unnatural.

The fourth and final section deals with common stylistic problems, a somewhat more subjective domain. Nonetheless, certain standards do exist and are important to uphold. Here we look at choppy sentences, overuse of the passive voice, unnecessary modification, imprecise vocabulary, and sexist language.

If you regularly have trouble with the areas discussed in this appendix, consult one of the following handbooks or style guides:

Barnet, Sylvan, and Marcia Stubbs. *Practical Guide to Writing.* 4th ed. Boston: Little, Brown, 1983.

Ebbitt, Wilma R., and David R. Ebbitt. *Writer's Guide and Index to English.* 7th ed. Glenview, IL: Scott, Foresman, 1982.

Government Printing Office. *Style Manual.* Washington, D.C.: U.S. Government Printing Office, 1982.

Lefcowitz, Allan B. *The Writer's Handbook.* Englewood Cliffs, NJ: Prentice-Hall, 1976.

Leggett, Glenn, et al. *Handbook for Writers.* 9th ed. Englewood Cliffs, NJ: Prentice-Hall, 1985.

Watkins, Floyd C., and William B. Dillingham. *Practical English Handbook.* 7th ed. Boston: Houghton Mifflin, 1986.

PUNCTUATION

APOSTROPHES ’

Apostrophes can be used

- to indicate possession.
- to indicate contraction.
- to indicate omission.
- to indicate certain plural forms.

Possession

Use *'s* to indicate the possessive form of singular nouns and plural nouns that do not end in *s*.

Singular: child's, engineer's, man's, schema's
Plural: children's, men's, schemata's

Use *'s* to indicate the possessive form of singular nouns that end in *s*. However, when the next word in the same sentence begins with *s*, use only the apostrophe.

Copernicus's theory that the earth revolved around the sun dramatically changed astronomy.
Copernicus' system laid the foundation for modern astronomy.

Use only the apostrophe to indicate the possessive form of plural nouns ending in *s*.

The committee reviewed the engineers' joint proposal.

Use *'s* to indicate the possessive form of indefinite pronouns.

anyone's, everyone's, nobody's, someone's

Use *'s* with the last noun in a pair or series to indicate the possessive form of that pair or series.

The professor and the student's case history will be published. (one case history written by two people)

The professor's and the student's case histories will be published. (two case histories)

Contraction

Use an apostrophe to indicate the contraction of two words into one. Contraction usually occurs with the negative form of a verb or with a form of the verb *to be*. Note that contractions are rarely acceptable in formal written documents.

aren't (are not), can't (cannot), didn't (did not), don't (do not), he's (he is), isn't (is not), it's (it is), she's (she is), they're (they are), we're (we are), won't (will not), you're (you are)

Omission

Use an apostrophe to indicate the omission of a letter (number) or letters (numbers) from a word or phrase.

The fifth-generation computers of the '90's will process information in ways unforeseeable today.

The meeting convened at five o'clock.

Pluralization

Use an apostrophe to indicate the plural form of named letters, numbers, and words.

The student received three A's and two B's this semester.

The gymnast scored perfect 10's in all four events.

Delete unnecessary *could's* and *might's* to make your presentation more assertive.

BRACKETS [ ]

Brackets can be used

- to indicate interpolations in quotations.
- to indicate parentheses within parentheses.
- to indicate instructions to a printer or speaker.

Interpolation

Use brackets to indicate an interpolation, or insertion, in a quotation. Use of brackets generally clarifies a quotation taken out of context. A bracketed [*sic*] placed after an error or misspelling indicates that the error appears in the quotation.

"[Transplanting of normal bone marrow] gives one more hope that gene therapy could correct defective genes." (The information in brackets replaces the pronoun *it.*)

Parentheses within Parentheses

Use brackets to enclose parenthetical information within a larger base of parenthetical information. (Try to avoid such complex structures whenever possible.)

With increasingly successful gene-splicing experimentation under way, the scientific community is enjoying greater acceptance of the potential benefits of such procedures. (Note: A new *Business Week*/Harris Poll indicates that most Americans support gene therapy [p. 85].)

Instructions

Use brackets to indicate instructions regarding the treatment of information within a document. These instructions are a form of commentary and do not appear in the final printed document.

Speaker: [Hold up cord.] The personal computer monitor is attached to the central processing unit by inserting this cord here. [Point to socket in back of central processing unit.]

COLONS :

Colons are used

- to separate main clauses.
- to introduce lists, appositives, or quotations.
- to end business letter salutations.

Separation of Main Clauses

Use a colon to separate two main clauses when the second clause explains the first.

A moving particle possesses kinetic energy: the faster it goes, the greater its kinetic energy.

Introduction

Use a colon to introduce a list or series of related items. (Note: Do not use a colon after a linking verb or preposition.)

The writer discusses three basic points for each category: effectiveness, construction complexity, and cost.

Use a colon to introduce a formal appositive, or word or phrase used as a noun that explains another word. (Note: The colon appears before such expressions as *namely* or *that is*.)

The engineer made a difficult decision: he would abandon the design.
The engineer made a difficult decision: namely, that he would abandon the design.

Use a colon to introduce a direct quotation.

The European DP manager commented that violence against computer centers is linked to social or military protest campaigns: "Our industry and our products are taken as a symbol of modern business."

Business Salutation

Use a colon to end the salutation in a business letter. (Note: A comma should be used to end the salutation of a personal letter.)

Dear Mr. West:

Your Commander XZ-74 computer has been repaired and is being returned to you with this letter. We regret any inconvenience . . .

COMMAS ,

Commas can be used

- to separate items in a series.
- to separate main clauses (with a conjunction).
- to set off nonrestrictive modifiers or appositives.
- to set off introductory phrases or clauses.
- to help avoid ambiguity.
- to punctuate conventions such as dates, places, titles and degrees, numerals, quotations, and direct address.

Separation in a Series

Use commas to separate individual items in a series.

Three criteria were used for this experiment: seedling size, root length, and color of leaves.

Separation of Main Clauses

Use commas *and a conjunction* to separate main clauses.

The design is quite elegant, and it should be adopted without difficulty.

Nonrestrictive Modifiers or Appositives

Use commas to set off nonrestrictive modifiers or appositives from the word or phrase described. (A nonrestrictive modifier is a word or phrase that adds information but is not essential to the meaning of the sentence. An appositive is a word, phrase, or clause used as a noun and placed beside another word to explain it.)

Computers, which are still viewed by some executives as an expense rather than as an investment, can play an increasingly important role in improving the competitiveness of American manufacturers.

The report in the appendix, *Tolerance of Plantago ovata to Three Herbicides,* pinpoints the degree candidate's experiments and findings.

Introductory Phrases or Clauses

Use a comma at the end of an introductory phrase or clause to separate it from the balance of the sentence.

Phrase: With approval of his proposal in hand, the construction engineer began rounding up his crews.

Clause: As soon as the committee approved his proposal, the construction engineer began rounding up his crews.

Avoiding Ambiguity

Use commas between adjectives in a series in which each adjective independently modifies the noun. (Note: Do not use commas between adjectives when they cumulatively modify a noun: *The large dark stain streaked the slide.*)

The exhibit displayed the efficient, customized, user-friendly menu system.

Use a comma to prevent a possible misreading.

After eating, the horse responded more favorably to the treatment.

Conventions

Use commas to separate elements in a date.

Neil Armstrong made history on July 20, 1969, when he became the first person to set foot on the moon.

Use commas to separate elements in the name of a place.

This experiment was first conducted at an IBM laboratory in Yorktown, New York, in 1985.

Use commas to set off degrees or titles when they appear after the name of a person.

James K. Kimberly, Ph.D., will chair the special task force investigating the pipeline explosion.

Dr. Suzanne Clarkson, Chief of Neurosurgery, performed the delicate operation in less than six hours.

Use commas in numerals of more than three digits.

The $17,000 computer graphics system will be self-amortizing as soon as the engineers begin using it instead of traditional drafting tools.

Use commas to set off direct quotations.

Edward F. Staiano, the senior vice president responsible for Motorola's cellular business, says the decision has finally put his company on an equal footing with the Japanese. "In an even fight," he says, "we're not at any disadvantage. In fact, I think our products are better than theirs."

Use commas to set off words in direct address.

John, you and I both know that we must make a decision soon, or we will lose our competitive advantage.

DASHES —

Dashes can be used

- to separate a summary from the balance of the sentence.
- to indicate interruptions.
- to set off asides, or parenthetical remarks.
- to provide special emphasis.

When typing, use two hyphens to indicate a single dash. Do not add space before or after the dash.

Summary

Use a dash to separate a summary from the balance of the sentence. (Note: When the summary comes at the end of the sentence, a colon is usually more appropriate.)

Multiprocessing, parallel processing, vector processing—all of these buzz words are mistakenly being used interchangeably.

Interruptions

Use a dash to indicate a sudden interruption or change of thought. In technical writing, such a technique is rarely necessary or appropriate, although it may be effective in informal memos.

Jim, I think this report will please you—it may even shock you.

Parenthetical Remarks

Use dashes before and after asides, or parenthetical remarks. In most cases, these remarks could otherwise be contained within parentheses.

Every aspect of the product life cycle—design, manufacture, delivery, installation, and typical utilization—undergoes extensive planning, testing, and evaluation.

Increased Emphasis

Use a dash to achieve additional emphasis for a qualifying remark. In a technical environment, this tactic is used infrequently, but it appears often in advertising copy.

The multiple personalities of MAGIC 8 help you face the challenges of programmers and nontechnical users alike—while *you* control costs.

ELLIPSES . . .

An ellipsis, indicated by three spaced periods, can be used to shorten quotations. (If the ellipsis comes at the end of a sentence, there should be no spacing between the last letter and the punctuation. The ellipsis is then indicated by three spaced periods following the end punctuation.)

Original: Boeing Computer Services has proposed a solution: TOP (Technical and Office Protocol), a group of data communications specifications based on the International Standards Organization's Open Systems Interconnection (OSI) model. If adopted as standards, these specifications will enable machines from any vendor, whether in the office or on the factory floor, to communicate with any other machine. Whether TOP will be successful or go the way of Esperanto, the international language that no one speaks, will depend on several factors, most notably vendor and user cooperation.

Shortened version: Boeing Computer Services has proposed a solution: TOP . . . based on the International Standards Organization's Open Systems Interconnection (OSI) model. . . . [T]hese specifications will enable machines from any vendor . . . to communicate with any other machine. Whether TOP will be successful . . . will depend on . . . vendor and user cooperation.

EXCLAMATION POINTS !

Exclamation points are used after a word, phrase, or sentence to indicate strong feeling. Since technical writing is rarely infused with strong feeling, exclamation points should be used sparingly, though they are often used in promotional writing. (Note that exclamation marks appear inside of quotation marks only when the quotation itself is of an exclamatory nature.)

"This is the most exciting software product to appear on the market in years!" say our leading users.

HYPHENS -

Hyphens can be used

- to indicate compound numbers and certain compound words.
- to indicate that two words act together as a single modifier.
- to indicate the continuation of a word that cannot be completely typed at the end of a line.

Compound Numbers and Words

Use a hyphen to indicate compound numbers when they are spelled out in a sentence.

Ninety-five researchers signed the petition before it was sent.
His is only the twenty-third patent owned by the company.

Use a hyphen to indicate certain compound words. There are no consistent rules for the hyphenation of compound words, as the following list demonstrates. When in doubt about whether a word should be spelled closed up or with a hyphen, consult your dictionary.

air-shed	aircraft	air mass
back-check	backcourt	back order
color-bearer	colorfast	color phase
day-tripper	daylight	day bed

Compound Modifiers

Use a hyphen to link together two or more words that act as a single modifier. (Note: Do not use a hyphen with adverbial modifiers.)

Decision-support software has to be user-friendly, or executives will not use it.

The half-finished Central Arizona Project (CAP) requires six times its original appropriation in order for it to be completed.

But: The partly finished experiment was abandoned.

Continuation

Use a hyphen to indicate the continuation of a word from one line to the next. Follow standard syllabication techniques for breaking a word. When in doubt, consult your dictionary.

Next to new cancer therapies, the quest for memory-improving
drugs is the hottest area of medical research. "The first com-
pany that comes out with something that is even modestly ef-
fective is going to be very successful," says one industry
consultant.

PARENTHESES ( )

Parentheses can be used

- to enclose additional, but not crucial, information.
- to direct readers to related material, such as figures.
- to set off numbers in a list.

Additional Information

Use parentheses to enclose information additional to the main idea of a phrase, clause, or sentence. Parenthetical information ranges from a single word to a phrase, clause, or complete sentence. (Note: If the parentheses enclose a complete sentence, be sure to place the period *inside* the parentheses.)

Since it launched its first commercial satellite four years ago, the European space agency Ariane has put twelve into orbit. NASA has launched thirty commercial satellites in that same period. . . . A satellite launched by shuttle costs \$25 million; Ariane charges \$24 million. They have cut the price as much as \$3 million to land some U.S. customers. (The prices do not include the cost of the satellites, which runs from \$40 million to \$100 million, depending on their size and complexity.)

Directions

Use parentheses to direct readers to related information, such as that found in figures, tables, appendices, or other items within the same document. Parenthetical comments can also be used when describing or discussing figures.

Depending on the complexity of the report, several levels of subordination may appear in the table of contents. (See Chapter 4 for an explanation of levels of subordination in decimal outlines.) For the sake of clarity, however, the level of detail should not be any more specific than the third level of subordination (Figure 13-5).

Caption for photograph: A Canadian satellite (top) floats out of the cargo bay of the U.S. space shuttle *Discovery* during a 1985 flight.

Lists

Use parentheses to enclose numbers or letters that segment items of information in a list or series.

Follow these four simple steps to a healthier lifestyle: (1) get at least six to eight hours of sleep each night; (2) make breakfast your biggest meal, lunch a medium-size meal, and dinner a light snack; (3) perform aerobic exercise at least four times a week for at least one half-hour each session; (4) do not smoke or chew tobacco in any form.

PERIODS .

Periods can be used

- to end sentences that make a statement or command.
- to indicate certain abbreviations.
- to indicate decimal points in numbers.

Sentences

Use a period to end a sentence that makes a statement.

Solar-heating systems are gaining popularity in the Sun Belt.

Use a period to end a sentence that expresses a command.

Insert the disk into slot A.

Abbreviations

Use a period after abbreviations such as *Mr.*, *Dr.*, *Ave.*, A.M., B.C., *Ph.D.*, *e.g.*, and others. Note that many governmental and international agencies do not use periods in the abbreviations of their names (*NASA*, *FCC*, *NATO*). Consult a dictionary to determine correct usage.

Decimals

Use a period to indicate a decimal point.

He approved the $398.75 expenditure yesterday.

The 51.4 percent majority vote was hardly a sign of overwhelming popularity for the measure.

QUESTION MARKS ?

Question marks can be used

- to indicate direct questions.
- to show that dates or figures are uncertain.

Questions

Use a question mark at the end of a direct question. A question mark should be enclosed within quotation marks only when the quotation itself poses a question. In the case of a title that poses a question, the question mark is also enclosed within quotation marks.

How many ways can the inside of a system be arranged without an external observer being aware that rearrangements have occurred?

In his book *The Second Law*, P. W. Atkins asks, "But what of Kelvin's statement of the Second Law: how is dispersal related to the conversion of heat to work?"

"Did Comets Kill the Dinosaurs?" was the feature article in *Time*.

But: Have you had a chance to read Harry's article, "How to Promote a Jet"?

Uncertainty

Use a question mark after a date or the name of a figure to indicate uncertain historical accuracy.

Amenhotep III was King of Egypt (1411?–1375 B.C.).

QUOTATION MARKS “ ”

Quotation marks can be used

- to indicate brief quotations.
- to indicate certain types of titles.

Brief Quotations

Use quotation marks at the beginning and end of brief quotations. If you are typing quotations of four or more lines, you should single-space and indent ten spaces. Do not use quotation marks with such blocked, longer quotations.

In *To Engineer Is Human,* Petroski states, "Celebrated writers and engineers alike learn more from the errors of their predecessors and contemporaries than they do from all the successes in the world."

However, if the quotation is longer:

Petroski discusses the benefits of making mistakes:

> Yet the process of successive revision is as common to both writing and engineering as it is to music composition and science, and it is a fair representation of the creative process in writing and in engineering to see the evolution of a book or a design as involving the successful elimination of faults and error.

Use a single quotation mark to indicate a quotation within a quotation.

As Petroski made these points to his students, he remarked, "These students were coming to the realization that T. H. Huxley expressed in his book, *On Medical Education:* 'There is the greatest practical benefit in making a few mistakes early in life.' "

Use the following punctuation conventions when quotation marks are used with other forms of punctuation:

- Place periods and commas inside quotation marks.

Petroski states, "The Kansas City Hyatt Regency Hotel became a synonym for the greatest structural tragedy in the United States."

"The Kansas City Hyatt Regency Hotel became a synonym for the greatest structural tragedy in the United States," according to Petroski.

• Place semicolons and colons outside closing quotation marks.

The NASA official calmly remarked, "We have a major malfunction"; the gravity of the situation was not immediately apparent.

• Place question marks or exclamation points inside quotation marks only when the quotation itself is an interrogative or exclamatory sentence.

Atkins asks, "What is the final state of the universe?"

Titles

Use quotation marks to enclose the title of articles, papers, essays, or chapters. They are also used for the title of television shows, songs, and poems.

The "Design as Revision" chapter of Petroski's book can be valuable to technical writers.

SEMICOLONS ;

Semicolons are used

- to separate main clauses.
- to separate items in a complex series.

Main Clauses

Use a semicolon between main clauses not connected by a coordinating conjunction.

The problem of verifying a cutoff can be divided into two subsidiary problems. The first is ensuring that significant quantities of fissile material are not diverted to weapons from legitimate activities; the second is ensuring that there are no significant clandestine facilities.

Lists

Use a semicolon between items in a series that include their own punctuation.

Some of the frustrations facing middle management include computers, to the extent that a manager feels unable to use them; fewer promotions, due to business cutbacks; an influx of bright, capable women with MBAs; young managers who want to be VPs in three years; the perception that there are limited resources to get the job done with; and limitations on the manager's decision-making authority, as decisions get pushed downward or pulled upward in the organization.

MECHANICS

ABBREVIATIONS

In technical writing, only standard abbreviations should be used. (See the lists of standard scientific and technical abbreviations and commonly used abbreviations, pages 524–526.) If you use an unfamiliar abbreviation, spell out the word the first time it is used in the text, followed by its abbreviation in parentheses. (Whether an abbreviation is considered unfamiliar will depend on the audience.) Explain such abbreviations in tables by means of footnotes; when they appear in illustrations, use captions to explain them. Once defined, an abbreviation may be used throughout the document, although it may be a good idea to repeat the definition in later sections of a manual or a long report that may not be read completely through.

Abbreviations for units of weight, measure, and time are the same in both singular and plural forms (*in., cm, kg, hr*). *No.* is the abbreviation for both *number* and *numbers.* Avoid the use of symbols to indicate inch ("), foot ('), pound (#), or number (#) in formal documents; however, such usage is appropriate in scientific notation.

Abbreviations spelled in capital letters should never be written in lower-case letters. Abbreviations spelled in lower-case letters (such as *etc., cont., no.*) should be changed to all capital letters when they are part of a heading that is all capitals. However, scientific and technical abbreviations (such as *mA, kHz, bpi, cpm*) should always appear in their standard form, regardless of the context in which they appear. Note as you look at the lists that some abbreviations have periods and some do not. Follow the conventions in your own discipline for these and any other abbreviations.

Scientific and Technical Abbreviations

alternating current	ac	binary digit	bit
ampere	amp	bits per inch	bpi
angstrom	A	bits per second	bps
ante meridiem	A.M.	British thermal unit	Btu
azimuth	az	calorie	cal

Celsius (Centigrade)	C	megahertz	MHz
centimeter	cm	meter	m
circumference	circ	miles per hour	mph
cubic centimeter	cm^3, cc	milliampere	mA
cubic inch	$in.^3$	milligram	mg
decibel	db	millimeter	mm
degree	deg,°	millisecond	ms
direct current	dc	milliwatt	mW
dozen	doz	minimum	min
external	ext	minute	min.
Fahrenheit	F	nanosecond(s)	ns
farad	f	number(s)	no.
feet per minute	fpm	pint	pt
fluid ounce	fl oz	post meridiem	P.M.
foot	ft	pound	lb
frequency modulation	FM	pounds per square foot	psf
gallon	gal	pounds per square inch	psi
gigahertz	Ghz	quart	qt
gram	g	radian	rad
grams per square meter	g/m^2	revision	rev.
hertz	Hz	revolutions per minute	rpm
horsepower	hp	revolutions per second	rps
hour	hr	roentgen	r
inch	in.	second	s, sec
inches per second	ips	square inch	$in.^2$
internal	int	square millimeter	mm^2
kilobyte	KB	temperature	temp
kilocalorie	kcal	tensile strength	ts
kilogram	kg	tolerance	tol
kilohertz	kHz	volt	V
kilovolt	kV	voltampere	VA
kilovoltampere	kVA	volts, alternating current	Vac
kilowatt	kW	volts, direct current	Vdc
latitude	lat	watt	W
liter	l	wave length	wl
logarithm	log	week	wk
longitude	long	words per minute	wpm
lumen	lm	yard	yd
maximum	max	year	yr

Commonly Used Abbreviations

about	c., ca.	and so forth	etc.
and others	et al.	anonymous	anon.

Commonly Used Abbreviations (continued)

appendix	app.	following page(s)	f., ff.
born	b.	illustrated, illustration	illus.
bulletin	bull.	line(s)	l., ll.
chapter	ch., chap.	manuscript(s)	ms., mss.
compare	cf.	namely	viz.
copyright	©	note	n.
died	d.	page(s)	p., pp.
dissertation	diss.	part(s)	pt., pts.
editor	ed.	section	sec.
especially	esp.	that is	i.e.
example	ex.	thus, so	sic
figure	fig.	translation	trans.
for example	e.g.	volume	vol.

ACRONYMS

Acronyms should be used only when their meanings have been defined (or are well known) to readers. Spell out the acronym the first time it is used in a document, followed by the acronym in parentheses. Redefine the acronym in later sections of a lengthy document. Acronyms should not contain, or be followed by, periods. Use "s" to form the plural of acronyms (e.g., *CPUs, CRTs*).

Use acronyms selectively; reliance on them will reduce your document to jargon. The less expert your audience, the fewer acronyms you should use.

CAPITALIZATION

Capitalize the following elements in your technical documents:

- The first word of a sentence or sentence fragment

Now is the time for all men to come to the aid of their country.

- The first word of a direct quotation

"Four score and twenty years ago . . ."

- The first word of a direct question

Are you going to the convention?

• Words in titles, with the exception of conjunctions, prepositions, and articles (unless these occur as the first word)

The Birds of Arizona

• The first word in the salutation and the complimentary close of a letter

Dear Mr. Roberts,
Yours truly,

• The first, middle, and last names of individuals

Paul William Roberts

• The names of places

Newark, New Jersey
The Dew Drop Inn

• Definite geographic regions but not general terms of direction

Enclosed please find my report on my trip to the Midwest.
But: He was driving west when he encountered the moose herd.

• The names of organizations, departments, bureaus, courts, and companies

United Nations, Rotary International, United States Court of Appeals for the Ninth Circuit, Texaco

• The names of ships, aircraft, and spacecraft (Note: These names should also be italicized.)

S.S. *Wisconsin, Spirit of St. Louis, Challenger*

• The names of peoples and their languages

English, Iroquois, Swedish

• Derivations from proper names

Newtonian physics, Darwinian evolution

- Titles of family relationship preceding the first name of that person

Uncle Henry, Cousin Mary

- Titles of a person preceding the first or last name of that person

President Ronald Reagan, Pope John Paul II, Professor Beltramini, Dr. Smith

- Formal epithets used instead of the name of a person or thing

Old Hickory, Old Ironsides, Old Faithful

- The pronoun *I*
- Days of the week, months of the year, and holidays

Tuesday, December, Thanksgiving

- The names of treaties, acts, and laws

Treaty of Versailles, Taft-Hartley Act

- Registered trademarks (if they are registered as capitalized)

Kleenex, Xerox

- Names of geological eras, periods, epochs, strata, and prehistoric divisions

Silurian period, Pleistocene epoch, Age of Reptiles, Stone Age

- Names of planets, constellations, asteroids, and groups of stars. Note: *Sun, earth,* and *moon* are not capitalized unless they are listed with other capitalized astronomical names.

Mars, Little Dipper, Sirius, Pleiades

- Genus names in binomial scientific classifications (specifically, zoology and botany); the names of species are not capitalized. (Note: These terms should also be italicized.)

Plantago ovata, Cynomys ludovicianus

- Latinate names of classes, families, and all groups above genera. However, derivative forms are not capitalized.

Gastropoda, gastropod

- Acronyms

COBOL, UNESCO

ITALICIZATION

Use underlining to indicate italics for any of the following circumstances:

- Titles of books (except the Bible), magazines, journals, newspapers, paintings, plays, and motion pictures. Do not underline spacing between words or intervening punctuation.

Time, *Wall Street Journal*, *Star Wars*

- Titles of ships and trains

Queen Elizabeth II, *Orient Express*

- Foreign words used in an English context

They took me to the *boma* (a thorn-fence enclosure) to milk the cows.

- Words or phrases being defined or discussed in a sentence

The phrase *math anxiety* has received a lot of attention lately.

- Special emphasis for a word, phrase, or statement. In technical writing, frequent use of this tactic is inappropriate.

"With fabric, form *must* follow function," says Burger.

- To distinguish a caption from the surrounding text

The inscription, right, from a tablet at Palenque tells the story of a coronation.

- A biographical or background note to an article or paper

Virginia Morrell writes extensively about anthropology and other sciences. She is currently writing a book for Simon and Schuster about her recent travels.

NUMBERS

Write out numbers in technical documents if:

- The number can be written in one or two words.

The professor said that thirty-four students attended the meeting.

- The number is the first word of a sentence. (Note: If this number cannot be written in one or two words, revise the sentence.)

Thirty-four students attended the meeting.
But: The meeting drew 412 participants.

- The number is approximate rather than actual.

At least three hundred orders came in for the book.

- The number precedes a second number.

The experiment used fourteen 8-oz beakers.

- The number refers to a street name lower than 100.

The museum is located on Fifty-second Street, but I live on First Avenue.

Use numerals in technical documents if:

- Precise figures, decimals, or technical measurements are required.

The room measures 4 ft × 12 ft.
The return on investment for 1985 was 17.3 percent.
The device used 84 Btus.

- The number includes a fraction.

The plant experiments yielded 6½ oz of seed.

- The number is larger than ten and cannot be expressed in a word or two.

We traveled 6,136 miles, round trip.

- The number follows a spelled-out number.

We have ordered sixteen 4-foot software cabinets for the computer lab.

- The number is a date.

They will be arriving on January 28.

- The number is a unit of measure.

The horse was clocked at 36 miles per hour.

- The number is an age.

John Strauss, 38, was elected to the post yesterday.

- The number is a percentage.

Only 32 per cent of the eligible voters went to the polls.

- The number is an exact time.

The astronomical portion of the experiment must begin at 6:17 P.M.

- The number reflects statistical data (such as census figures).

The sample interviewed 50 male line workers at the Detroit assembly plant.

- The number is a sum of money.

We paid $4,922.86 for the new surveying equipment.

- The number is a page or line number.

Turn to page 12, line 14, to see the passage.

Use *both* written-out numbers and numerals (in parentheses) in legal documents such as contracts and agreements.

The COMPANY agrees to pay the CONTRACTOR one hundred fifty thousand dollars ($150,000) upon the satisfactory completion of the structure. . . .

SPELLING

Writers of technical documents are often faced with especially difficult words. Nonetheless, it is every writer's responsibility to spell correctly. Invest in a good dictionary and keep a list of words with which you have difficulty handy for quick reference.

COMMON SENTENCE ERRORS

SENTENCE FRAGMENTS [frag]

A sentence fragment is a part of a sentence that is punctuated as a complete sentence. These fragments are incomplete thoughts and do not make sense in and of themselves. Although promotional writing and advertising make frequent use of sentence fragments for dramatic effect, technical writing standards do not accommodate such usage. Eliminate these sentence errors by completing the thought.

INCORRECT Gene-splicing experiments being successful.

CORRECT Recently, gene-splicing experiments have been successful.

INCORRECT The solution is a reliable, consistent, computer system. A computer system such as the Commander X-74.

CORRECT The solution is a reliable, consistent computer system, such as the Commander X-74.

COMMA SPLICES [cs]

A comma splice is the joining of two or more main clauses by using a comma without a coordinating conjunction (*and, but, or, nor, for, yet, so*). Eliminate this error by using either a semicolon or a comma with the coordinating conjunction.

INCORRECT The remarkable capability of ZSM-5 to make gasoline molecules and nothing else was not deliberately designed, it just happened to be there.

CORRECT The remarkable capability of ZSM-5 to make gasoline molecules and nothing else was not deliberately designed; it just happened to be there.

INCORRECT Catalyst design remains enormously complex, many catalytic reactions are still poorly understood.

CORRECT Catalyst design remains enormously complex, and many catalytic reactions are still poorly understood.

RUN-ON SENTENCES [ro]

A run-on, or fused, sentence is one in which two or more main ideas are joined without benefit of punctuation. The result can be very confusing or long-winded. Eliminate this error by supplying the necessary punctuation or by breaking the sentence into two or more complete sentences.

INCORRECT The company's limitation on health benefits is $300,000 and the patient's treatment cost is nearly $500,000.

CORRECT The company's limitation on health benefits is $300,000, and the patient's treatment cost is nearly $500,000.

INCORRECT The computer system supporting the needs of scientists and engineers must be efficient, reliable, and secure and it must offer a range of processing capabilities and applications to meet the diverse workloads these specialists undertake.

CORRECT The computer system supporting the needs of scientists and engineers must be efficient, reliable, and secure. It must offer a range of processing capabilities and applications to meet the diverse workloads these specialists undertake.

INCORRECT SUBJECT-VERB AGREEMENT [agr sv]

Incorrect subject-verb agreement occurs when a single noun or pronoun is used with a plural verb, or when a plural noun or pronoun is used with a singular verb. Eliminate this error by changing the verb form to match the noun or pronoun status. Here are some useful guidelines.

- Use a plural verb with a compound subject that includes the conjunction *and*.

INCORRECT Civil and mechanical engineering is equally complex.

CORRECT Civil and mechanical engineering are equally complex.

INCORRECT Anticipated tension and stress determines construction design specifications.

CORRECT Anticipated tension and stress determine construction design specifications.

• Use a verb that agrees in number with the nearest part of a compound subject when that subject includes the conjunctions *or, nor, either/or, neither/nor, not/but.*

INCORRECT Neither the vice president nor the project manager are confident that the proposal will be accepted.

CORRECT Neither the vice president nor the project manager is confident that the proposal will be accepted.

INCORRECT The spark plugs or the piston ring are the source of the problem.

CORRECT The spark plugs or the piston ring is the source of the problem.

• Use a verb that agrees in number with its subject even when intervening phrases or clauses occur in the sentence.

INCORRECT The booster rocket, as well as the propellant jets, were responsible for the explosion.

CORRECT The booster rocket, as well as the propellant jets, was responsible for the explosion.

• Use a singular verb with indefinite pronouns (*anybody, anyone, each, either, everybody, everyone, neither, nobody, no one, one, somebody, someone*).

INCORRECT Neither of the experiments were successful.

CORRECT Neither of the experiments was successful.

• Use a verb that matches the noun or pronoun in a phrase whose subject indicates a part or portion (e.g., *all, half, none, part, quarter, some, third*).

INCORRECT Some of the distillate accumulate at the base of the beaker.

CORRECT Some of the distillate accumulates at the base of the breaker.

• Use a singular verb with the phrase *the number.*

INCORRECT The number of surviving bacteria were surprising.

CORRECT The number of surviving bacteria was surprising.

• Use a plural verb with the phrase *a number* when it means the plural of *some* (usually the case).

INCORRECT A number of scientists was opposed to the use of fissile materials.

CORRECT A number of scientists were opposed to the use of fissile materials.

• Use a verb that matches its subject in sentences beginning with *there* or *here*.

INCORRECT There was sparks flying from the tail of the booster.
CORRECT There were sparks flying from the tail of the booster.

Note: *There is* may sometimes be used to introduce a compound subject when the first noun in that subject is singular.

CORRECT There is a mortar and pestle in the cabinet.

• Use a verb that agrees with its subject rather than with any other part of the sentence.

INCORRECT His report and proposal is the result of diligent research and creativity.
CORRECT His report and proposal are the result of diligent research and creativity.

• Use a singular verb when a title is the subject, even when that title contains plural words or ideas.

INCORRECT *The Birds of Arizona* are a comprehensive survey of the varied and unusual birds of that state.
CORRECT *The Birds of Arizona* is a comprehensive survey of the varied and unusual birds of that state.

INCORRECT PRONOUN-REFERENT AGREEMENT [agr p]

A pronoun must agree in number with the noun it replaces (its referent). In other words, a singular pronoun must refer to a singular noun, and a plural pronoun must refer to a plural noun. Here are some guidelines for correct usage.

• Use a singular pronoun to replace a singular noun. Use a plural pronoun to replace a plural noun.

INCORRECT The scientist is looking for a way to speed up the chemical reactions in their experiments.

CORRECT The scientist is looking for a way to speed up the chemical reactions in his experiments.

INCORRECT Because enzymes are generally fragile and work only under precise conditions, it is replaced by human-made catalysts.

CORRECT Because enzymes are generally fragile and work only under precise conditions, they are replaced by human-made catalysts.

- Use a plural pronoun when it refers to a compound noun that uses the conjunction *and*.

INCORRECT Molecular structure and function are closely related. It can be affected by catalytic action.

CORRECT Molecular structure and function are closely related. They can be affected by catalytic action.

- Use a pronoun that agrees in number with the closest word of the compound noun it replaces when that compound noun uses conjunctions such as *or, nor, either/or, neither/nor, but also*.

INCORRECT Neither the student nor the professor completed their experiment.

CORRECT Neither the student nor the professor completed his experiment.

INCORRECT Either the professor or the students will represent his university in the competition.

CORRECT Either the professor or the students will represent their university in the competition.

- Use a singular pronoun to replace a collective noun when the members of that group are treated as a unit. Use a plural pronoun to replace a collective noun when the members of that group are treated individually.

INCORRECT The special task force presented their findings.

CORRECT The special task force presented its findings.

INCORRECT The special task force took its seats for the press conference.

CORRECT The special task force took their seats for the press conference.

- Use a singular pronoun to replace indefinite pronouns (*anybody, anyone, each, either, everybody, everyone, neither, nobody, no one, one, somebody, someone*).

INCORRECT Not one of the doctors felt that they had a fair schedule.
CORRECT Not one of the doctors felt that he or she had a fair schedule.

UNCLEAR PRONOUN REFERENCES [ref]

Unclear pronoun reference occurs when a previous sentence, clause, or phrase contains more than one noun for which the pronoun might be substituted. Eliminate this error by supplying the appropriate noun or by recasting the sentence(s) to clarify the relationship between noun and pronoun.

INCORRECT The researchers believe that every 26 million years a rain of comets bombards the earth. The impact of some of the larger comets spews enough debris into the atmosphere to block the sun for months. This could be responsible for the extinction of dinosaurs.

CORRECT The researchers believe that every 26 million years a rain of comets bombards the earth. The impact of some of the larger comets spews enough debris into the atmosphere to block the sun for months. This darkening of the earth could be responsible for the extinction of dinosaurs.

INCORRECT PRONOUN CASES [c]

Errors occur in pronoun case, or form, when you do not consider the way in which the pronoun is used in a sentence. Here are some guidelines for proper usage.

- Use the subjective case of a pronoun when it is used either as the subject of a sentence or as the subjective complement.

INCORRECT Dr. Alvarez and me will be speaking at the conference.
CORRECT Dr. Alvarez and I will be speaking at the conference.

INCORRECT The speakers at the conference will be Dr. Alvarez and me.
CORRECT The speakers at the conference will be Dr. Alvarez and I.

- Use the objective case of a pronoun when it is used as the object of a preposition.

INCORRECT The competition was between he and I.
CORRECT The competition was between him and me.

INCORRECT A few of we researchers were chosen for the project.
CORRECT A few of us researchers were chosen for the project.

• Use the objective case of a pronoun when it is the subject or object of an infinitive.

INCORRECT The panel considers we to be the primary candidates for the Nobel Prize in chemistry.
CORRECT The panel considers us to be the primary candidates for the Nobel Prize in chemistry.

• Use the same case for both an appositive and the pronoun that refers to it.

INCORRECT Two theorists, Dr. Alvarez and me, were chosen to speak at the conference.
CORRECT Two theorists, Dr. Alvarez and I, were chosen to speak at the conference.
INCORRECT The committee chose two theorists, Dr. Alvarez and I, to speak at the conference.
CORRECT The committee chose two theorists, Dr. Alvarez and me, to speak at the conference.

• Use the same case for a pronoun that follows the words *than* or *as* in an incomplete (elliptical) clause that you would use if the clause were complete.

INCORRECT No one else at the forum was as knowledgeable as him.
CORRECT No one else at the forum was as knowledgeable as he.
INCORRECT The student respected no one else as much as he.
CORRECT The student respected no one else as much as him.

• Do not use an apostrophe with the possessive form of personal pronouns (*hers, his, its, ours, theirs, yours*). Do use an apostrophe with the possessive form of indefinite pronouns (*anybody's, anyone's, everybody's, everyone's, nobody's, no one's, one's, somebody's, someone's*).

• Use *who* and *whoever* when the subjective case of a pronoun is required in a clause. Use *whom* and *whomever* when the objective case of a pronoun is required.

INCORRECT Whom was in charge of the quality review board?
CORRECT Who was in charge of the quality review board?

INCORRECT Whoever the panel selects will be the Nobel Prize winner.
CORRECT Whomever the panel selects will be the Nobel Prize winner.

DANGLING MODIFIERS [dg]

A dangling modifier is a verbal phrase occurring at the beginning of a sentence that does not modify, or qualify, the correct subject. Eliminate this error by supplying the appropriate subject immediately following the phrase.

INCORRECT After mixing the solution, the chemicals were put away.
CORRECT After mixing the solution, the laboratory assistant put the chemicals away.

INCORRECT To clean the countertop, the chemicals were put away.
CORRECT To clean the countertop, the laboratory assistant put the chemicals away.

MISPLACED MODIFIERS [mm]

A misplaced modifier is any adjective clause that has an intervening word, phrase, or clause between it and the word it modifies. Eliminate this error by placing the modifier and the word it describes as close together in a sentence as possible.

INCORRECT Some food additives are still used in processed foods that are suspected of being carcinogenic.
CORRECT Some food additives that are suspected of being carcinogenic are still used in processed foods.

FAULTY PARALLELISM [par]

Faulty parallelism occurs when parallel thoughts are not expressed in parallel grammatical constructions. Phrases should be matched with phrases, clauses with clauses, nouns with nouns, verbs with verbs.

INCORRECT The catalysts were effective and acted quickly.
CORRECT The catalysts were effective and fast-acting.

COMMON STYLISTIC ERRORS

CHOPPY SENTENCES

Choppy sentences result from overdependence on simple sentences or from overuse of short independent clauses linked by coordinating conjunctions. They result in a boring, telegraphic style that can cause readers to overlook important information because there are insufficient transitions between ideas. Eliminate this stylistic error by using subordinating conjunctions to form complex sentences.

INCORRECT Some scientists link dinosaur extinction to volcanoes. This theory is based on the explosion of Krakatoa in 1883. This particular volcano is between Java and Sumatra. It darkened the skies for weeks. It triggered a giant tidal wave. The tidal wave drowned 36,000 people.

CORRECT Some scientists link dinosaur extinction to volcanoes, basing their theories on the effects of the explosion of Krakatoa (between Java and Sumatra) in 1883, which darkened the skies for weeks and triggered a giant tidal wave that drowned 36,000 people.

OVERRELIANCE ON PASSIVE VOICE

The passive voice uses a sentence pattern of recipient/action/actor. This construction emphasizes the thing that is acted upon rather than the actor itself. Use it selectively, for overreliance on the passive voice can distort emphasis, create unnecessarily wordy sentences, and bore readers. In most writing situations, use the active voice.

WEAK The investigators were frustrated by the lack of evidence.

PREFERRED The lack of evidence frustrated the investigators.

WEAK The houses on Wainuenue Street are endangered by the lava flow.

PREFERRED The lava flow endangers the houses on Wainuenue Street.

OVERRELIANCE ON LINKING VERBS

Linking verbs express condition rather than action. The most common linking verbs are forms of the verb *to be* (*is, were, am,* etc.). Other linking verbs include *appear, become, feel, look, seem, smell, sound,* and *taste.* An overreliance on linking verbs can result in wordy sentences that convey little or no sense of action.

WEAK Active verbs are preferred in most writing situations.

PREFERRED In most writing situations, use active verbs.

WEAK Active verbs are lively in technical documents, and they are more interesting to readers.

PREFERRED Active verbs enliven technical documents and interest readers.

UNNECESSARY MODIFIERS

Unnecessary modifiers are descriptive words or phrases that add little or no information to the sentence. These elements can be struck from the sentence without jeopardizing the point being made. Strengthen the remaining verbs, adjectives, and adverbs as appropriate.

WEAK Very few people realize that red and violet, which appear to be somewhat similar in color, are actually at very opposite ends of the color spectrum.

PREFERRED Few people realize that red and violet, seemingly similar colors, are at opposite ends of the spectrum.

WEAK Purpurin ($C_{14}H_8O_5$) is used to create a very, very red dye.

PREFERRED Purpurin ($C_{14}H_8O_5$) is used to create a deep red dye.

WORDINESS

Wordiness results from the use of unnecessary words, phrases, and clauses that add little or no new information. These elements can be deleted from your sentence without changing their meaning. Here are some tips for achieving a more concise style.

- Avoid redundancy, or unnecessary repetition of words, ideas, or information, by using synonyms and pronouns and cutting words wherever possible.

WEAK The fact that the experiment failed does not mean that it was a complete failure, for the experiment had some successes.

PREFERRED Although the overall experiment failed, we gleaned some valuable information from it.

WEAK The sides of the canal were fortified with gabions to strengthen them.

PREFERRED The sides of the canal were fortified with gabions.

• Use repetition of a word or phrase only to create logical connections between sentences and paragraphs or to reinforce key ideas. In other cases, repetition is unnecessary and frequently tedious. Compress related ideas into complex sentences.

WEAK A *quipu* is a counting device that was used by the ancient Inca Indians. These Indians used the device to calculate figures by means of a series of different-colored strings. These strings were hung from a main cord, and groups of knots were tied along the strings. Each string color, knot size, and knot location signified a different numeric value.

PREFERRED A *quipu* is a calculating device used by the ancient Inca Indians. It is made up of a main cord and a series of colored strings, in which groups of various-sized knots were tied at different locations to signify specific numeric values.

• Do not overuse prepositional phrases. They can be used effectively to indicate relationships between a subject and an object, such as possession, location, time, or condition. However, too great a reliance on prepositions creates wordy sentences that can be difficult to follow. Eliminate this stylistic error by deleting unnecessary propositional phrases and substituting more precise words.

WEAK From the answers on the questionnaire, it was found that the average amount of use that these units will receive is 100 hours per month.

PREFERRED Questionnaire answers indicate that these units will be used approximately 100 hours per month.

• Replace relative clauses with a single word or phrase unless they are used to introduce information that distinguishes the subject or object in a sentence.

WEAK The suggestion was submitted by someone who did not sign a name.

PREFERRED The suggestion was submitted anonymously.

WEAK The panel defines the criteria that are satisfactory.

PREFERRED The panel defines the satisfactory criteria.

• Eliminate expletives, valueless words or phrases that add nothing to the meaning of a sentence. Common examples include *there are, there is, it is felt*. These phrases are often combined with unnecessary relative clauses. Avoid these stylistic excesses by paring down your sentences.

WEAK There are certain kinds of fungi and lichens that produce nonmotile sperm cells by means of a spermagonium, which is a flasklike structure.

PREFERRED Certain kinds of fungi and lichens use their spermagonium (a flasklike structure) to produce nonmotile sperm cells.

WEAK It is felt that the increase in water usage will deplete the city water supply.

PREFERRED The increase in water usage may deplete the city water supply.

JARGON

Jargon is specialized vocabulary, generally understood only by those schooled in a particular discipline. It is a useful shorthand in technical documents addressed to experts in a field. However, it should not be used unnecessarily or with nontechnical audiences. Eliminate this stylistic error by using field-specific terminology only when it is essential to do so.

WEAK The journalization of yesterday's operations can be called from memory and accessed from the CRT.

PREFERRED Yesterday's operations records can be retrieved from the computer's storage facilities and displayed on the terminal screen.

WEAK Expunge periphrasis to make writing more precise.

PREFERRED Remove roundabout words and phrases to make writing more precise.

IMPRECISE VOCABULARY

Imprecise vocabulary results from using general rather than specific words and vague rather than concrete statements. Eliminate this stylistic error by selecting precise, descriptive language.

WEAK The gold coins of ancient Greece were fancy.

PREFERRED The gold coins of ancient Greece were elaborately decorated with twisted wreathes, furling banners, and political seals.

WEAK These things make me a strong candidate for the job.

PREFERRED The courses and part-time jobs I have described in my résumé make me a strong candidate for the job of junior technical writer.

WORDS WITH STRONG CONNOTATIONS

The connotation of a word is the idea or mood suggested by or associated with it. In most technical writing situations, words with strongly negative or strongly positive connotations are inappropriate because readers expect objectivity. Eliminate this stylistic error by using precise and factual language that does not convey personal opinion.

WEAK The city council discussed the problem of bums sleeping in the park.

PREFERRED The city council discussed the problem of transients sleeping in the park.

WEAK The famous arbitrator was manipulated by the mob of demonstrators.

PREFERRED The well-known arbitrator was influenced by the large group of demonstrators.

SEXIST LANGUAGE

Sexist language is biased, frequently demeaning, vocabulary that ascribes positions of authority to men and subservient positions to women. Eliminate sexist language by using neutral terms, such as those in the following list. Minimize the use of the sometimes awkward phrase *he or she* (*him or her*) by using plural nouns and pronouns if possible.

Sexist Term	Neutral Term
authoress	author
businessman	businessperson
chairman	chair, chairperson
housewife	homemaker
manhours	hours
mankind	humankind
manpower	human resources
mailman	mail carrier
policeman	police officer
foreman	supervisor

WEAK Each doctor wants his patients to be prompt.

PREFERRED Doctors want their patients to be prompt.

Continuation of copyright page.

Text credits
7 Reprinted by permission of Morris, Clester, Abegglen & Associates.

21, 22, 132, 162, 306, 407 Reprinted with the permission of Honeywell Information Systems Inc., Large Computer Products Division, Phoenix, AZ.

36 From *Robert Bentley's Volkswagen Rabbit/Jetta Diesel Service Manual: 1977–1984*. Reprinted by permission of Robert Bentley, Inc., Cambridge, MA.

114 From Gene Bylinsky, "The Magic of Designer Catalysts," *Fortune,* May 27, 1985. Reprinted by permission.

166 From Benjamin F. Miller and Claire B. Keane, *Encyclopedia and Dictionary of Medicine and Nursing,* Philadelphia, W.B. Saunders Co., 1972. Reprinted by permission.

179 Reprinted from *The New Science of Skin and Scuba Diving,* 6th Edition. Copyright 1985 Council for National Cooperation in Aquatics. Reprinted by permission of New Century Publishers, Piscataway, NJ.

218 From Natalie Angier with Robert Buderi and Christine Gorman, "Did Comets Kill the Dinosaurs?" *Time,* 125 (May 6, 1985). Copyright 1985 Time Inc. All rights reserved. Reprinted by permission from *Time.*

344 From "Educational Programs for Women in the Construction Industry" by Jacqueline Waite Fleury. Reprinted by permission.

483 From "Tolerance of *Plantago ovata* to Three Herbicides" by Peggy Specht. Reprinted by permission.

INDEX